AF750926

Lecture Notes in Geoinformation and Cartography

Series editors

William Cartwright, Melbourne, Australia
Georg Gartner, Wien, Austria
Liqiu Meng, München, Germany
Michael P. Peterson, Omaha, USA

The Lecture Notes in Geoinformation and Cartography series provides a contemporary view of current research and development in Geoinformation and Cartography, including GIS and Geographic Information Science. Publications with associated electronic media examine areas of development and current technology. Editors from multiple continents, in association with national and international organizations and societies bring together the most comprehensive forum for Geoinformation and Cartography.

The scope of Lecture Notes in Geoinformation and Cartography spans the range of interdisciplinary topics in a variety of research and application fields. The type of material published traditionally includes:

- proceedings that are peer-reviewed and published in association with a conference;
- post-proceedings consisting of thoroughly revised final papers; and
- research monographs that may be based on individual research projects.

The Lecture Notes in Geoinformation and Cartography series also includes various other publications, including:

- tutorials or collections of lectures for advanced courses;
- contemporary surveys that offer an objective summary of a current topic of interest; and
- emerging areas of research directed at a broad community of practitioners.

More information about this series at http://www.springer.com/series/7418

Henrik Hargitai
Editor

Planetary Cartography and GIS

Springer

Editor
Henrik Hargitai
NASA Ames Research Center/NPP
Moffett Field, CA, USA

Eötvös Loránd University
Budapest, Hungary

ISSN 1863-2246 ISSN 1863-2351 (electronic)
Lecture Notes in Geoinformation and Cartography
ISBN 978-3-319-62848-6 ISBN 978-3-319-62849-3 (eBook)
https://doi.org/10.1007/978-3-319-62849-3

Library of Congress Control Number: 2018965443

© Springer Nature Switzerland AG 2019
This work is subject to copyright. All rights are reserved by the Publisher, whether the whole or part of the material is concerned, specifically the rights of translation, reprinting, reuse of illustrations, recitation, broadcasting, reproduction on microfilms or in any other physical way, and transmission or information storage and retrieval, electronic adaptation, computer software, or by similar or dissimilar methodology now known or hereafter developed.
The use of general descriptive names, registered names, trademarks, service marks, etc. in this publication does not imply, even in the absence of a specific statement, that such names are exempt from the relevant protective laws and regulations and therefore free for general use.
The publisher, the authors and the editors are safe to assume that the advice and information in this book are believed to be true and accurate at the date of publication. Neither the publisher nor the authors or the editors give a warranty, express or implied, with respect to the material contained herein or for any errors or omissions that may have been made. The publisher remains neutral with regard to jurisdictional claims in published maps and institutional affiliations.

This Springer imprint is published by the registered company Springer Nature Switzerland AG
The registered company address is: Gewerbestrasse 11, 6330 Cham, Switzerland

Preface

This book is created for students, planetary scientists, and cartographers who are involved in one way or another in making maps of planetary bodies. Planetary mapping involves expensive and complex infrastructure, from building and remotely operating cameras to spacecraft platforms that carry them, and to the Deep Space Network that receives their signals. In a complex process, image and different raw data are transformed into calibrated data that can be read by scientific software including GIS and remote-sensing applications. The actual production of a particular planetary map, for example, a geologic map, may take months to years of intensive work. This book is written for those whose academic or commercial projects involve working with planetary spatial data. These include university students studying Earth and Planetary Sciences or Cartography; early career scientists who study the surfaces of planets and moons; professional cartographers and GIS experts who wish to add an extraterrestrial perspective to their cartographic portfolio; and graphic artists who wish to visualize other worlds.

However, this is not a conventional textbook. Each of the chapters is written by an expert in that particular field, giving a review of both conceptual background and practical details on how to create good and usable planetary maps.

We begin with the basics: the role of planetary cartography, its history and development, the planetary toponyms, and the fundamental geophysical frameworks that provide the backbone of the planetary spatial infrastructure. We continue with thematic mapping chapters, giving details on geologic and topographic mapping, landing site selection, and a planetary specialty: irregular bodies. The next chapters deal with different cartographic approaches: mapping by teams, map design, Atlas making, and maps for human and automatic surface operations. Various mapping methods are presented afterwards: feature databases, grid mapping, analysis of distribution and topographic roughness, maps that visualize extraterrestrial 3D circulation models. The next chapter looks into the future, the final frontier, preparing for the mapping of exoplanets. The book ends with a roster of many currently active facilities working in the field of planetary cartography internationally.

I am very grateful for the continued support of Trent Hare and Andrea Naß throughout the editorial works of this project that started in 2015 at NASA Ames in Moffett Field. Authors contributed to this book internationally, from Canada, China, the Czech Republic, Germany, Hungary, Italy, Poland, Russia, Sweden, the UK, and the USA, reflecting the increasingly international activities in planetary mapping where many of the authors work together with the future generation of planetary mappers.

This book is dedicated to the memory of Kira Shingareva who was my mentor when I created my first map of Mars.

Moffett Field, USA/Budapest, Hungary Henrik Hargitai

Contents

Part I
The Basics

Cartography: Its Role and Interdisciplinary Character in Planetary Science

Andrea Naß, Henrik Hargitai and Manfred Buchroithner

Abstract Cartography is the science, technique, and art of filtering and compiling spatial data into map information and to communicate complex spatial relationships and interdependences by advanced visualization techniques. In this context, Cartography provides the whole environment and necessary analysis toolsets to derive mapping results and produce maps. This chapter gives a description about what planetary cartography is about and how planetary cartographic products are produced, the process from data, via information, to knowledge and understanding, and also discussing the recent technical and conceptual transition from hardcopy end products to interactive, dynamic digital databases.

Keywords Cartography · Planetary cartography · Maps · GIS · Standards · Sustainability

1 Introduction

Maps are one of the most important tools for communicating geospatial information between producers and receivers. Geospatial data, tools, contributions in geospatial sciences, and the communication of information and transmission of knowledge are a matter of ongoing cartographic research. Thus, it covers a wide area of different information and knowledge levels, and ranges from a technical and engineering focus to a scientific abstraction. This applies to all topics and objects located on Earth or on any other body in our Solar System. In planetary science, cartography

A. Naß (✉)
German Aerospace Center (DLR), Rutherfordstr. 2, Berlin 12489, Germany
e-mail: andrea.nass@dlr.de

H. Hargitai
Eötvös Loránd University, Múzeum krt 6-8, 1088 Budapest, Hungary
e-mail: hargitaih@caesar.elte.hu

M. Buchroithner
Technical University Dresden, Helmholtzstr. 10, Dresden 01069, Germany

© Springer Nature Switzerland AG 2019

H. Hargitai (ed.), *Planetary Cartography and GIS*, Lecture Notes in Geoinformation and Cartography, https://doi.org/10.1007/978-3-319-62849-3_1

and mapping have a history dating back to the roots of telescopic space exploration and are now facing new technological and organizational challenges with the rise of new missions, new global initiatives, organizations, and opening research markets.

The aim of the present chapter is to provide an informative overview of what is planetary cartography about. This section gives a short review about the general definition of cartography, and the distillation process of gaining abstracted information in general, and in cartography in particular. Section 2 focuses on cartographic issues in planetary science in particular. How the terms *cartography* and *mapping* are currently used will conclude the chapter. Section 3 deals with the cartographic workflow in planetary sciences and, thus, the chronological process from *data*, via *information*, to the desired *knowledge* and, ultimately, *understanding*. Section 4 provides an overview of planetary map *products*. Section 5 sums up with *tasks and challenges* planetary cartography has to face today and in the near future.

1.1 Cartography—What is it About?

As stated by the International Cartographic Association (ICA), "a map is a symbolised image of geographical reality, representing selected features or characteristics, resulting from the creative effort of its author's execution of choices, and is designed for use when spatial relationships are of primary relevance" (http://icaci.org/strategic-plan). This definition describes a map entirely and with its all facets. Kraak and Fabrikant (2017) currently presented a more general definition and described a map as a "visual representation of an environment" (Kraak and Fabrikant 2017, p 6). In comparison to previous definitions, this updated version explains a map (1) less specified to include also temporal and dynamic nature, and (2) broad enough to cover also new developments in geographic information sciences and technology. A historical overview of map definitions is given by Andrews (1996).

Turning to the discipline where a map is directly dedicated to the field of cartography is *"the art, science, and technology of making and using maps"* (Fig. 1), first defined by de Brommer in 1959 (Ormeling 1987, p 20). This definition persisted through the last decades (Meynen 1973), although not without debates (Krygier 1995), and is also mentioned in the strategic plan 2003–2011 and 2011–2019 of International Cartographic Association (ICA) (http://icaci.org/strategic-plan).

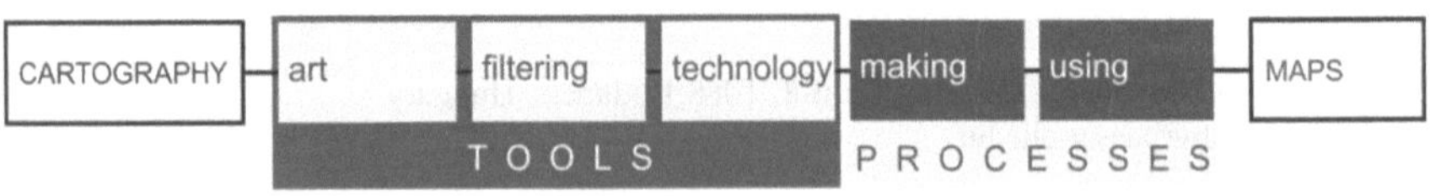

Fig. 1 Definition and relations of the fields between and within cartography and maps (extracted and slightly modified from http://icaci.org/strategic-plan)

More specifically, one can say that cartography is the art of filtering and compiling spatial data into map information and communicating complex spatial relationships and interdependences by advanced visualization techniques. Considering the components in more detail, it can easily be seen that the meanings of the different terms are very variable and depend on current technical developments and scientific cognition in this fields.

1.2 *Distillation, Abstraction, and Visualization of Information*

Visualization of data in general, and visualization of research data in particular, represents a simplified view on the real world, covering complex situations as well as the relationship between them (Ware 2004; Mazza 2009). The process to accomplish this can generally be divided into four parts: (1) data preprocessing and transformation, (2) visual mapping, (3) generation of views, and (4) perception/cognition by the user. This workflow is, e.g., described by Haber and McNabb (1990) and Carpendale (2003) and they called it *visualization pipeline* (Fig. 2).

This process is independent of actual production techniques, methods, and from the field where the visualization will be realized. If this pipeline will be transferred to the cartographic visualization process, the steps of the workflow are describable as *acquisition and filtering of raw data* (input), *abstraction and generalization of information* (distillation), and *rendering, i.e., visualization of results* (output) (Fig. 2).

This workflow clearly shows parallels to the *data–information–knowledge–wisdom* hierarchy (e.g., Ackoff 1989), retrieving its bases from the field of information sciences and knowledge management. More recent discussions about this are shown, e.g., by Rowley (2007).

Geologic and geomorphologic mapping of physical surfaces, whether located on Earth or on any other planetary body, represent a special way of spatial data visualization within its scientific context. It does not show an objective visualization of a situation, but rather a subjective distinction of individual surface structures, units, and objects (see also Haber and McNabb 1990; Hargitai et al. 2015a). This interpretation will generally be conducted by experts in their topical field whose expertise allows isolating, differentiating, and describing individual planetary

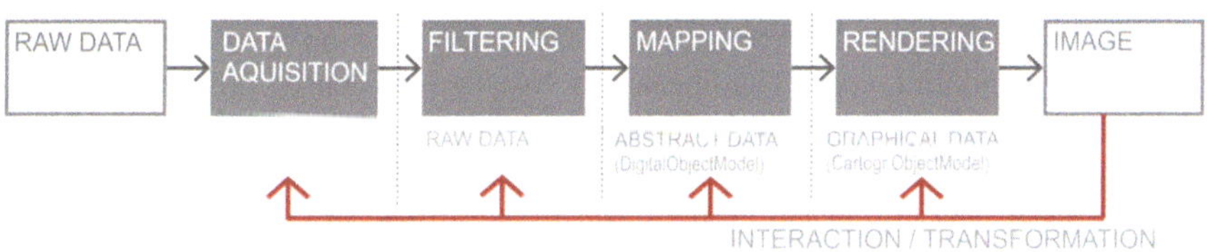

Fig. 2 Visualization pipeline of data in general and is transferable to spatial data (modified by Haber and McNabb 1990; Carpendale, 2003)

objects based on remote sensing data to reconstruct processes which shaped the current character of a surface. Any geologist or geomorphologist, utilizing the same standardized procedure of mapping, would produce similar, although not identical, output. The mappers are credited in such maps, which is typically not the case for elevation maps or image mosaic products that are less sensitive to a few individual's decisions.

2 Intersection of Planetary Cartography and Mapping

2.1 From Past to Present

The history of planetary cartography as extraterrestrial mapping dates back to shortly after the invention of the telescope at the beginning of the seventeenth century which marked a milestone in planetary exploration. The most comprehensive review on all aspects of planetary cartography was published by Snyder (1982, 1987), and Greeley and Batson (1990). For detailed summaries on the development and evolution of planetary cartography, the reader is referred to Shevchenko et al. (2016) for the history of Soviet and Russian planetary cartography, and to Jin (2014) for Chinese Lunar mapping results. The history of planetary mapping is discussed in, e.g., Kopal and Carder (1974), and Morton (2002) and recent planetary cartographic techniques and tools are reviewed in Beyer (2015) and Hare et al. (2017a). A detailed historical overview of planetary cartography is given in Hargitai and Naß (2019).

2.2 Cartography and Mapping Today—and Their Different Meaning

When the scientific community talks about planetary cartography, they frequently focus on topics like generating reference systems by geodetic models, data calibration, and the transfer of cartographic projections. However, conducting cartographic products in terms of maps also all other components like scale, legend, grid, color ramp, and labeling have to handle.

The primary goal of planetary mapping is to create and provide scientifically sound products after the successful termination of a planetary mission by *distilling data into (image and thematic) maps*. That means, to visualize condensed information out of specific base data at different scale. This provides the basis to support planning (e.g., landing site selection, observation from orbit, and traverse planning) and to facilitate mission operations during the lifetime of a mission (e.g., observation tracking and hazard avoidance; iterative in-mission geologic mapping, Williams 2016a, 2016b). After a mission's lifetime, information is stored in data archives—and eventually compiled into maps and higher-level data products—to form a basis for research and new scientific and engineering studies. The complexity of such tasks increases with every new dataset that is put onto this stack of information, and in the same way, as the complexity of autonomous probes increases, also tools that support these challenges require new levels of sophistication.

The major components of the planetary cartographic portfolio are topographic and remote sensing image maps and data-synthesizing geologic maps that latter are probably one of the highest-order spatial data products: A geologic map is "a visual representation of the distribution and sequence of rock types and other geologic information" (Yingst et al. 2014) that synthesize morphologic, compositional, brightness (albedo), and geophysical information. The end product of what is called "geologic mapping" is not just a geologic map, though. For extraterrestrial planetary bodies, "geologic mapping is an investigative process that seeks to understand the evolution of planetary surfaces" (Williams et al. 2014). A detailed description of geological mapping in planetary science is given by Hauber et al. (2019).

2.3 Transition from Analog to Digital Mapping

The emergence of electronic platforms changed the meaning of what a map fundamentally is. Today, we see a clear transition from aesthetically composed map sheets to georeferenced data files. The latter ones can be directly used for quantitative analysis and be combined (Hargitai 2016) with other maps (as "layers") in various GIS, WebGIS, or *WebMapping Services* (WMS) platforms. These include spatially referenced image mosaics in complete, downloadable single-band image files, multiband image cubes, and online Web *Map Tile Services* (Hare et al. 2017b). Map layers within Web-mapping services range from global to local scales and include thematic raster or vector layers. It is debatable, though, whether such data layers can be regarded as *complete* "maps," or just intermediate spatial data products. However, if these products are separated into thematic layers in GIS-compatible format, include metadata, permanent digital identification (e.g., DOI), and "perpetual" access in online repositories, similar to paper maps stored in archives and libraries, these products are useful for further analysis. In a GIS-based research environment, a printed map and even a digital simple image file such as a *.jpeg, or a one-layer *.pdf, is a static product, where data is inseparably merged into one final, unchangeable view. Complex static maps with marginalia are long-term visual and aesthetically valuable cartographic documentation of previous mapping efforts. Hardcopies provide a complex "analog" overview of the mapped area that cannot be experienced on a screen, but these cannot be used directly for further, computer-based analysis, because that requires integratable thematic data layers. While the well-composed completeness was considered the attribute of the highest level product, today a final (un-importable) map image product has limited use. Interactive digital environments should facilitate the placement of different types of spatial data into the *map platform* as the user wishes, but, as much as possible, confined to aesthetically pleasing and scientifically accurate possibilities, where the ad hoc on-screen map image, a combination of layers linked by georeferencing, can be comparable any well-crafted paper map. This transition from static, complete, hardcopy "map-images" to layers of digital spatial data, taking place in the 2010s, is comparable to how word processors changed typography in the 1990s. Well-designed printed maps and atlases might remain relevant in education and public outreach (EPO), but electronic maps and virtual globes are replacing paper maps in EPO, too.

3 Cartographic Work Process

The mapping process in planetary cartography is comparable to established processes commonly employed in Earth-bound cartographic workflows. The GIS-based workflow has the benefit that it delivers a digital object model (ger. Digitales Objekt Model (DOM)) (Hake et al. 2002), which groups the objects by classes. Thus, additional attributes can refer to the objects and objects layers, and that they permit the user to perform analysis based on spatial relations but also on thematic attributes.

The individual tasks within the planetary mapping process refer to the single steps within the *visualization pipeline* (Fig. 2). Through the missing ground truth (except very local investigations in predominantly lunar and Mars exploration) scientific interpretation, mapping is mainly based on remote sensing data. The following paragraphs describe briefly the different levels (Fig. 3).

3.1 Input—Data

Independent of the investigation target and mapping purpose, the first step of every mapping process is to establish an individual basis of fundamental, i.e., primary

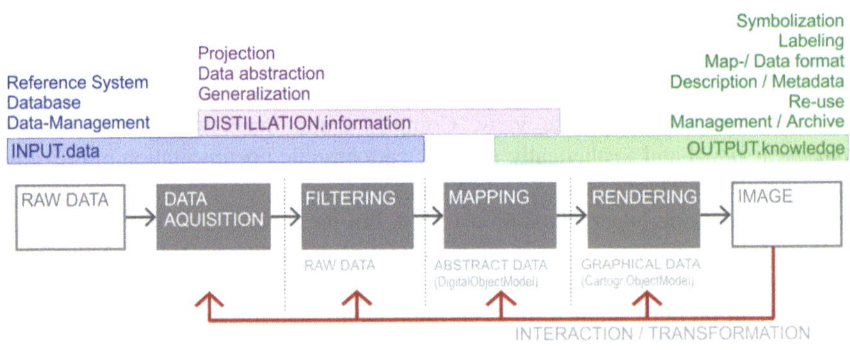

Fig. 3 Visualization pipeline linked to both the process during planetary mapping and the *data–information–knowledge–wisdom* hierarchy (cf. Fig. 2)

data. In planetary science, the community has access (from *open* to *restrict*) to different digital data archives and connected portals which serve mostly raster, but also a selection of vector-oriented data. The variety of primary data reaches from images (visible, infrared, multispectral, and radar wavelength), terrain data (such as laser altimeter) to other derived data, e.g., digital terrain models (DTMs).

The best-known planetary data archives are: Planetary Data System—PDS/NASA, Planetary Science Archive—PSA/ESA, United States Geological Survey, Astrogeology—USGS, Lunar and Planetary Institute—LPI, and Data Archives and Transmission System—DARTS/JAXA. In addition to these digital data archives and portals, there are still a large number of analog map sheets available, which contain already derived information, and can valuably completed the mapping process.

Data selection for the individual mapping process is primarily based on the availability of base data sets covering the investigation area, but at the same time highly influenced by the individual project and mapping goal. That means, while the first data selection the available datasets from different missions and measurement instruments will be filtered (by spatial request), in the next step a qualitative selection will follow. During this second step the usable data have to be filtered which depends on 1. is the dataset technically correct, and 2. does the dataset really covers the necessary area, and show the expected information for the thematic mapping purpose. As soon as these two-level selection for input data could be done the mapper has the valuable basis for the multi-layered mapping process.

3.2 *Distillation—Information*

After the acquisition and compilation of the primary data, they have to be preprocessed in order to make them useable for the further processing, e.g., in a GIS environment. The data itself generally contain all spatial information required for georeferencing (parameters of the reference body and of the image location itself) within their file header. How these parameters are to be defined in a uniform way is determined by the *IAU Working Group on Cartographic Coordinates and Rotational Elements* (Archinal et al. 2011). The preparation starts with several processing steps such as radiometric, geometric, and photometric control. Nowadays, most of the data portals support the community with both raw and corrected datasets. Thus, the users generally do not have to handle these complex processes and can easily download the corrected data. An exemplary overview for digital planetary data processing and mapping is discussed in Frigeri et al. (2011), and Hare et al. (2017a, b).

The structure of project data, i.e., the organization of spatial data in *separate* and *related* data and mapping layers, depends on the purpose of mapping. Furthermore, the mapping *process* also depends on map *content*, and output *format*, i.e., how the data should be accessible and modifiable after the mapping is finished. As already known from terrestrial workflows (e.g., Gustavsson et al. 2006, 2007), there are two classical methods on how GIS-compatible mapping data can be structured: The data are organized (1) in different map layers, and within them, all datasets are stored based on their geometry type (point, line, or polygon). (2) Within an individual and predefined object model, this includes classes, hierarchies, and relations. The data structure is not dominated by the geometry type of the data; rather the individual classification of the object attributes and properties is determinative. Object models like this can be used and modified for many purposes. There exist also standardized models that describe the usable object layers and their relations (e.g., INSPIRE 2012). Both methods are applicable for vector and raster, but also for hybrid datasets. The second method represents the properties of the spatial objects and the whole planetary surface in the most realistic way (Gustavsson et al. 2006). Within such a logical data structure, queries and spatial analysis are also possible. One first implementation for the integration of planetary object data within a GIS-based data model was realized by van Gasselt and Naß (2011).

Based on these preliminary considerations and preparations, i.e., having corrected and georeferenced primary data and the definition of a data structure, the GIS-based mapping process begins. The users can choose between desktop systems (such as ArcGIS™[1]—proprietary, *Quantum GIS,*[2] *Saga GIS*[3] or *GrassGIS*[4]—open source), or Web-based mapping environments (*geoportal,*[5] *map-a-planet,*[6] *JMARS,*[7] and *openplanetary*[8]). For more detailed descriptions of the application of GIS in planetary science, see Hare et al. (2017a).

All software systems are based on the structures of spatial databases, either on the most popular open-source database *PostgreSQL*[9] with the extension *Postgis* or on proprietary solutions with individual environments. Further databases handling the spatial context are, e.g., *Oracle*[10] *(Oracle spatial)*, *MySQL*[11] *(MySQL spatial)*, and *neo4j (neo4j spatial).*[12]

The mapping process itself, i.e., the spatial and substantial distinction of particular spatial objects, represents the first part of the interpretative mapping process and is dominated by multi-level comparisons of the different primary data. This process is described in detail in Hauber et al. (2019), and will nowadays, result in a digital object model (DOM) that is not specifically visualized yet.

---

[1]www.arcgis.com.

[2]qgis.org.

[3]www.saga-gis.org.

[4]grass.osgeo.org.

[5]cartsrv.mexlab.ru/geoportal/.

[6]astrogeology.usgs.gov/tools/map-a-planet-2.

[7]jmars.asu.edu.

[8]openplanetary.co.

[9]www.postgresql.org.

[10]www.oracle.com.

[11]www.mysql.com.

[12]neo4j.com.

3.3 Output—Knowledge

The second part of the surface feature interpretation process is followed by the contextual and generic classification of the individual objects. These attributes are stored within the attribute tables of the DOM and could subsequently be used for further analysis and calculations. Features of planetary surfaces may be classified based on their shape (morphology), origin (geology), or evolution that, combined, result in particular landscapes. An example to shape-based classification is the system of descriptor terms used in the planetary nomenclature,[13] while planetary landforms may also be classified based on their formative processes inferred from modeling or potential terrestrial analogs if there is any (Hargitai and Kereszturi 2015; Hargitai et al. 2015a).

Some maps are less sensitive to interpretation and conceptual approach. Their production requires mainly a *technical workflow*. The process producing a DEM dataset is based on clear goals and tools, where visualization of the data may not even be included. Producing an outreach map or geological map, however, can be done in many ways and requires a *conceptual workflow*. Standard procedures and symbology developed over decades simplify the actual mapmaking by limiting the possible cartographic decisions, but it also limits the cartographer's creativity. Thematic geological maps in journal articles may illustrate specific aspects of a given research. Outreach maps can include more creative solutions. For both scientific and outreach maps, the mapper does not only choose the topic but also the *focus of the map*, the problems or characteristics to be communicated. Therefore, the decisions on which and how cartographic elements should be included, emphasized, or excluded depend on the purpose of the map, or the map's implicit story. These conceptual decisions will define the cartographic tools used: scale, size, style, technical generalization, symbols, colors, tones, labels, annotations, etc.

For the cartographic visualization, observation-/ and interpretation-derived attributes are required to distinguish and merge different object classes with the most representative cartographic symbolization. In this context, the most representative means are the commonly used symbols for different objects and object classes, and it may be necessary to generate an individual catalog, which links the objects or object types to individual symbols. Data visualization can be compared to spoken languages, and in both cases, the involved individuals have to have the same lowest common basis. The establishment of these symbols should follow similar rules like vocabulary, spelling, and grammar. In the world of data visualization in general, and in cartography in particular, these rules are transferable to the type and characteristics of the cartographic symbols (i.e., for colors, points, lines, polygons, and pattern). "These rules have been defined to ensure that the map content is correctly communicated between mapper/cartographer and map reader. A classic example of such rules is the displacement guidelines for cartographic generalization. In order to (a) implement such rules and guidelines in a user-friendly

[13]planetarynames.wr.usgs.gov.

way, (b) for a (technically) correct visualization, and (c) to ensure comparability between various map products (analog and digital domains), the most commonly used spatial objects are predefined by individual symbols" (Naß et al. 2011, p. 2).

In order to find a single graphical language for geological mapping (for Earth and other planetary surfaces), a standard should be used, which describes a large number of potential map objects specifically in earth and planetary sciences (see FGDC 2006). Naß et al. (2011) discuss how this symbol standard can be directly implemented into GIS and linked to an underlying data model. The final task within the mapping process is the textual and attributive description of the interpretation and analysis results within the GIS-based mapping data.

Besides the graphical representation of the spatial objects, one major part in cartography is the *nomenclature* that is shown on the map as labels. All information about the planetary nomenclature, on how names for individual objects are adopted, which rules the community has to follow during the naming process, and which features are already named (the gazetteer) is contained in the IAU[14] webpage, and also discussed in Strobell and Masursky (1990), Hargitai (2006), and Hunter et al. (2019). Bruno and Ruban (2017) discuss the problems and challenges in naming planetary surface objects.

The last step of the cartographic visualization process is the presentation of the results of analyses and interpretation on a digital or analog map sheet. This step contains the definition of the layout and style of the map sheet, its marginalia elements (including scale, grid, and legend), and the (optional or obligatory) information about the mapping content by defining the *metadata* description. This last point is necessary for the digital archiving of the map contents within a map sheet document or as GIS-compatible project data. For metadata description, there exist already several standard recommendations. The most prominent ones are the *Geographic information—Metadata* (ISO19115, ISO 2014a), the *Geographic information—Metadata—XML schema implementation* (ISO19139, ISO 2014b), and the *Content Standard for Digital Geospatial Metadata* (CSDGM) (FGDC 1998, 2000). Beside this, nowadays the *eXtensible Markup Language* (XML) becomes very popular for a system-independent description of (meta)data. This format is implemented in many GIS environments and allows a hierarchical text-based encoding system which is ideal for structured data queries.

A general overview about metadata in planetary science is given by Hare (2011) and Hare et al. (2011). Naß et al. (2010) discuss, how metadata can be defined for planetary mapping results, thus implying how the individual character of the geological objects and the overall mapping project is described in the most efficient way.

[14]www.iau.org.

4 Types of Map Products

In this and the following sections, we present two classifications of planetary map products. These terms are used in ICA Commission on Planetary Cartography (CPC) documents for classifying historic and contemporary planetary map products (Hargitai and Pitura 2018) (Fig. 4).

4.1 Map Type Based on Motivation

Based on the specific motivation of their generation, we distinguish four types of mapping projects: investigative *science-driven (1)* maps, *reconnaissance* (2) maps, exploratory *outreach (3)* maps, and *citizen* (4) maps.

The *first* type of planetary map is created through a *scientific hypothesis-driven investigation*, which aims at answering a specific science question related to the mapping region and where the mappers typically include planetary scientists.

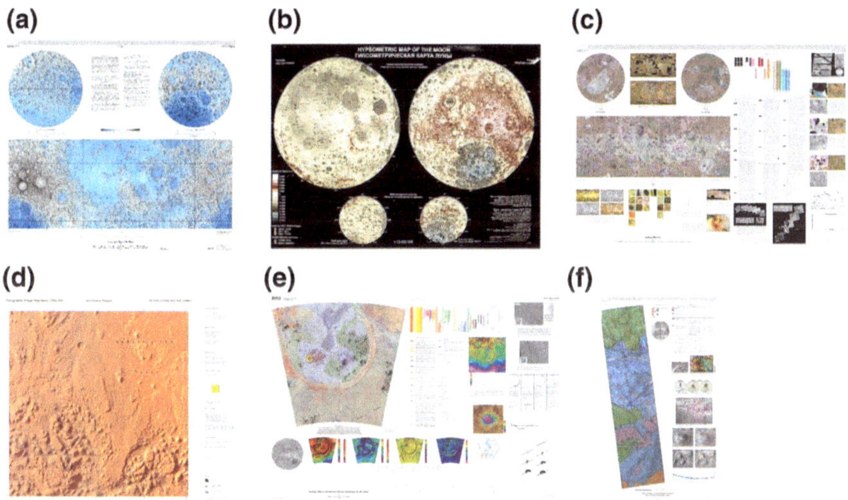

Fig. 4 A selection of recently published printed scientific and outreach planetary map layouts. **a**. *Outreach* type topographic map of the Moon (USGS, 1:10M, LRO WAC data, Polar Stereographicand Mercator projections, Hare et al. 2015 (b); **b**. *Hypsometric* map of the Moon (SAI/MSU/MIIGAiK, 1:13M, LOLA data, Lambert Azimuthal Equal Area projection and Polar Stereographic projection, Grishakina et al. 2014); **c**. Global *geologic* map of Io (USGS, 1:15M, Voyager–Galileo data, Polar Stereographic and Mercator projections, Williams et al. 2011). **d**. *Topographic image* map of Iani Chaos, Mars (© Image Data: ESA/DLR/FU Berlin (G. Neukum) © Map Compilation: Technische Universität Berlin, 2006), 1:200k, HRSC data, Sinusoidal projection); **e**. Regional *geologic* map of the Artemis Chasma quadrangle, Venus (USGS, 1:5M, Bannister and Hansen 2010); **f**. Local (small-scale) *bedrock geologic* and *structural* map through the western Candor Colles region of Mars (USGS, 1:18k, based on HiRISE data, Transverse Mercator Projection, Okubo 2014)

In these projects, mapping is a necessary step in the workflow, and the end product is one or a series of research papers, even if eventually a stand-alone map is also published. This map may be a geological map, or maps of specific features, composition, etc. Geologic maps, especially the *USGS Scientific Investigations Map* (SIM) series that systematically maps planetary bodies, have a special place in planetary cartography, because these are treated with the full range of traditional cartographic procedures, from uniform design and editing to a rigorous review. They are generated over a several-year-long period and published both in full-resolution print and online GIS database versions (Tanaka et al. 2011; for historical production procedure, see Wilhelms 1972). Probably, the highest number of planetary maps, however, is published as figures in research papers and scientific reports in conferences. Some journals allow the publication of these figures as supplementary GIS-ready files. Specialized journals publish large-format static maps that provide a new platform from and for international planetary mappers. Global topographic mapping efforts are frequently *directly* driven by the need to define a reference surface for a body to control any future spatial products, but are also fundamental parts of any geological investigations. *Local* topographic mapping is often done in ad hoc (science-driven) DEM production.

General or feature-specific geologic and geomorphic mapping processes can also generate spatial databases that may be visualized on a map, but originally exist as data tables (e.g., crater or other feature catalogs).

A *second* type of planetary map enables or facilitates future scientific investigations (NASA 2017), such as the historic topographic shaded relief *mission planning maps*. Although these maps also answer the science question "what is the terrain like within the mapping area?" this type of mapping does not result immediately in research papers and is therefore less popular for proposal-based research projects. Today, the scientifically *justified need* for mapping (Williams 2016a, 2016b) or naming a feature (Blagg and Muller 1935) is emphasized; however, the role of *reconnaissance* mapping is also getting recognition (Baker 2017 and Skinner 2015). In addition, renovation (i.e., the generation of GIS maps from previously published geologic feature catalogs, hardcopy maps in various scales and formats; or merging quadrangles) makes it possible to effectively use the maps published before the GIS era (Hargitai 2016).

The *third* type of planetary map is usually called an *outreach* map. For terrestrial geography, educational maps that are used in formal, classroom education also belong to this category, but since astrogeology is not part of formal education curricula, there is no commercial demand for the production of educational planetary maps. However, outreach maps that target the interested public are included in popular science publications, applications, exhibitions, etc. Outreach mapmaking, as a mapping process, does not involve traditional scientific investigation. Instead, it focuses on the creative and easily accessible visualization and synthesis of already existing data. While outreach maps have no research question, they do have a focus, and tell a specific story determined by the cartographer-editor. Cartographic aesthetic aspects are also important elements of outreach products. Outreach maps may also serve prestige purposes for both, producers and users. It may use the same tools as those utilized in any investigative mapping project. When speaking of outreach cartography, it is important to note some differences from terrestrial maps. (1) Planetary maps are typically not used for navigation, which is the main purpose of map use on Earth, (2) planetary maps only display geologic/geomorphologic features (no administrative units, roads, vegetation, etc.), (3) the most common features on planetary maps are similar-shaped craters, and (4) planetary maps show only land surface and no hydrological features (except Titan, and Mars paleomaps). Consequently, there are much fewer visual cartographic "landmarks" for finding a position and create a sense of the spatial scales (Gede and Hargitai 2015, 2017). The general absence of linear map objects on most planetary topographic and image maps is especially important, because prominent lines are the backbones of a map and they draw immediate attention of map readers (Ooms et al. 2014, Albert et al. 2017).

Outreach maps may be produced by professional planetary scientists (Hare et al. 2015), amateur astronomers, educators, artist or designers and cartographic companies (Hargitai and Pitura 2018). Perhaps the most popular hardcopy outreach products are the amateur visual observer's maps of the Moon (e.g., Rükl 2012, originally published in 1976) that are specifically designed to locate or identify features through a telescope and were perhaps the most popular in 1969 during the first Apollo landing. Popular outreach products further include Web cartographic tools such as the freely accessible Google Mars and Moon (Hancher et al. 2009) or the wall maps published by the National Geography Society, produced by "Earth" cartographers (e.g., National Geographic Society 1969) famous for its font faces and characteristic arrows.

We note that neither popular astronomy books nor world atlases include planetary maps, with very few exceptions (such as the Swiss World Atlas, ETH 2017)—perhaps simply because their publishers are not aware of the planetary cartographic resources. In these platforms, "exploratory" planetary maps could serve the same purpose as seafloor or Antarctic maps where the mapped regions are also typically inaccessible for the reader. Special types of exploratory outreach maps include multilingual maps (e.g., Buchroithner 1999; Buchroithner et al. 1999; Shingareva et al. 2001, 2002, 2003, 2007; Shingareva et al. 2007), topographic as well as 3D maps representing the Martian surface (Lehmann et al. 1997; Dorrer and Zhou 1998; Buchroithner and Wälder 2003; Albertz et al. 2005; Gehrke et al. 2005), and planetary maps designed for children (Hargitai et al. 2015b).

Outreach map design is especially sensitive to map-reading issues (Albert et al. 2017). While shaded relief maps directly communicate surface relief, and landscape information, the geologic information is hidden on these displays. Furthermore, reading the abstract symbology of geologic maps requires training—both for understanding the symbols and the concepts behind the symbols. Outreach maps should find a balance between direct or iconic visual data and abstract symbols that are familiar from school atlases or everyday online navigation tools.

Finally, a *fourth type* of planetary map includes enthusiast (*citizen*) maps. In outreach maps it is the scientific or cartographic community that reaches out to the general public, whereas in this type, citizens produce maps, "peer to peer," visualizing scientific datasets typically with graphic design software and sophisticated but not scientific image processing methods. These maps represent a new trend arising at the end of the 2010s. While these maps can be useful to visualize planetary surfaces, the scientific use of citizen maps is limited and typically cannot be used for measurements or analysis.

At the intersection of enthusiast, maps and science-driven maps lie the crowd-sourced (citizen scientist) maps, which are produced by combining microtasks of several (up to thousands of) individual non-scientists in a science-ready online platform developed by researchers for answering science questions.

4.2 Map Type Based on Production and Content

Based on the *production process* and *map content,* we distinguish the following map types. This distinction is to make it clear what we can call a map today, in a period when visual spatial information is ubiquitous.

Maps offer more than a "realistic" representation of the surface. Maps are more than a photograph because they are based on analysis and/or interpretation (see Sect. 3.2); maps have to simplify (generalize) or emphasize surface characteristics. At the same time, maps are less than a photograph because they omit the details from the "infinite" complexity of the real world in order to make it "legible" for the human map reader.

Maps are a composite product of different human activities, on the intersection of science, technology, and arts. Each component is based on a variety of human decisions that may have no *objective* "right" or "wrong" choices. Its components are similar to a book that has content (the text for books—spatial data for maps), visualization tools (the typefaces—symbology), organization (layout—projection), the data-bearing object for hard copies and the interface for digital ones. The harmonic (subjective) interplay of these components together results in the "map experience."

Drawings of planetary bodies and planetary *images (photographic, radar, or other)* record and document the appearance of a planetary body at a given moment.

1. *Image maps* are georeferenced raster data. They are characterized by geometric resolution, given in reference size/data pixel (meter/pixel) or data pixel number/ 1° reference surface units. Image maps are frequently used as basemaps for higher-degree cartographic products.

   1.1. A *photomosaic* is a georeferenced mosaic of images without maginalia and visual cartographic elements. Photomosaics typically show surface reflectance as measured by the camera or sensor (photometric value, radar echo). Source images are typically processed to optimally show details and data gaps may be filled.

   1.2. A *photomap* is a map-projected photomosaic with marginalia, grid, and other cartographic elements.

   1.3. *Data maps* are georeferenced data that form an image. Data maps show quantitative, calculated data (albedo, thermal inertia, emission, elevation, mineralogy from spectra, etc.). Data maps may contain marginalia.

2. We introduce the term *metamap* for the data behind the maps. These data are also representations of the surface in that they contain spatial information but without defining visual attributes such as line width, symbol type, color, and map projection. Such datasets should contain the coordinates and one or several feature attributes (metadata) for each data point or data object. Examples include feature catalogs and DEMs. Their formats range from spreadsheets to shapefiles. Metamaps may contain marginalia as metadata.
3. *Cartographic maps* are usually produced using image or data maps and are higher-order cartographic products. Additional to standard map elements such as nomenclature, grid and marginalia (including legend), cartographic maps also contain, e.g., linework, graphic work (e.g., human-produced airbrush, and drawing), symbols and are made using various cartographic techniques such as generalization. Cartographic maps may be composed of multiple data maps.

Cartographic maps may be digital static maps (pdf/jpg), online/digital dynamic maps (GIS, WMS), or printed products. They are typically characterized by their scale (ratio of hardcopy or screen display size to reference size, 1:x). Examples include geologic maps, geomorphologic maps, thematic maps, traverse maps, feature maps, outreach maps, shaded relief maps (e.g., airbrush), contour maps, and sketch maps.

The quality and accuracy of cartographic maps can only be ensured when map reviewers and cartographers (professional, specialized-content, and graphics editors) are involved in the process. It must be noted that both at USGS and National Geographic—the two most important publishers for scientific and outreach planetary maps, respectively—cartographic planetary map production are part of regular "Earth" mapmaking and involves cartographers. Planetary geologic results are often visualized through maps in research papers. These thematic maps are produced almost exclusively by planetary scientists, and are typically not reviewed by cartographers before publication. There is a serious need for closer cooperation between the planetary and cartographic communities and the inclusion of cartography in the curricula where planetary scientists are trained. Vice versa, GIS specialists and cartographers are included very rarely in planetary research teams. Instead, planetary scientists become self-educated cartographers.

Image and data maps are mostly objective scientific products where the workflow is standardized, while cartographic maps are produced using a substantial number of aesthetic choices and artistic elements, such as choices of color(s), line style, typeface, grid density, symbol type (even if standardized), the positioning of labels, line and point elements. The choice of layers is unique. The surface representation methods, visual style and layout may be defined by the cartographer, the user, informal or formal standards, or a combination of thereof. We note that in cartographic map series, individual maps are produced using a standardized pipeline, visual design, layout and symbology developed for that particular series to ensure that the data on map sheets are scientifically comparable and visually matched at the borders. However, in series spanning over long time periods, mapping groups may use different methods, symbols or color schemes, depending of map focus and the mapped region's characteristics. The basic cartographic and mapping rules, described in e.g., Tanaka et al. (2011), should be uniform. Scientific and visual quality and comparability is ensured at USGS by the applied standardized production pipeline across all products, which is not the case for ad hoc research papers.

Besides these three basic map types, we describe some additional terms used in planetary cartography:

The term *albedo map* may be used specifically (Vincendon et al. 2011) or generally for a surface image map where shadows are not apparent and represents surface brightness (or reflectance).

On Earth, *topographic maps* are large-scale reference maps showing contour lines, natural and artificial surface features and detailed nomenclature. However, in planetary usage, the term "topographic map" is used for elevation (altitude) maps at any scale. Topographic planetary maps are also called hypsometric maps in Russia. For Magellan Venus mapping, "topographic maps" show shaded relief with contour lines and color-coded topography, while "altimetric radar image maps" are constructed from a color-coded DTM overlaying a radar image mosaic. The term "topographic image map" (Alberz et al. 2004) or "topophotomap" *(Schimerman* 1975*)* refers to large-scale (approx. 1:10k–1:400k) photomaps also showing contour lines, which is the closest analog to terrestrial topographic maps.

Talking about *geologic maps*, the term "geologic" is an umbrella term in planetary cartography. Recently, the term "geoscience map" is also used for maps that delineate and describe units of terrains (Skinner et al. 2018). The units on planetary geologic maps are determined differently from how terrestrial rock units are identified. The type of rock is usually not identified; instead, formative process is a key identifier. Ages of units are inferred from crater counting. Planetary geologic maps are primarily based on the remote sensing analysis of surface geomorphology (relief, texture), stratigraphic relations, and where available, supporting data may also include information derived from spectral data (albedo, mineralogy, thermal inertia) or radar properties. Depending on the focus of the particular mapping project, these maps may be named differently. The basic themes of most planetary geologic maps could be described as "morpho-stratigraphic", where mappers try to reconstruct the geologic evolution from the stratigraphic relations of morphology-based units (Galluzzi 2019). Several Russian/Soviet publications (MIIGAiK) are "geologo-morphologic" maps, and a recent German mapping project (Bernhardt et al. 2016) is titled "photogeologic", referring to the method and limitation of that geologic investigation. Other themes within the broader geologic framework include paleotectonic, paleoerosional, "bedrock and structural geologic", and geomorphic maps.

Nomenclature reference maps are special reference maps to show adopted place names, published online at USGS by the IAU Working Group for Planetary System Nomenclature.

Feature maps are thematic maps, either cartographic or metamaps, which contain spatial information concerning a single or a selected set of geomorphologic, geologic, etc., feature types. It is recommended to generate an ID for each mapped object, for their identification in research papers and further catalogs.

Renovation maps are digital, and GIS-ready maps recreated digitally, based on maps previously published as printed (hardcopy) products.

A quadrangle (also called quad, section, and tile) is a standardized subdivision unit of a planetary surface (e.g. Greeley and Batson 1990) where the entire surface is divided into roughly similar area portions for mapping purposes. Typically, each quadrangle is mapped separately, but in certain cases, quadrangles are combined in one map, such as the 1:1M Mars MTM series.

A *sheet* is a map canvas or a subdivision of a single map publication into separate "pages" that display different parts of the surface or different themes in a complementary way.

An *atlas* is the systematic collection of maps, drawn up according to the general procedure as a complete product (Salishev 1982, Karachevtseva et al. 2019). There exist several understandings of this term. An atlas may be a single cartographic publication where maps at various scales and themes are collected together. Several planetary atlases are, however, books that contain a collection of images (photographs) with accompanying text and figures. USGS publications use the term atlas for a *series* of map sheets with uniform theme and scale, produced and published over years or even decades.

3D maps include analog and digitally animated maps that show the terrain in perspective view. The novel *lenticular foil* technology allows the generation of flip-image effects, short animations, and true 3D displays in hardcopy form (Buchroithner et al. 2005 and Hargitai and Naß 2019).

5 Challenges and Future Tasks

Standardization of cartographic methods and data products is critical for accurate analysis and scientific reporting. This is more relevant today than ever before, since researchers have comparatively easy access to a wide variety of digital data as well as to the tools to process and analyze these various products, and planetary mapping centers are now distributed globally. The life cycle of cartographic products can be short and standardized descriptions are needed to keep track of different developments. One of our aims herein was to subdivide the processes of planetary cartography and to define, describe, and present the overall mapping process all the way through its successive segments (see Fig. 3).

Processes related to the **Input** segments cover all aspects that allow not only to produce higher-level products but also to create a basis for their stable representation and reusability. One of the major future issues will be to establish an international map database by digitizing analog maps and by establishing a uniform structure to describe existing data which can then be easily be queried and accessed (see Hargitai and Pitura 2018 for an attempt to establish such an internationally complete database). For digital map products, a metadata description (i.e., a digital equivalent known from map legends and additional information related to the map content), along with validation tools and platforms capable of providing access to archiving, distribution and querying, needs to be established. Standards for metadata exist already partially on a national level and some of the older higher-level map data products are currently transferred to fit into such schemes. USGS includes planetary geologic maps in the digital platform for their terrestrial geologic maps; a database of planetary geologic maps is maintained on the USGS Astrogeology portal[15]; the Lunar and Planetary Institute (LPI) maintains an archive of historic topographic, shaded relief and image maps,[16] and the ICA CPC has a catalog for historic and recent international planetary maps.[17] However, many non-standardized map products exist outside these national institutions all around the world and are distributed across different institutes. One task will be to review such products and to establish a methodological repertoire to transfer maps, to establish a common metadata scheme, and to provide a common semantical basis.

Within the **Distillation** process, the core issues are the abstraction of data, the (carto)graphic visualization, and GIS-based management of derived data. The three major tasks that are necessary to accomplish this are as follows: (1) the definition and setup of *rules* and recommendations for GIS-based mapping processes (cf. Tanaka et al. 2011); (2) advocacy of the GIS-based implementation and distribution of international cartographic *symbol* standards; (3) generation of generic, modular *data models* for GIS-based mapping, which could be used by the mappers to fill in their individual mapping data and scientific results. Currently, efforts focus on creating a template-based framework for the evaluation and optimization of existing map templates. In particular, the short lifetime of products during ongoing missions represents a considerable challenge when creating such models and putting them into operational use. Furthermore, recent work focuses on revising recommendations for cartographic symbols for geological mapping. This encompasses critical review and updating of existing standards for planetary geological symbols (FGDC 2006).

The Web-based mapping systems will also play an important role in the future, because the scientists will have the possibility to easily exchange their mapping approaches and results in joint investigation and analysis projects. As all these systems are relatively new developments in Earth and also planetary science, different topics like usability (i.e., graphical user interface and handling), data exchange and management as well as cartographic visualization have to provide an efficient use of these software systems in the near future.

One first attempt to organize planetary maps into a Web-based interface was done by the USGS (Hare and Tanaka 2001). More recent implementations are *JMARS/JVesta*[18] project led by *Arizona State University (ASU),* or the Web Map

[15]astrogeology.usgs.gov.

[16]www.lpi.usra.edu/resources/collections/.

[17]planetarymapping.wordpress.com.

[18]jmars.asu.edu.

Tile Service *Vesta/MarsTrek*[19] developed at NASA's *Jet Propulsion Laboratory* (JPL) and *NASA's Ames Research Center.*

Output processes cover all aspects of publishing and archiving mapping results, independent of the output format whether digital or analog. The results should be easily accessible and searchable in archives, intuitive online interfaces, platforms and metadatabases that connect the contents of existing databases. One method to achieve this is to incorporate already published maps along with their metadata into a universally accessible international digital map archive. This includes digitized analog maps, digital maps, and mapping products in comparable formats and builds on existing definitions that benefit from existing validation tools. The NASA *Planetary Data System* (PDS 2009), e.g., provides a flexible toolset to accomplish parts of this task in cooperation with USGS/ACS. Existing efforts covering this topic of metadata are, e.g., described by Hare et al. (2011), Hare (2011) and Naß et al. (2010). The existing archives like the *PDS/NASA*, the *PSA/ESA* (Besse et al. 2017), or *DARTS/JAXA* could be extended to include digital maps.

The last issue of this segment covers aspects of interoperability and exchange of map projects between different mapping and database systems. It is now common that the same datasets are available through different and often incompatible platforms, from Web-mapping servers to raw data download sites. Since different research institutes and individuals use different tools for mapping and data storage, procedures have to be established to allow conversions and also collaborative mapping in the future. Lawrence et al. (2016), Laura et al. (2017), and Hare et al. (2017b) discuss how this interoperability could be improved and implemented for planetary primary data, i.e., the substantial amount and diversity of mission data (a kind of *big data*). However, also for secondary data products, i.e., for the derived scientific mapping results containing valuable scientific analysis results, the issue of *sustainable availability and access* has to be solved in the near future.

6 Conclusion

Since 1960, more than two thousand planetary maps have been produced and published during various different framework programs and projects. Therein, different mapping efforts exist, either on national level or as collaboration between groups participating as investigators in mapping missions. However, the coordination of such tasks does not end with the compilation and publication of a set of maps. Coordination may only be considered successful when mapping products are preserved for the upcoming generations of researchers and mappers to allow efficient reuse into new sustainable databases. This may be the ultimate goal for planetary geologic or other thematic maps produced by different groups. Thus, they should use comparable principles in data collection, analysis, and display.

[19]marstrek.jpl.nasa.gov/index.html.

In order to accomplish this, mapping infrastructure, workflows, communication paths, and validation tools have to be developed and made available. Activities dealing with this and the status of planetary cartography are described by Pędzich and Latuszek (2014), Kirk (2016), Naß et al. (2017), and Laura et al. (2017).

References

Ackoff R (1989) From data to wisdom. J Appl Syst Anal 16:3–9

Albert G, Ilyés V, Szigeti C, Kis D, Várkonyi D (2017) How hard is it to design maps for beginners, intermediates and experts? In: 28th International cartographic conference (ICC). Washington, D.C., USA, https://doi.org/10.1007/978-3-319-57336-6_17

Albertz J, Gehrke S, Wählisch M, Lehmann H, Schumacher T, Neukum G (2004) Digital cartography with HRSC on mars express. In: XXth Congress ISPRS, Istanbul, Turkey, 12–23 July 2004

Albertz J, Attwenger M, Barrett J, Casley S, Dorninger P, Dorrer E, Ebner H, Gehrke S, Giese B, Gwinner K, Heipke C, Howington-Klaus E, Kirk RL, Lehmenn H, Mayer H, Muller J-P, Oberst J, Ostrovskiy A, Renter J, Reynik S, Schmidt R, Scholten F, Spiegel M, Stilla U, Wählisch M, Neukum G, the HRSC Col-Team (2005) HRSC on mars express—photogrammetric and cartographic research. Photogram Eng Remote Sens (PE&RS) 71(10):1153–1166

Andrews JH (1996) What was a map? The lexicographers reply. Cartographica: Int J Geogr Inf Geovis 33(4):1–12

Archinal BA, A'Hearn MF, Bowell E, Conrad A, Consolmagno GJ, Courtin R, Fukushima T, Hestroffer D, Hilton JL, Krasinsky GA, Neumann G, Oberst J, Seidelmann PK, Stooke P, Tholen DJ, Thomas PC, Wiliams IP (2011) Report of the IAU working group on cartographic coordinates and rotational elements: 2009. Celest Mech Dyn Astron 109(2):101–135. https://doi.org/10.1007/s10569-010-9320-4

Baker VR (2017) Debates—hypothesis testing in hydrology: pursuing certainty versus pursuing uberty. Water Resour Res 53:1770–1778. https://doi.org/10.1002/2016wr020078

Bannister RA, Hansen V (2010) Geologic map of the Artemis Chasma quadrangle (V-48), Venus. U.S. Geological Survey Scientific Investigations Map 3099

Bernhardt H, Hiesinger H, Ivanov MA, Ruesch O, Erkeling G, Reiss D (2016) Photogeologic mapping and the geologic history of the Hellas basin floor, Mars. Icarus 264:407–442

Besse S, Vallat C, Barthelemy M, Coia D, Costa M, De Marchi G, Fraga D, Grotheer E, Heather D, Lim T, Martinez S, Arviset S, Barbarisi I, Docasal R, Macfarlane A, Rios C, Saiz J, Vallejo F (2017) ESA's planetary science archive: preserve and present reliable scientific data sets. Planet Space Sci 1–10. https://doi.org/10.1016/j.pss.2017.07.013

Beyer RA (2015) An introduction to the data and tools of planetary geomorphology. Geomorphology 240:137–145

Blagg MA, Müller K (1935) Named lunar formations. In: International astronomical union, percy lund, p 196. Humphries and Co Ltd, London

Bruno DE, Ruban DA (2017) Something more than boulders: a geological comment on the nomenclature of megaclasts on extraterrestrial bodies. Planet Space Sci (PSS) 135:37–42. https://doi.org/10.1016/j.pss.2016.11.006

Buchroithner MF (1999) Mars map: the first of the series of multilingual relief maps of terrestrial planets and their moon. In: Proceedings 19th international cartographic conference (ICC), Ottawa

Buchroithner M, Gründemann T, Kirk RL, Habermann K (2005) Three in one: multiscale Hardcopy depiction of the Mars Surface in true-3D. Photogram Eng Remote Sens 71 (10):1105–1108

Buchroithner MF, Shingareva K, Krasnopevtseva B (1999) The Series of Multilingual Maps for Terrestrial Planets and their Moons. Abstract Book Moscow Symposium on Planetary Cartography, p 126–127

Buchroithner MF, Wälder O (2003) True 3D presentation of the mars surface. In: 21st International cartographic conference, Durban, South Africa, pp 1540–1543

Carpendale MST (2003) Considering Visual Variables as a Basis for Information Visualization. Research report 2001-693-16, Department of Computer Science, University of Calgary, Calgary, Canada

Dorrer E, Zhou X (1998) Towards optimal relief representation from mars imagery by combination of DEM and shape from shading. Int Arch Photogram Remote Sens 32(4):156–161

ETH (2017) Schweizer Weltatlas. ETH Zürich, Institut für Kartografie und Geoinformation

FGDC (1998) Content Standard for Digital Geospatial Metadata. FGDC-STD-001-1998, Federal Geographic Data Committee, Washington, D.C

FGDC (2000) Content Standard for Digital Geospatial Metadata Workbook. Version 2.0. For Use with FGDC-STD-001-1998, Federal Geographic Data Committee

FGDC (2006) Digital Cartographic Standard for Geologic Map Symbolization. FGDC-STD-013-2006, Federal Geographic Data Committee, Washington, D.C

Frigeri A, Hare T, Neteler M, Coradini A, Federico C, Orosei R (2011) A working environment for digital planetary data processing and mapping using ISIS and GRASS GIS. Planet Space Sci (PSS) 59:1265–1272, Special Issue: Planetary Mapping, https://doi.org/10.1016/j.pss.2010.12.008

Galluzzi V (2019) Multi-mapper projects: collaborative mercury mapping. In: Hargitai H, Wolak J (eds) Planetary Cartography—Concepts, tools, and methods, Lecture Notes in Geoinformation and Cartography. Springer

Gede M, Hargitai H (2015) Country movers—an extraterrestrial geographical application. In: Reyes NJ (eds) Cartography beyond the ordinary world joint ICA symposium, pp 178–183

Gede M, Hargitai H (2017) An online planetary exploration tool: “Country Movers”. Acta Astronautica 137. https://doi.org/10.1016/j.actaastro.2017.04.028

Gehrke S, Neukum G, das HRSC Co-Investigator Team (2005) Das kartographische Softwarepaket. Planetary Image Mapper (PIMap), Photogrammetrie—Fernerkundung—Geoinformation (PFG), 5/2005, pp 417–422

Greeley R, Batson RM (1990) Planetary mapping. Cambridge University Press, Cambridge

Grishakina EA, Lazarev EN, Rodionova JF, Shevchenko VV (2014) Hypsometric map of the Moon, Gipsometritseskaya Karta Luny 1:13M. Sternberg Astronomical Institute and the Moscow State University of Geodesy and Cartography (MIIGAiK), Moscow

Gustavsson M, Kolstrup E, Seijmonsbergen A (2006) A new symbol-and GIS-based detailed geomorphological Mapping System: renewal of a scientific Discipline for understanding Landscape Development. Geomorphology 77:90–111, https://doi.org/10.1016/j.geomorph.2006.01.026

Gustavsson M, Seijmonsbergen A, Kolstrup E (2007) Structure and Contents of a new geomorphological GIS Database linked to a geomorphological map—with an example from Liden, central Sweden. Geomorphology 95(3-4):335–349. https://doi.org/10.1016/j.geomorph.2007.06.014

Haber RB, McNabb DA (1990) Visualization idioms: a conceptual model for scientific visualization systems. In: Shriver B, Neilson GM, Rosenblum L (eds) Visualization in Scientific Computing, pp 74–93. IEEE Computer Society Press

Hake G, Grünreich D, Meng L (2002) Kartographie: visualisierung raum-zeitlicher Informationen. 8., vollst. neu bearb. und erw. Aufl., Berlin New York, de Gruyter

Hancher MD, Beyer R, Broxton M, Gorelick N, Kolb E, Weiss-Malik M (2009) Visualizing Mars data and imagery with Google Earth. In: 40th Lunar and planetary science conference (LPSC), LPI No. 1903, #2308

Hare TM (2011) Standards-based collation tools for geospatial metadata in support of the planetary domain. (M.A. thesis), Northern Arizona University, Flagstaff, AZ

Hare TM, Pio Rossi A, Frigeri A, Marmo C (2017b) Interoperability in planetary research for geospatial data analysis. Planet Space Sci (PSS). https://doi.org/10.1016/j.pss.2017.04.004

Hare TM, Skinner JA, Fortezzo CM, Bailen MS (2011) FGDC geospatial metadata for the planetary domain. In: 42th Lunar and planetary science conference (LPSC), LPI No. 1903, #2154

Hare TM, Skinner JA, Fortezzo CM, Gaddis LR (2015) Planetary GIS at the U.S. geological survey astrogeology science center. In: 2nd planetary data workshop, Held June 8–11, 2015 in Flagstaff, Arizona. LPI No. 1846, #7005

Hare TM, Skinner JA, Kirk RL (2017a) Cartography tools. In: Rossi A-P, van Gasselt S (eds) Planetary Geology. Springer, Berlin, p 540

Hare TM, Tanaka KL (2001) Planetary interactive GIS-on-the-web analyzable database (PIGWAD). In: 20th International Cartographic Conference (ICC), Beijing, China

Hargitai H (2006) Planetary maps: visualization and nomenclature. Cartographica 41(2):149–167. https://doi.org/10.3138/9862-21ju-4021-72m3

Hargitai H (2016) Metacatalog of Planetary Surface Features for Multicriteria Evaluation of Surface Evolution: the Integrated Planetary Feature Database. 48th Devision of Planetary Science (DPS)/11th European Planetary Science conference (EPSC), #426.23

Hargitai H, Gede M, Zimbelman J et al (2015b) Multilingual narrative planetary maps for children. In: Robbi SC, Madureira Cruz CB, Leal de Menezes PM (eds) Cartography—maps connecting the world. Lecture Notes in Geoinformation and Cartography. Springer, Heidelberg, pp 17–30

Hargitai H, Naß A (2019) Planetary mapping: a historical overview. In: Hargitai H, Wolak J (eds) Planetary Cartography—Concepts, tools, and methods, Lecture Notes in Geoinformation and Cartography. Springer

Hargitai H, Kereszturi A (eds) (2015) Encyclopedia of planetary landforms. Springer, New York. https://doi.org/10.1007/978-1-46149213-9

Hargitai H, Page D, Canón-Tapia E, Rodrigue CM (2015a) Classification and characterization of planetary landforms. In: Hargitai H, Kereszturi Á (eds) Encyclopedia of planetary landforms. https://doi.org/10.1007/978-1-4614-3134-3

Hargitai H, Pitura M (2018) International Catalog of Planetary Maps 1600–2017. 49th Lunar and planetary science conference (LPSC), LPI No. 1903, #2083

Hauber E, Naß A, Skinner J, Huff A (2019) Planetary geological mapping. In: Hargitai H, Wolak J (eds) Planetary cartography—concepts, tools, and methods. Lecture Notes in Geoinformation and Cartography. Springer

Hunter M, Hayward R, Hare T (2019) Planetary nomenclature. In: Hargitai H, Wolak J (eds) Planetary Cartography—Concepts, tools, and methods, Lecture Notes in Geoinformation and Cartography. Springer

INSPIRE (2012) Data Specification on Geology. Draft Guidelines.2.8.II/III.4_v3.0 rc2,—Thematic Working Group Geology

ISO (2014a) Geographic Information—Metadata. ISO 19115, ISO/TC 211, Geographic Information/Geoinformatics

ISO (2014b) Geographic information—Metadata—XML schema implementation. ISO19139, ISO/TC 211, Geographic Information/Geoinformatics

Jin S (2014) Planetary geodesy and remote sensing. CRC Press, Boca Raton

Karachevtseva I, Kokhanov A, Rodionova Zh (2019) Atlas planetary mapping: phobos case. In: Hargitai H, Wolak J (eds) Planetary Cartography—Concepts, tools, and methods, Lecture Notes in Geoinformation and Cartography. Springer

Kirk R (2016) Planetary cartography: what, how, and why begin with where. In: 47th Lunar and planetary science conference (LPSC), LPI No. 1903, #2151

Kopal Z, Carder RW (1974) Mapping of the moon. Springer Science + Business Media, Dordrecht

Kraak M-J, Fabrikant SF (2017) Of maps, cartography and the geography of the International Cartographic Association. Int J Cartogr 1–23. https://doi.org/10.1080/23729333.2017.1288535

Krygier JB (1995) Cartography as an art and a science? Cartogr J 32(6):3–10

Laura JR, Hare, TM, Gaddis LR, Fergason RL, Skinner JA, Hagerty JJ, Archinal BA (2017) Towards a planetary spatial data infrastructure. ISPRS Int J Geo-Inf 6:181, https://doi.org/10.3390/ijgi6060181

Lawrence S, Hagerty J, Gaddis LR, Archinal BA, Radebaugh J, Byrne S, Sutton S, DellaGiustina D, Thomson B, Mazarico E, Williams D, Skinner J, Hare T, Fergason R, Laura J (2016) The Mapping and planetary spatial infrastructure team (MAPSIT): addressing strategic planning needs for planetary cartography. In: 47th Lunar and planetary science conference (LPSC), LPI No. 1903, #1710

Lehmann H, Scholten F, Albertz J, Wählisch M, Neukum G (1997) Mapping a whole planet—the new topographic image map series 1: 200,000 for Planet Mars. In: 18th International cartographic conference (ICC), Stockholm, Sweden

Mazza R (2009) Introduction to information visualization. Springer, London

Meynen E (1973) Multilingual dictionary of technical terms in cartography. International Cartographic Association, Stuttgart

Morton O (2002) Mapping mars: science, imagination, and the birth of a world. Picador, New York

NASA (2017) Planetary Data Archiving, Restoration, and Tools. NASA Research Announcement Solicitation: NNH17ZDA001 N-PDART

Naß A, Di K, Elgner S, van Gasselt S, Hare T, Hargitai H, Karachevtseva I, Kersten E, Manaud N, Roatsch T, Rossi AP, Skinner J, Wählisch M (2017) Planetary Cartgraphy and mapping: where we are today, and where we are heading for? Int Arch Photogram Remote Sens, Hong Kong, vol XLII-3/W1. https://doi.org/10.5194/isprs-archives-xlii-3-w1-105-2017

Naß A, van Gasselt S, Jaumann R (2010) Map description and management by spatial metadata: requirements for digital map legend for planetary geological and geomorphological mapping. In: AutoCarto, symposium on computer-based Cartography and GIScience, #1457, Orlando, Florida

Naß A, van Gasselt S, Jaumann R, Asche H (2011) Implementation of cartographic symbols for planetary mapping in geographic information systems. Planet Space Sci (PSS) 59:1255–1264, Special Issue: Planetary Mapping, https://doi.org/10.1016/j.pss.2010.08.022

National Geographic Society (1969) The earth's moon. Chamberlin W, Grazzini AD (eds) Supplement to National Geographic, 135, 2

Ooms K, De Maeyer P, Fack V (2014) Study of the attentive behavior of novice and expert map users using eye tracking. Cartogr Geogr Inf Sci 41(1):37–54

Ormeling FJ Sr (1987) ICA 1959-1984. The first twenty-five years of the International Cartographic Association. Enchede, International Cartographic Association, p 20

PDS (2009) Planetary Data System Standard Reference. Technical Report, Jet Propulsion Laboratory, California Institute of Technology, California

Pędzich P, Latuszek K (2014) Planetary cartography—sample publications, cartographic projections, new challenges. Pol Cartogr Rev 46(4):388–396

Rowley J (2007) The wisdom hierarchy: representations of the DIKW hierarchy. J Inf Sci 33 (2):163–180. https://doi.org/10.1177/0165551506070706

Rükl A (2012) A Hold Atlasza (Atlas of the Moon, Hungarian edition) Geobook Hungary, Szentendre

Salishev KA (1982) Kartovedeniye (Cartography). 3rd edn, M, 400 p. (In Russian)

Schimerman LA, (1975) Lunar Cartographic Dossier. Section 4.3.7

Shevchenko V, Rodionova Z, Michael G (2016) Lunar and planetary cartography in Russia. Springer, Heidelberg. https://doi.org/10.1007/978-3-319-21039-1

Shingareva KB, Krasnopevtseva BV, Buchroithner MF (2002) Moon map. a new map out of the series of multilingual relief maps of terrestrial planets and their moons. In: Proceedings InterCarto 8, Helsinki—St. Petersburg, pp 392–395

Shingareva KB, Krasnopevtseva BV, Buchroithner MF (2001) Venus map (The Series of Multilingual Maps for Terrestrial Planets and their Moons). In: 20th International Cartography conference (ICC), Beijing, China, pp 3279–3284

Shingareva KB, Krasnopevtseva BV, Leonenko SM, Buchroithner M, Wälder O (2003) Mercury—a new map in the series of multilingual relief maps of terrestrial planets and their moons. In: 21st international cartographic conference (ICC), Durban, South Africa, pp 1551–1555
Shingareva KB, Krasnopevtseva BV, Zimbelman JR, Pérez Gómez R, Vázquez Hoehne A, Buchroithner MF, Dorrer E (2007) A new version of the multilingual glossary of planetary cartography. In: 23rd international cartographic conference (ICC), Moscow, Russia, 12 p. CD-ROM
Skinner JA Jr (2015) The Challenges of Standardized Planetary Geologic Mapping. Second Planetary Data Workshop, #7071
Skinner JA Jr, Fortezzo CM, Gaither TA, Hare TM, Huff AE, Hunter MA (2018) The USGS-NASA planetary geologic mapping program: status, process, and future plans. In: 49th Lunar and planetary science conference (LPSC), LPI No. 1903, #2083
Snyder JP (1982) Map projections used by the U.S. Geological Survey. U.S. Geological Survey Bulletin 1532
Snyder JP (1987) Map projections used by the U.S. Geological Survey. U.S. Geological Survey Professional paper 1395
Strobell ME, Masursky H (1990) Planetary nomenclature. In: Greeley R, Batson RM (eds) Planetary mapping. Cambridge University Press, New York, pp 96–140
Tanaka KL, Skinner JA Jr, Hare TM (2011) Planetary Geologic Mapping Handbook—2011. USGS
van Gasselt S, Naß A (2011) Planetary mapping: the datamodel's perspective and GIS framework. Planet Space Sci (PSS) 59:1231–1242, Special Issue: Planetary Mapping. https://doi.org/10.1016/j.pss.2010.09.012
Vincendon M, Audouard J, Langevin Y, Poulet F, Bibring J, Gondet B, (2011) OMEGA albedo map of Mars. American Geophysical Union (AGU) Fall Meeting, #P23A-1699
Ware C (2004) Information visualization—perception for design, 2nd edn. Elsevier Morgan Kaufmann Publisher, San Francisco
Wilhelms DE (1972) Geologic mapping of the Second Planet. Interagency Report: Astrogeology 55 Prepared under NASA Contract W-13,204
Williams D (2016a) Cartographic needs for geologic mapping during active orbital planetary missions. In: 47th Lunar and planetary science conference (LPSC), LPI No. 1903, #1588
Williams D (2016b) NASA's planetary geologic mapping program: overview. *Int Arch Photogramm Remote Sens Spatial Inf Sci* XLI-B4:519–520. https://doi.org/10.5194/isprs-archives-xli-b4-519-2016
Williams DA, Keszthelyi LP, Crown DA, Yff JA, Jaeger WL, Schenk PM, Geissler PE, Becker TL (2011) Geologic map of Io: U.S. Geological Survey Scientific Investigations Map 3168, scale 1:15,000,000, 25 p. https://pubs.usgs.gov/sim/3168/
Williams DA, Yingst, RA, Garry WB (2014) Introduction: the geologic mapping of Vesta. Icarus 244:1–12
Yingst RA, Mest SC, Berman DC, Garry WB, Williams DA, Buczkowski D, Jaumann R, Pieters CM, De Sanctis MC, Frigeri A, Le Corre L, Preusker F, Raymond CA, Reddy V, Russell CT, Roatsch T, Schenk PM (2014) Geologic mapping of Vesta. Planet Space Sci 2–23. https://doi.org/10.1016/j.pss.2013.12.014

Planetary Mapping: A Historical Overview

Henrik Hargitai and Andrea Naß

Abstract The development of the methods of visualization, control, and content of planetary maps goes in parallel with terrestrial ones. Both reflect technological, scientific, sociopolitical, and graphic design changes. However, while terrestrial maps are ubiquitous and show abstract or iconic representations of the Earth features, planetary surfaces are much more frequently represented with uninterpreted images, despite the wealth of planetary spatial data. In this paper, we highlight the key maps and map series made before the space age and the new cartographic methods introduced in the early 1960s when rectified, geologic and airbrush maps, and space-borne planetary photography, revolutionized the way we can look at planetary surfaces. This chapter also highlights the most recent novel approaches in planetary cartography.

Keywords Planetary cartography · History of cartography · Mapping · Geologic maps · Nomenclature

1 Introduction

We can distinguish the following major time periods in planetary cartography: the era of Earth-based visual observations, that of the photographic observations, and the era of digital spectral and topographic observations from space.

The *visual era* started with naked eye observations (with only one example in 1600), followed by astronomers' observations using their own telescopes (selenographers and areographers, 1610–ca. 1960), and astronomers using a large observatory (ca. 1900s–1960s). In *the photographic era*, astronomers used

H. Hargitai (✉)
Eötvös Loránd University, Múzeum Krt 6-8, Budapest 1088, Hungary
e-mail: hargitaih@caesar.elte.hu

A. Naß
Institute of Planetary Research, Planetary Geology, German Aerospace Center (DLR), Rutherfordstr. 2, 12489, Berlin, Germany

© Springer Nature Switzerland AG 2019

H. Hargitai (ed.), *Planetary Cartography and GIS*, Lecture Notes in Geoinformation and Cartography, https://doi.org/10.1007/978-3-319-62849-3_2

photographic plates collected from different sources for mapping that were typically supplemented by *visual observations*. The era of "armchair astronomers" started by the International Astronomical Union (IAU) and their lunar reference map that was produced without the personal visual observations of the mappers in 1935. In the late 1950s, all major observatories' best available photographs were combined for creating planetary atlases to support both IAU's lunar mapping agenda and America's entry into the space age. However, it also became evident that even the best photographic plates by itself are unable to show the geology behind the landscape.

Planetary maps at the end of the 1950s with the new approaches were created using Earth-based images but knowing that this will change soon: Future maps would be created using space-based cameras, and it would soon become possible to start studying the geology of the Moon in situ, ending the era of speculations. The introduction of the new mapping methods signaled a preparation for the space age (Almar I., personal communication 2018).

Understanding the lunar geology was essential for choosing the appropriate landing sites for the Apollo and automatic missions. Planetary mapping moved into adulthood with the application of *geologic mapping* methods to planetary surfaces. This, advanced, form of manual mapping is based on the same types of images that were used in previous mappings, and it also included morphologic–structural details as before. The new element on these maps was the lines for material unit contacts, combined with the analysis of these units' stratigraphic, and consequently, age relations. Instead of studying a heterogenous image, a geologic map reduces the complexity into comprehensible units (Williams 2016). Planetary mapping was decoupled from astronomy, and now it supported the newly born discipline of *astro*geology.

In the space age, sensors onboard orbiting or flyby platforms provide source material for mapping with additional data from landed missions. The survey of the relief of planetary surfaces, without theodolites, was made possible by applying the new cartographic technologies, such as stereophotogrammetry, laser and radar altimetry, which have become the source for topographic data since the 1960s.

Regarding the end product, paper maps began to be replaced by GIS databases in 2011, since when all new planetary maps at USGS are produced in GIS format while the most significant previous (historic) maps and atlases are also being digitized ("renovated").

When speaking of the history of planetary mapping, it is important to distinguish maps from other graphical depictions of a landscape. Many astronomers created drawings of planets that were not maps. We may distinguish *drawings*–(popular or scientific)–from *maps* using various criteria. We list some of these in the followings, noting that maps may utilize any or several of these criteria. Most importantly, (1) *maps* offer a uniform view across the planetary body both in geometry and theme. Uniformity is expressed in that the image has a map projection (including orthographic view, e.g., for the Moon), and features are shown under similar conditions, e.g., illumination. Furthermore, (2) maps show the surface in a synthetic way compiled from multiple observations, showing *inherent* characteristics and not the actual expression of these qualities at a particular moment, as much as available data make it possible; (3) maps display designations of the represented features, preferably a systematic, standardized nomenclature with a hierarchical system of local and regional names, taken from a well-defined pool of specific elements and descriptor terms; and this is displayed in a systematic typographic way; (4) maps use a coordinate system, and positions of the surface features and markings may be linked to the coordinate frame by controlled measurements; (5) the map (or atlas) of an entire body may be split into sheets or quads, and may have multiple, thematic views of the same regions; (6) maps represent the surface in consistent ways, including stylized (iconic) or symbolic (abstract) visual representations (symbology); and (7) maps include metadata (data about the data, e.g., title, legend, information on the scale, projection, production, error, source of data, etc.); some of this information may appear as marginalia on paper or static maps in the form of graticules, north arrow, scale bar, etc.; or as a supplementary file in digital maps.

Maps provide higher fidelity view of planetary surfaces than photographs in many ways. During image-based mapping, we turn individual observations (single views or photographs) into a representation of the surface that accurately shows real spatial relations. These maps may use either a photographic or photorealistic (synthetic) representation of the surface, or a more abstract, symbolic, generalized cartographic representation. This cartographic representation communicates "deeper" (analyzed, distilled) information than an image map where this *information* may be included but in hidden, "raw" *data* form (Naß et al. 2019). Maps could be regarded as a type of "augmented reality" in that they combine multiresolution, multitemporal data, their analysis, and annotation, into a single, often customizable, view.

2 Milestones in the History of Planetary Cartography

2.1 Before the Space Age

The history of the planetary cartography (also called *extraterrestrial mapping*) dates back to shortly after the invention of the telescope at the beginning of the seventeenth century. This event marked a milestone in planetary exploration, because observational evidence could have started replacing reasoning-based natural philosophy. Additionally, the telescopic view of the Moon was not only scientifically exciting but it was also appealing for the general public. Numerous map manuscripts were created in the next centuries that were reproduced in popular books and encyclopedias in the forms of better or worse quality engravings. Claude Mellan's engravings of the Moon, made in 1635, were remarkably realistic. Besides detailed maps, these early drawings of the disk of the Moon served the same purpose as photographs today, so they are not real maps. However, they *are* maps because the illustrator had to use the mapping principle of generalization and made decisions of classifying the surface features and using iconic representations as part of the drawing, for repeated features such as craters.

The first detailed *maps of the Moon* showed different visual and toponymic approaches to represent the lunar landscape. *M. van Langern* (also known as *Langrenus* (Krogt and Ormeling n.d.)) in 1645 used an iconic, simplified representation of the surface; introduced the mare/terra distinction, as well as the concept of assigning personal names to craters, abstract concepts to terrae, and terrestrial descriptor terms (Montes, Lacus, etc.).

J. Hevelius in 1647 created two Moon displays: a realistic full Moon engraving and a symbolic, highly interpretive map showing the Moon almost as a region on Earth, in both visuals and labels. In this map, he used the "termite hill"-style, then standard, representation of mountain ranges for crater rims (that were called "ring mountains" in the eighteenth century), and clearly marked the coastlines of maria that he thought to be water bodies with islands. Accordingly, he assigned geographic names from the Mediterranean to these features with similar arrangement to that on Earth.

G. B. Riccioli and F. M. Grimaldi in 1651 created a third nomenclature and an objective visual representation, reflecting the telescopic view even if the features seen could not be explained. Riccioli produced the nomenclature, taking van Langern's descriptors, but changed specifics of crater names from dignitaries to ancient and modern astronomers. He also changed the names of mare to weather-related terms. The Riccioli nomenclature became the basis of today's planetary nomenclature scheme. For centuries, however, Hevelius' and Grimaldi's maps (both the visuals and nomenclature) were used in parallel, with some Moon maps showing both versions (e.g., Dopplemayr 1742).

Many improvements were introduced during the upcoming centuries, and extraterrestrial mapping became a scientific discipline. Map design was determined by the artistic talent of the observing astronomer or his financial status to hire good engravers, while the level of details was determined by the used telescope.

A milestone in planetary mapping is the map of "the *first* selenographer," T. J. Mayer (1748), who first used control points, measured at the telescope, which became a standard procedure afterward. He established the lunar coordinate system with equator and prime meridian.

The title of the "*last* visual selenographer" goes either to P. J. H. Fauth or H. P. Wilkins. Fauth used his own observatory, and his 342-cm-diameter map was completed by his son, H. Fauth, in 1964. The addition of details in drawing-based selenographic maps culminated in the 300-in. (7.62 m diameter) map of Wilkins (1946). This map was divided into 25 regular sheets. Additional special sheets showed the libration zones on farside-centered stereographic projection, and two polar stereographic views completed this work with actual data for the nearside portion only. These latter map views became common only after the space age. Wilkings also attempted something unprecedented: to draw the "probable appearance" of the farside. Data for the marginal regions were taken from libration observations. He traced nearside crater rays back to the farside and added speculative maria (Wilkins 1953).

Early maps showed the Moon north-up, as seen by the naked eye but standard lunar maps later became oriented south-up until the 1960s. The projection, understandably, remained sub-Earth-centered orthographic throughout the telescopic era: Maps were produced for users of telescopes, showing the view as seen through a telescope. One of the last projects using this style was the map supplement of the System of Lunar Craters catalog (Arthur et al. 1963).

Mapping the Moon and Mars required entirely different approaches until the space age, where available data types became more uniform. *Mars,* unlike the Moon, could be mapped globally as it rotated. Surprisingly, the earliest maps of Mars (Herschel 1784; Beer and Mädler 1841) used polar projections. The first map, by *Herschel* (1784, his Fig. 25; Stooke 2012) is also the first map made by image *mosaicking* technique. Herschel combined the drawings of his successive observations of the Martian disk in a south pole-centered petal-like mosaic. This process resulted in a single figure, allowed Herschel to correct single-observation distortions, and showed a view of Mars that can never be seen from Earth: a triumph for cartography. Mercator projection for Mars was first used by a geologist, *Philips* (1865). Mercator and polar stereographic projections together first appeared in *Green's* map (1879) and were also used by Schiaparelli, whereas popular reproductions of the map manuscripts of Schiaparelli and Flammarion were reprojected to two-hemisphere Mollweide projection.

The Venus map of Bianchini (1728) shows Venus in equirectangular projection. The next example is Kaiser's Mars map (1864). Equirectangular projection is commonly used today for computer-generated maps (Hargitai et al. 2019).

Telescopes provided details for 1:10 M lunar maps in the 17th century and for 1:3.5 M lunar maps in the 19th century (e.g., Wilhelm Lohrmann's 1824 maps). As telescopes became more powerful, it began to be possible to map the albedo features of Mars (*Beer and Mädler* 1841), during the oppositions every two years. Mars, unlike the Moon, displayed no shadows, and only albedo markings could be mapped that, showing another difference, significantly changed in time. The mapping of the low-contrast albedo features seen on the small diameter Martian disk (Fig. 1) through variable seeing conditions soon resulted in surprisingly detailed and highly controversial maps that showed a network of linear features called canals, starting with *G. Schiaparelli*'s maps that accompanied his very detailed descriptions of the surface features he named. On these maps, canal features first appeared as sharp, wide, curved bands (1877 opposition), then lines (1879), and parallel-running ("twin") straight lines (1881) (Tucci 1998).

The canal controversy had major implications on Mars science and cartography. During the visual observer era, maps were drawn after the observers' dynamic, ever-changing personal visual experiences. In addition, what the maps showed were filtered through the cognition process of the observer, i.e., they were subjective. At the turn of the nineteenth–twentieth century, two observer schools existed, one that had seen and drawn canals and another that had not.

The most detailed canal maps were those made by *Lowell* (1906) at his personal observatory in Flagstaff. Lowell based his theory of intelligent life on Mars on the "cognitively observed" canal properties. This theory survived until the space age despite canalists failed in its first test of objectivity. It was demonstrated that canals were likely mere optical illusions (Evans and Manunder 1903). However, for the public, with the effective help of journalists, canal maps served as a true and

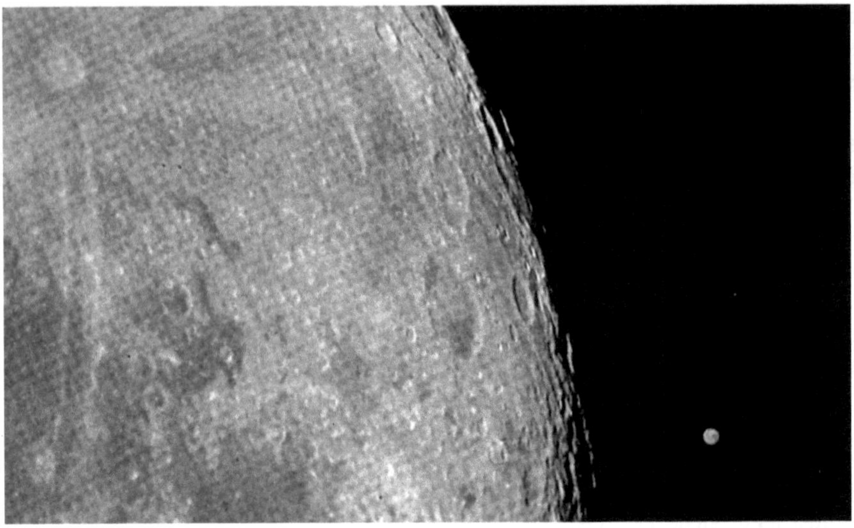

Fig. 1. Telescopic view of the Moon (left) with Mars (right) The picture was taken when Mars appears larger than usual, around opposition. *Photograph*, 12.04.1911. (Plate III in Slipher 1962a)

exciting representation of Mars that was generally believed to be inhabited. The debate triggered new, photographic techniques (Markley 2005), and canal maps remained authoritative sources of the Martian geography for the interested public.

However, the difference is not just the content and level of details but also the type of representation. Mars mapping at this time was about seeing and recording the surface details during the sometimes seconds-long favorable seeing conditions. Sharp lines could result from extremely good conditions or eyesight—or wishful seeing. The brain's visual data reduction system automatically connects random spots to more meaningful shapes, as the author of this chapter could have experienced during geomorphologic field mapping. Linear objects are the most prominent and best remembered elements on any map (Ooms et al. 2014; Albert et al. 2017). Lines on a map usually represent an abstract concept, artificial object or are *symbolic/iconic* representations. In contrast, these Mars maps claimed to be true representations of Mars: hand-drawn maps with "photorealistic" details but sharper than what photographic techniques could have achieved (Fig. 2). For the canalists, a realistic, pictorial representation may have been unconsciously confused with the urge to create a better, that is, map-like (cartographic) representation of Mars. They appeared to be cartographic maps, but in fact, they were (imagined) image maps (Fig. 3) (the problem of the clear distinction between the representation and the represented also appeared in the arts: Magritte's famous *Ceci n'est pas une pipe* image was painted around this time, in 1928–29). As Lane (2011) notes, these early maps of Mars lent the planet a fundamentally geographic identity, while previously Mars was an astronomical object. Maps competed for the visual authority of depicting the real Mars. This "representation war" began at the 1876–77 opposition.

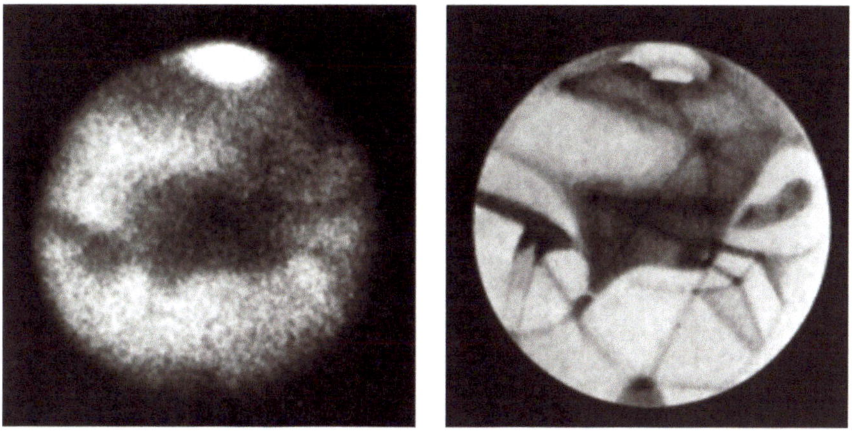

Fig. 2. Two views of Mars on the same night (Sept. 2. 1924) by R. Trumpler. A photograph and a drawing that are intended to faithfully represent what was seen through the telescope. Even if we consider occasionally excellent seeing conditions, the resulting map goes beyond the available contrast and spatial resolution and somewhat overinterprets the low-quality data (Trumpler 1924: Plate XX)

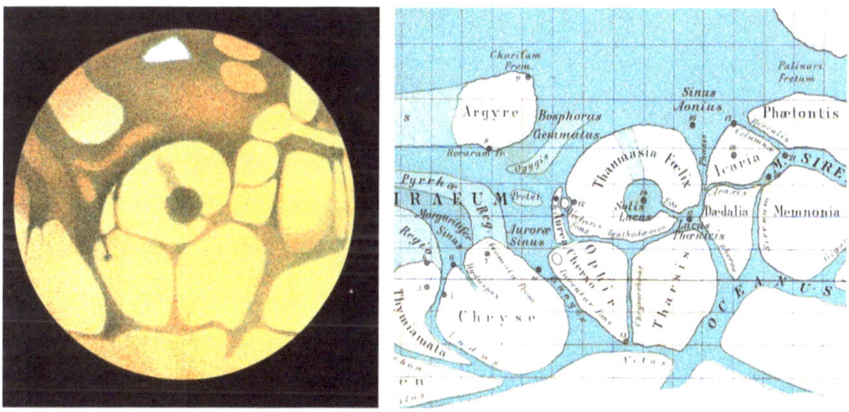

Fig. 3. A drawing that represents what the observer saw through the telescope, and a map representing features on that drawing. Argyre, corresponding to Argyre Basin, is the roughly circular white feature on the upper left of the map (Schiaparelli 1878)

The seemingly more accurate maps were those that showed prominent, sharp outlines (e.g., Schiaparelli 1878, Fig. 4 and also by Flammarion, from 1876), providing a familiar, "cartographic" visualization that the readers could recall from their school atlases. In contrast, less accurate-looking, definitely *less map-like*, maps displayed blurred spots and subtle colors (Green's map of 1879, Fig. 4, and the maps of Antoniadi), much closer to the actual telescopic view. We would today call it an image map. Photographic maps are rarely used for Earth and are difficult to

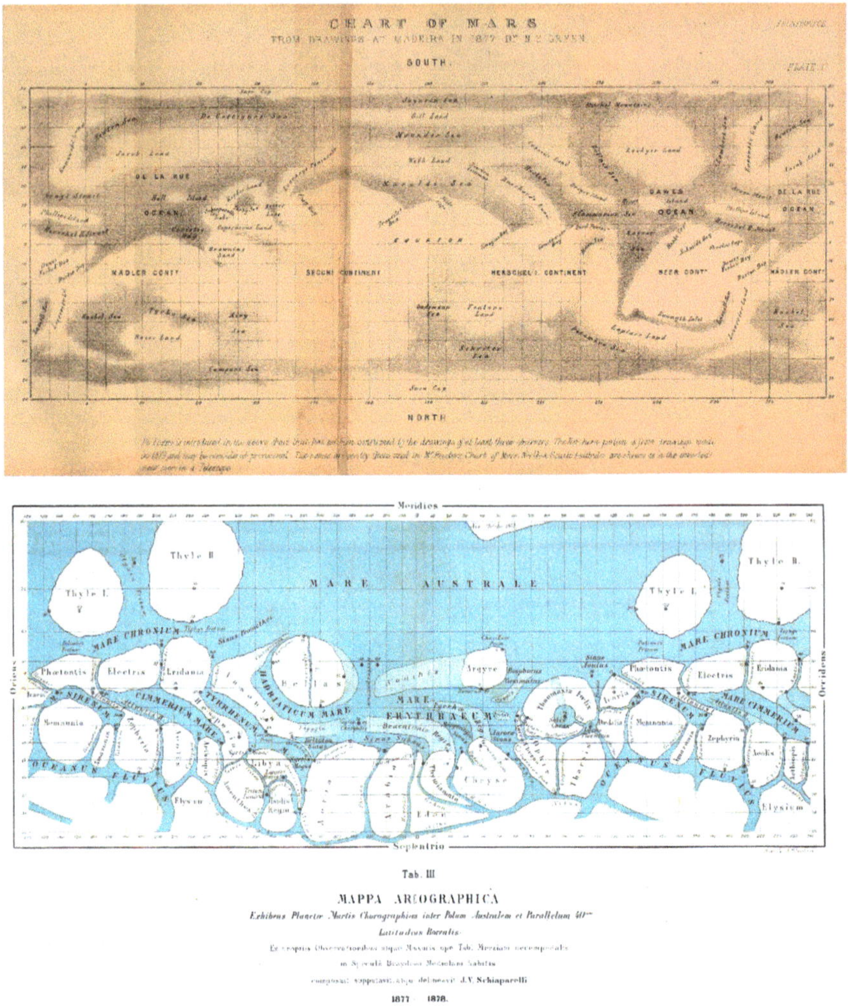

Fig. 4. Two different visual representations of the same planet, Mars, both drawn in 1877. Top: map by N.E. Green, who was a painter and painted Mars in pale yellow-brown colors in the original version (Ledger 1882) Bottom: map by Schiaparelli (1878). Note that central longitudes are different and the upper map duplicates the marginal zones. South is up

understand because of their uninterpreted, raw content. The maps with subtle, blurred spots communicated to the general public that these maps are *less transparent* and *less detailed* than the familiar cartographic ones with strong outlines, and razor sharp straight lines. Both map types were updated almost at each apparition, adding new details every second year. The sharp lines eventually evolved into a system of double canals and oases, tied to the fast-developing implications for life on Mars.

After decades of uncertainty, canals and their names were eventually removed on IAU's official nomenclature reference map, made by de Mottoni at Pic du Midi (IAU 1960, plate I–II, Dollfus 1961) despite Lowell meticulously named all his canals and oases. Interestingly, mission planning maps of the first NASA Mars project did show the canal features (Slipher 1962b; JPL 1967) until the Mariner 6, 7, and 9 missions sent back the first close-up photographs and straight canals were permanently removed from subsequent planning charts (Roth and de Vaucouleurs 1971) and Mars' geology (Sagan and Fox 1975; Moore 1977). While Lowell triggered a decades-long public debate about extraterrestrial geology and biology, many scientists today regard Lowell's contribution to planetary mapping as damaging that created distrust in Mars science. An ironic twist in this story is that most of the dark markings are geologically relatively insignificant ephemeral aeolian features.

2.2 *The Photographic Era*

Since the beginning of the telescopic era, individual astronomers mapped the Moon and later Mars, as parts of "personal space missions" (MacDonald 2017). In the first half of the twentieth century, planetary exploration moved to the largest observatories.

For lunar mapping, atlases with large-sized photographic plates provided a new way to show and study the surface, using Earth-based telescopic views. The first major project of this kind was that of the *Paris Observatory* (Loewy and Puiseux 1896) and simultaneously another at *Lick Observatory* (Holden 1896). However, these and the subsequent photographic atlases were systematic collections of reference photographs, not cartographic *maps*. In 1910, the British *W. Goodacre* combined photographs from two major observatories with his visual observations in making his line art Moon map. At around the same time, M. A. Blagg and W. H. Wesley entirely abandoned visual observations for their line-drawing Moon map made in 1911 and 1922 for IAU (Blagg and Müller 1935). As for Mars, the new, canal-free IAU map of Mars, presented in 1958 (IAU 1960), was based on both visual and photographic observations (Dollfus 1961) at the Pic du Midi Observatory.

2.3 *Transition to Modern Planetary Mapping*

Transitional periods are the most exciting in history. These nodes in time define future practices. The technology, organization, and the place of mapping all have changed fundamentally by the beginning of the 1960s when the world entered the space age, marking the transition from traditional to modern planetary mapping. Changes in mapping practices were accompanied by a significantly advanced understanding of the geologic processes that shaped the mapped surface.

At this time, static, but still Earth-based *photographs* provided a solid base for mapping instead of the ever-changing and personal telescopic view. For a short period in the 1960s, decades after the first photographic lunar atlases were published, lunar cartographic products (atlases, image mosaics, contour maps) were based on carefully selected photographs that had been taken at the largest observatories. These projects used the best available data, which could be compared to the invention of writing that decouples the text from limitations in time and space: Observations from different times and places were combined for the first time.

At the end of the 1950s, IAU, an *international organization*, began new mapping projects that were based on the photographs of the largest telescopes, and supplementary visual observations, for both Mars and the Moon. Finally, the focal point of planetary mapping activities moved from Europe to the *USA* (and also the *Soviet Union*) where, now entering the Space Race, *systematic* planetary mapping was organized by the military and civilian *government-sponsored agencies.* This was the scene when the first spacecraft were launched.

This change of technology coincided with a change in the organization of map production. Teams of cartographers, planetary scientists, and graphic artists, and later mission team members, produced maps instead of individual observing astronomers, and typically by government-sponsored agencies that operated space missions.

The *Photographic Lunar Atlas* (PLA) (1960) project was led by G. Kuiper at Yerkes Observatory of the University of Chicago. Plans for this Atlas were first discussed by IAU in 1955. The Atlas was revolutionary in that it contained the best available photographs of not one, but five large observatories. It showed the nearside of the Moon, in several different illuminations, as seen from Earth. Photographs were selected in 1959 (Kuiper 1959). The photographs in the Atlas were oriented south-up, for use at the telescope. Its No. 2 supplement, the *Rectified Lunar Atlas* (1963) (Fig. 5), used a novel projection method: Earth-based photographs were projected onto a physical globe (Whitaker et al. 1963;

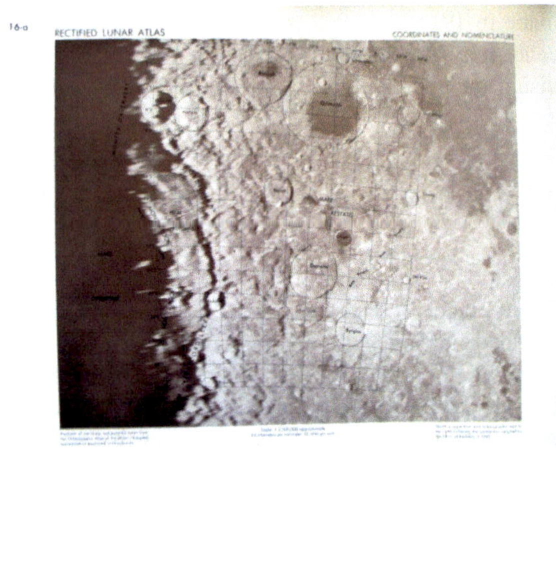

Fig. 5. "Discovery image" of the Orientale impact basin. Parts of Mare Orientale were previously known as individual marginal mountains. Photograph of the Rectified Lunar Atlas, Table 16-a (Whitaker et al. 1963)

Spradley 1962). The physical globe with the projected images was then photographed by directly looking "down" to the globe from the vantage point of an imagined spacecraft orbiting around the globe. This process generated photographs in orthographic projection, as seen from above the center of the photograph. This project produced the first rectified images of the limb regions; however, the analog photorectification method was originally developed around this time for the production of the LAC series at ACIC. Photographs there were rectified with an instrument called "Variable Perspective Projector." The resulting image was rephotographed and used as a mapping base (Carder 1962). Rectification technique abandoned Earth-bound viewing angles and provided an early preview of the space age orbital views of planetary surfaces. Just as digital mosaics today, this atlas was used for studying the Moon, including crater counting (Baldwin 1964).

By the time these atlases were published, the United States government also have had become interested in lunar mapping, as part of its Space Race politics. During these years, planetary mapping was taken over by mapping teams of professional (civilian and military) cartographers utilizing cutting-edge cartographic and instrumental techniques, to prepare for human lunar missions.

At this time, three options were considered in the USA and probably in the Soviet Union also (Myler 1957), for the use of the Moon as a resource in the Space Race: dropping a nuclear bomb on the Moon as a globally visible demonstration, establishing a permanent military base, and sending men to the Moon for a short trip. Each of these projects were fundamentally political missions. The US Air Force involved Kuiper who led lunar mapping at Yerkes (e.g., Ulivi 2004:44; US Army 1959). However, the United Nations forbid testing nuclear and other weapons and establishing military bases on the Moon in 1967 (UN 1967). Although lunar military plans were already dropped by that time, these early military mapping procedures defined the framework for subsequent planetary mapping. Nuclear bombs also played a role in the birth of modern astrogeology, also during this time, when nuclear explosion craters, observed by E. M. Shoemaker in 1959, provided evidence for the impact (explosion) origin of lunar craters and experimental data for the development of the method of crater counting (Shoemaker et al. 1963). Shoemaker created the prototype of lunar geologic maps in the following year, in 1960, of the Copernicus crater region.

Two independent lunar mapping programs started at the military: one at the *Air Force Aeronautical Chart and Information Center* (ACIC, St. Louis, MO) and another at the *US Army Map Service* (AMS, Washington, D.C.) in 1957 and 1958, respectively (Weir 2009).

AMS investigated the requirements for establishing a subsurface military outpost on the Moon. This project was called *Horizon* (US Army 1959). AMS focused on topographic mapping: It began works on a 1:5,000,000 topographic map of the nearside of the Moon in 1961, using stereo-photogrammetric technique on photographs from the large observatories. It is the first lunar topographic map that used photogrammetry, and also the first in stereographic projection (1-AMS 1963) (Fig. 6). Another, 1:1,000,000-scale series was planned to be completed between 1959 and 1962, based on future "earth-orbiting telescope camera systems," and

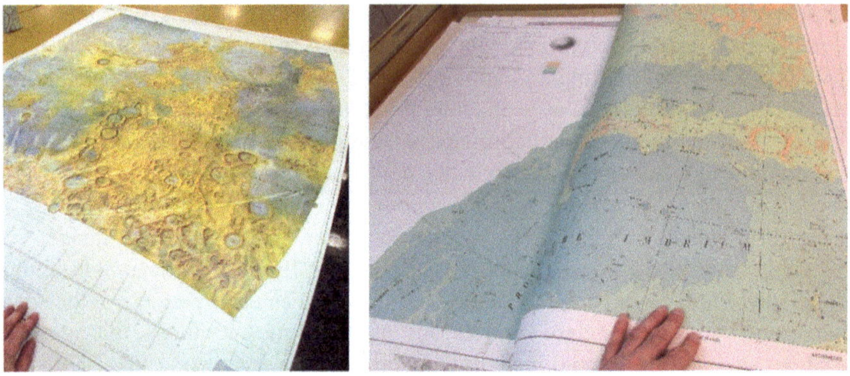

Fig. 6. Two sheets from the AMS topographic lunar maps (1962). Left: shaded relief version, right: gradient tint version (*Photograph* by H.H.)

"lunar circumnavigating and satellite vehicles." According to the plans, by 1964, 1:10,000 maps would have supported the landing site selection (US Army 1959).

ACIC also began mapping the Moon and planned to obtain far side images for creating topographic maps in cooperation with the newly (1958) established NASA. This mapping project was to support a planned human mission to the Moon within its proposed *Lunex* program (USAF 1961). This project continued as the Lunex and Horizon projects were dropped and the Apollo program was born. ACIC published the 1:5,000,000 Lunar Reference Mosaic (LEM-1) map in 1960 and was available by the time J.F. Kennedy announced the Apollo program. ACIC focused on the novel *airbrush hillshading* technique in its *Lunar Astronautical Chart* (LAC) (Carder 1962) for which ACIC used its existing terrestrial World Aeronautical Chart series used for navigation by pilots as a model (Weir 2009; Kopal and Carder 1974:115). LAC was the first lunar mission planning airbrush map series supporting the US Space Program, utilizing photographs from several observatories (similar to PLA) with supplementary visual observations first at Yerkes Observatory (with the help of Kuiper, Arthur, and Whitaker) (Carder 1962) and at the newly established Air Force Observation Unit at Lowell Observatory in Flagstaff (1:500,000 Apollo Intermediate Chart (AIC) series, 1:1,000,000 LAC series). ACIC put in enormous efforts into determining elevations, and while AMS experimented with stereogrammetry, ACIC chose the traditional shadow measurement technique. Photographs were taken every 20 seconds of the Moon at Pic du Midi with the direction of Zdenek Kopal, amassing 12 thousand images over 2 years and measured shadow lengths using a microdensitometer (Carder 1962).

Following this short period dominated by the US military, planetary mapping became coordinated by a civilian US government agency, the *United States Geological Survey* (USGS), where now astrogeologists took over planetary mapping. Mapping had a well-defined aim: to support future NASA missions and analyze the returned space-borne images. In Flagstaff, home of the Lowell Observatory, USGS began an institution-based systematic planetary survey program using NASA mission data (USGS Planetary Geologic Mapping Program, Tanaka et al. 2011). Another positive consequence of this move from private to government-funded mapping is that USGS maps and NASA's planetary image data are in the public domain, freely accessible to everyone.

Over the next decades, photographic atlases became common platforms for publishing planetary maps in easily accessible form. These books reproduced photomosaic or shaded relief NASA maps or newly produced photomosaics with nomenclature overlay with a collection of spectacular images (Batson et al. 1984; Davies et al. 1978; Greeley 1994; Greeley and Batson 1997; Bussey and Spudis 2004; Schenk 2010; for the new Indian Mars Orbiter Mission images: ISRO 2015). However, in addition to image mosaics and topographic maps, a third kind of planetary map was born: geologic maps.

2.4 The Beginnings of Astrogeologic Mapping

Until the 1960s, lunar cartography was developed by adding more and more minuscule details in the maps, as telescopes became more powerful, in accordance with a descriptive geographic approach (referred to as "selenography" for the Moon and "areography" for Mars). It became evident, though, that a new approach is needed to unravel the processes that shaped the landscape. This new conceptual approach was introduced with the foundation of the USGS *Brach of Astrogeology*. Instead of showing the surface morphology (or "physiography") in ever higher detail, and artistic realism, new planetary maps showed surface units classified by their geologic properties. Two novel photogeology-based approaches were proposed in 1960, which both required laborious manual investigation: one by Robert J. Hackman and Arnold C. Mason at AMS, and another led by Eugene Shoemaker at ACIC/USGS.

Mason and Hackman (1962, first print 1960) created a nearside map trio (Wilhelms 1993:37-40), still in traditional Earth-based orthographic projection, but North-up. Hackman's *photogeologic map* of the Moon defined rock formations based on their inferred age derived from *stratigraphic* relationships (pre-maria, maria, post-maria). This mapping required the delineation of the major stratigraphic units (enclosed by contact lines) across the entire surface. Hackman used stereopairs of photographs from different observatories enlarged to approx. 1:5,000,000 scale for the mapping, obtained with the help of G. Kuiper (Wilhelms 1993:38) (AMS used stereogrammetry to produce a contour line lunar map at that time). Another map, showing *physiographic divisions,* delineated and classified surface regions by *morphology* into lowlands, highlands, and crater provinces, and named them following geographic traditions (e.g., "Northern Lowlands"). This map followed a broader, landscape-scale categorization of the relief, where terrains (landscapes) were assumed to have resulted from a long and complex surface evolution (divisions based on "degree of similarity or difference in type of surface features, extent of preservation, type of modification, type of surface material, elevations, slopes, and structural disturbances"). Morphology, materials, features, and, perhaps most importantly, evaluation of landing, surface movement and in situ material excavation ("construction") possibilities, were described for each region. Additionally, a thematic feature map of the lunar rays completed this series of the "Engineer Special Study of the Surface of the Moon." Mason and Hackman's newly proposed nomenclature, physiographic (terrain) and chronologic division, was not accepted by the community. Physiographic analysis was replaced by a geologic approach.

A concurrent and independent photogeologic project in 1960 utilized the established method of terrestrial *geologic mapping* delineating *material units* based on stratigraphic analysis, for just a relatively small region around Copernicus crater (Shoemaker and Hackman 1962), at ACIC (drawing, printing, airbrush base chart) and USGS (where Shoemaker established the Branch of Astrogeology at this time), for NASA. This map departed from the telescopic-view-defined orthographic projection and used Lambert Conformal Conic projection. Geologic units were grouped according to their relative age (derived from superposition) and assumed material. Surface materials were inferred from morphology, which suggested a particular formation mechanism. The test map included a description of the observed morphology of each unit, their assumed material and formation process. The map also included a geologic cross-sectional view. This work also established the lunar timescale. This mapping approach immediately became the prototype, as intended (Shoemaker and Hackman 1961), and standard procedure for systematic planetary geologic mapping (Shoemaker and Hackman 1961, LPC 58 Chart, 1:1,000,000) (Portree 2013). Hackman also participated in the series, mapping the Kepler crater region (the first map in the ACIC series, Carder 1962), also signifying the end of the telescopic era in that he used his own "visual telescopic observations at the Leander McCormick Observatory" in 1960–'61 (USGS map I-355), perhaps the only planetary geologic map based (partially) on visual observations. This project also produced the first research paper on planetary stratigraphic analysis (Shoemaker and Hackman 1962). It was established that major periods in the geologic time scale can be defined by a type locality, or an age-indicator feature (e.g., crater rays).

Shoemaker and Hackman (1962) recognized first that "the geological law of superposition is as valid for the Moon as it is for Earth." With the introduction of geologic mapping in planetary science, it became possible to understand and reconstruct the evolution (chronology) of planetary surfaces. Interestingly, most methods and techniques used in planetary geologic mapping were laid down (Steno 1669) much before *planetary* geologic mapping was "invented." Terrestrial geologic mapping standards were already implemented at the USGS after its 1879 foundation (US Geological Survey 2006). The construction of a relative surface chronology is based on both stratigraphy and crater counting. Unlike morphology, crater counting can be used for *quantitative* age determination of geologic units. Crater counting was a new element in planetary geologic mapping, developed during the 1960s. The first attempts to connect relative ages to absolute ones used terrestrial analog regions to estimate the impactor flux over time (Öpik 1960; Shoemaker et al. 1963; Baldwin 1964; Hartman 1965). Just a few years later,

Apollo astronauts collected and returned lunar samples that could be used for accurate radiometric dating that in turn could be linked to crater density.

The discovery that craters are also common on Mars (Hartman 1972), and that they are the most common planetary landforms, enabled the usage of this technique across almost all solid surface solar system bodies, as proposed by Shoemaker et al. (1963).

Mars's surface morphology was first globally mapped by Mariner 9 (1971). The first geologic map of Mars used the same technique and conventions as for the Moon earlier, and *geologic* interpretation was based on *topographic* features that were "highly diagnostic of their origin" (Carr et al. 1973). Mariner 9-based stratigraphic mapping was formalized in the 1:25,000,000 geologic map (Scott and Carr 1978), which was later updated using Viking Orbiter (1976–80) imagery and crater density values were also assigned to units (Tanaka 1986). These maps were drawn over manually airbrushed shaded relief maps, while the most recent global geologic map of Mars is drafted over MOLA and THEMIS mosaics, similarly using photogeologic methods (Tanaka et al. 2014). Radar provided new geologic information as Earth-based and orbiting imaging radar data showed cm-scale surface roughness and radar reflectance properties, especially for mapping Venus.

The delineation of geologic material units is based on a standard procedure. On Earth, material properties are objective, descriptive parts of geologic maps. However, planetary material units are commonly delineated and properties *inferred* using macroscopic *morphologic or spectral* properties and not direct geochemical/elemental or microscopic analysis, or core sampling. Relative unit ages can be confidently determined through stratigraphy. Crater counting of homogenous map units and the application of crater chronology systems can provide cratering model ages that can be utilized to globally correlate spatially distant units, providing an "absolute" reference age (Williams 2016). New, higher resolution, or different wavelength image data, together with new impact flux models and statistical methods, may result in a different delineation of the units in updated geological maps (Fig. 7).

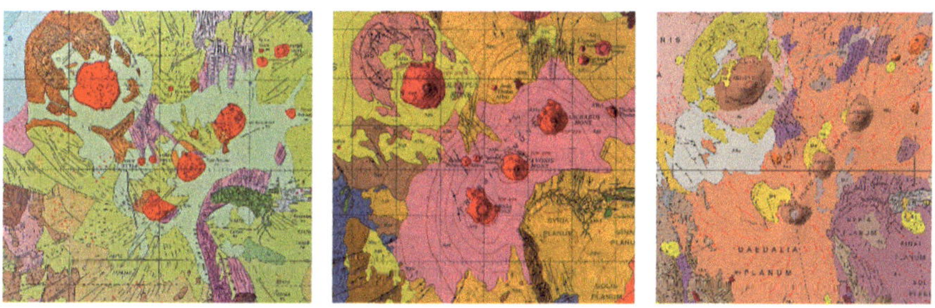

Fig. 7. Map cutouts of the Tharsis Montes region, Mars. Left: geomorphologic map, Makarova et al. (1978). 1:20 M; middle: geologic map, Scott and Carr (1978). 1:25 M; right: geologic map, Tanaka et al. (2014) (paper version), 1:20 M

2.5 New Mapping Techniques During the Space Age

Despite successful Earth-based telescopic and radar observations of Venus, Mars, and the accessible lunar nearside, topographic features of the planets and their satellites could only be mapped in detail from observation platforms onboard spacecraft.

The era of *space-borne mapping* started with the first set of pictures received from the far side of the Moon (Luna 3; Babarashov et al. 1960) and Mars (Mariner 6; Davies et al. 1970). Ranger VII–VIII photography was transformed into a series of different scale (1:1 M-1:10 k) airbrush maps (*Ranger Lunar Charts*) in 1964–66. Just a few years later, *Lunar Orbiter* images were used directly in the photomosaic maps of the equatorial zone of the Moon (LEMC), compiled manually (US Army 1968); and a global photographic reference atlas was also published (Bowker and Hughes 1971).

The first image-mosaic-based globes were produced, manually, from Mariner 9 Mars photographs glued to a very large globe (Staff 1973). Large-scale (1:10–1:50 k) "topophotomaps" during the mid-1970s were produced from Apollo 15 and 17 high-resolution panoramic photography with red 10- to 20-m-interval contour line overlay.

In *topographic mapping*, shadow length measurements (Schröter 1791) mainly provided local elevation data: peak elevations and crater rim profiles. The first attempt to create a topographic contour map of the Moon is that of Franz (1899). These techniques were superseded by stereophotogrammetry (AMS 1963), laser altimetry (Wollenhaupt and Siogren 1972), Earth-based and space-based radar altimetry (Pettengill et al. 1969, 1980), and radar interferometry (Zisk 1972). The photogrammetric control point technique developed by M. Davies at RAND Corporation was introduced to provide a basic reference framework for controlled planetary maps (Davies and Berg 1971) and served as a basis for stereophotogrammetry and bundle adjustment techniques (Wu 1978). Digital Elevation Models in 2018 are available for Mercury, Venus, the Moon, Mars, Ceres, Vesta, and the encounter hemispheres of Pluto and Charon. Regional topographic maps are generated from stereo- and photoclinometry for several other bodies.

New photometric technologies made it possible to objectively measure *radiance* data per pixel, first only indirectly, on photographic plates, and later with direct photoelectric telescopic observations (Pohn et al. 1970). Data from space-borne platforms all were digital that needs calibration.

While Mars was initially mapped visually as a disk with blurred dark markings, new, Earth-based techniques allowed *indirect mapping* of planetary surface markings before spacecraft could have visited and resolved them into multipixel images. Crude maps of surface albedo can be derived using the light curve inversion technique that analyzes the change in the photometric record of a body's light curve due to planetary rotation, occultation, or transits. Using light curve data collected over many decades, Iapetus was revealed to have two contrasting albedo hemispheres (Morrison et al. 1975) (Fig. 8b). Occultation light curves provided data for the first

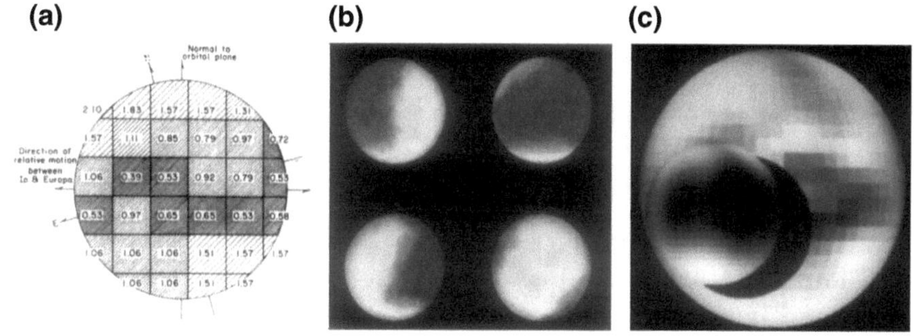

Fig. 8. Albedo maps derived from light curve analysis. Left: map of Europa, centered on 324°W longitude, showing relative reflectivity values on a scale normalized to 1.00 for the whole disk, visualizing reflectivity bins with cross-hatching in a 6 × 6-grid resolution (Vermilion et al. 1974); center: "artistic representation" of Iapetus based on a computed albedo distribution model centered on 0, 90, 180, 270° longitudes (Morrisson et al. 1975); right: Modeled albedo image of the transit of Charon across Pluto determined from right curve inversion (Buie et al. 1992)

map of Europa (Vermilion et al. 1974) (Fig. 8a). Pluto's surface map before the New Horizons' visit was calculated from mutual event light curve data, using a "disco ball" visualization (Buie et al. 1992, 1997; Young et al. 2001) (Fig. 8c).

2.6 *Planetary Mapping and Maps in the Digital Era*

Parallel to terrestrial developments, since the mid-1990s, digital cartographic techniques using vector- and raster-based graphic software arose. In the map *production* process, a major step was the introduction of the requirement that USGS geologic maps be produced in a *Geographic Information System* (GIS) *platform*. Astrogeologic maps are produced and published in GIS format at USGS since 1996, exclusively since 2011 (Tanaka et al. 2011).

NASA-produced digital web mapping tools, such as the *Mars Global Surveyor Interactive Data Maps,* were already available in 1999 and became quickly popular (Gulick and Deardorff 2003), almost a decade before Google released its planetary products. Today the production and display of planetary maps are available in diverse GIS and WebGIS/Web Mapping Service (WMS) platforms that support multilayer, multiple projection, multiple style, and multiscale viewing (Dobinson et al. 2005). GIS technologies dominate in planetary mapping and cartographic representation (e.g., Hare and Tanaka 2001; Hare et al. 2009; van Gasselt and Nass 2011; Nass et al. 2011; Frigeri 2011; Hare et al. 2015b).

Digital technologies have not only changed the mapping interface for both map creators and readers, but it is changing how map data are collected. New, supervised and unsupervised classification and machine learning techniques are emerging and are capable of automatically mapping surface features. After appropriate training, the user interface is not a map, but a content-based image search (e.g., “show images with > 30% coverage of dunes *and* fresh craters”). Neural networks capable of “mapping” millions of images (Wagstaff et al. 2018) and research papers (Wagstaff et al. 2016) could revolutionize planetary mapping the production and use of planetary maps.

2.7 *Planetary Cartography in the Soviet Union and East Asia*

During the Space Race time period, both the USA and the Soviet Union produced their lunar map series using manual hillshading and albedo, initially both involving military mapping facilities. However, in time, USGS maps became more and more detailed (from 1:5 M to 1:5 k) while Soviet mapping moved toward large-sized representative, thick-framed 1:10 M global maps of the Moon printed in 10,000 copies (Rodionova et al. 1985). USGS produced 173 geologic planetary maps between 1961 and 1990, during which time the Soviet Union produced 15, and other nations none (Hargitai and Pitura 2018).

The seven-sheet 1:1 M central nearside lunar map series (Lipsky 1968) and the nine-sheet 1:5 M “Complete map of the Moon” produced at the Sternberg Astronomical Institute with updated editions between 1967 and 1989 using the results of Soviet and American missions use similar visualization principles (but different colors) to the American airbrush maps (Shevchenko et al. 2016). Only few sheets of 1:5 M Mars maps were published, using images of the Soviet Mars 5 probe and covering small regions (Tefinim and Krestnikova 1977). In 1987–88, Venus radar maps of the Venera 15–16 Synthetic Aperture Radar images were published in both separate sheets and atlas format in the Soviet Union. Exceptionally, a “US/USSR Joint Working Group on the Solar System Exploration” produced a three-sheet (topographic, shaded relief, radar image) map from these Soviet radar data to support the American Magellan mission planning (USGS 1989, Basilevsky et al. 1990). Instead of geologic material unit mapping, Soviet scientists produced the “Tectonic map of the Moon” (Kozlov et al. 1969) and geomorphologic, structural geomorphologic (Makarova et al. 1978), and geologo-morphologic maps of Mars (Bugaevsky et al. 1992), and the USGS-published geomorphic/geologic map of Venus (Sukhanov et al. 1989) that latter show what Basilevsky et al. (1989) calls the “Soviet interpretation of the geology”. A unique product of Soviet-era planetary cartography was the first comprehensive cartographic planetary atlas. It was produced in over a decade of cartographic work at MIIGAiK, Moscow, and contained global thematic maps compiled using conventional map drawing techniques, and pencil-drawn shaded relief representations of the terrestrial planets and their moons along with a large number diagrams, photographs and thumbnail reproductions of historic and modern planetary maps (Bugaevsky et al. 1992[1]). Lunar maps from this

[1]http://planetmaps.ru/atlas.html.

Atlas were also published as a school atlas (Shingareva and Krasnopevtseva 2011). The 1970–1980s in the Soviet Union marked the peak in shaded relief map production of Mars, Phobos, the Moon, and radar maps and atlases of Venus (GUGK 1988). Today Russian planetary cartographic centers produce a variety of outreach-type global topographic maps and globes of terrestrial planets and the Moon (e.g., Grishakina et al. 2014; Lazarev et al. 2012) with a signature two-hemisphere layout. MIIGAiK's MExLab developed its own Web mapping service called Planetary Geoportal (Garov et al. 2015). Following their traditions in atlas cartography, the comprehensive atlas of Phobos offers a detailed view of this irregular body (Karachevtseva et al. 2012, 2019).

China recently joined the countries with a complete planetary mapping infrastructure from data acquisition to publication. Results from the *Chang'E* lunar probe series were published in photomosaic and topographic map, atlas and globe formats (e.g., Compiling Committee 2013, Mu et al. 2019), and also in an interactive WMS for Chinese audience. In *Japan*, the perspective views of lunar landscapes from the Japanese Kaguya mission's HDTV camera were published in a pictorial atlas format (Shiao and Wood 2011).

3 The Short History of Specific Cartographic Tools

3.1 Generating Synthetic Views: Concepts

Lunar telescopic observations reveal two "faces" of the Moon: one during high-sun (noon) and another representation reflecting the visual experience during low-sun (sunset/sunrise). The high-sun Moon shows albedo features: subtle tone variations, bright rays around craters, and dark and bright spots of different sizes and shapes. The low-sun Moon shows sharp shadows, which are cast to the east or west, depending on the actual local Moon time. While the high-sun (albedo) view of the Moon had not directly implied any geologic explanation before geologic mapping began (the water vs. land hypothesis was not considered very seriously in the last centuries), low-sun shadows and shading reflected topography, a key to geology. Later, Mars showed only the high-sun albedo view, while Venus revealed none of its surface optically.

For the lunar mapper, the basic problems in visualizing the lunar surface are how to synthesize the two characters of the lunar disk (brightness or albedo, and topography) and how to derive a single picture of the relief from the shading and shadows that change length, shape, and direction with the movement of the terminator line. The first mappers either produced one map view that combined albedo and sharp shadows, or separated the two themes. A combined-theme map assumes two different illumination conditions at the same time. Albedo is derived from a realistic full Moon view, showing the lunar disk during sub-Earth noon (or local noon views). The topographic view applies shading for the entire lunar disk with a single illumination angle, as if local times would be similar (near sunset or sunrise) on the entire disk. Maps may combine these two views: the uniform-illumination shading and albedo. The reader who is used to seeing planar surfaces with uniform illumination does not find anything unusual in a map view where all craters cast similar-oriented shadows. It is not realized consciously that here a sphere is projected on a plane and this synthetic view of normalized shadows never occurs in nature. Seller (1700), for good reason, called these views "Phasis Lunae Naturalis" and "Phasis Lunae Artificialis." Moreover, shadows and shading do not exist as materials do. Shading itself is not an inherent characteristic of the surface materials. Hillshading as a separate map theme is only used to create the mental image of the 3D relief in the map reader's mind, using a 2D technique, without perspective view.

The problem of showing relief was approached with different visual methods and techniques, from line drawing to hachure to airbrush hillshading, but the problem always remained how to create a synthetic, uniform reflectance and relief view of the surface as it never appears in real life. Generating composite maps usually needs some compromise. A topographic shading overlay on an albedo map makes albedo information ambiguous because both are expressed in a change of tone. Adding a low-resolution color overlay on a high-resolution panchromatic photomosaic is a common method of enhancing the map; however, if the image mosaic is in color, colors cannot be used to express elevation.

Modern technologies allow the creation of uniform-illumination maps by mosaicking multiple photographic observations, or synthetically by the generation of hillshading effect from elevation data. Albedo (reflectance) maps can be produced directly by combining noontime visual observations: DEMs have no albedo information. On the other hand, new computational techniques allow the reconstruction of surface albedo from low-sun images, by removing all effects of shading from the scene using DEM (Nefian et al. 2013).

Today, generic maps can be produced from the combination of several base raster layers: for example, DEM (coloration), reflectance (monochromatic tone that shows visual or infrared albedo), and shaded relief (Table 1).

3.2 Topographic, Brightness (Albedo), and Physiographic Maps: Examples

As for general (reference) visualization of a planetary surface, the images seen over a telescope were replicated by hand drawing and line engravings for printing. Albedo, relief, and nomenclature can be separated or combined in many ways.

In 1647, *Hevelius* produced three views: a realistic full Moon engraving (equivalent of a photograph), a symbolic line drawing with termite hills and names (the cartographic map), and a simple sketch map. *Grimaldi and Riccioli* made an engraved picture without names and a similar view with octant grid and carefully typographed names.

Table 1. A list of map themes and their source data

Map theme	Origin	Type
Albedo, visual	High-sun photographic or visual observation	Observed
Color, true or false	Multispectral observation	Observed
Relief: shadows	Low-sun photographic or visual observation or DEM	Observed or reconstructed (calculated)
Relief: shading	Low-sun photography or DEM	Observed, manually created or reconstructed (calculated)
Chronology: position (relative age)	Contact lines (manual determination units)	Derived from superposition (stratigraphic relations)
Chronology: absolute age	Crater counting, impact flux functions, radiometric ages of lunar samples	Derived from crater size frequency curves
Material	Morphology or spectral data (albedo, hyperspectral)	Inferred
Formative process	Morphology (shape, relief, pattern), albedo, or material	Inferred
Nomenclature	May be proposed by mission team. Assigned by IAU. Descriptor term is based on morphology.	Descriptor element is derived; specific element is not geology related
Surface roughness	cm-dm scale: radar echo, km-scale: from DEM	Observed or derived
Thermal inertia	IR observations (day and night)	Calculated from observation
Relief: elevation raster/ contours	Radar or laser altimetry, stereogrammetry, radar interferometry, shape from shading	Calculated from observation
Morphology/ structural feature type	Shadows, shading, DTM	Visual inspection or automated machine learning

Cassini's Moon map, engraved by *J. Patigny* (1679) (Whitaker 1999:140), shows albedo combined with a fine hillshading with a superior 3D effect.

The map of *J. Russell* (1805) combined pictures from his 40 years of observations separating albedo and topography and published two drawings of the Moon: a contrast-enhanced full moon (with "the rays of the sun falling oblique upon [the surface]") and one showing topographic details only (with "rays falling perpendicular to it") (Whitaker 1999:99–100). This separation of albedo and relief reappears regularly, both on Mars and Moon maps.

In finding ways to depict the lunar surface realistically, J. Nasmyth produced 3D plaster terrain models and photographed it in grazing light, instead of using actual photographs (Koeberl 2001). These pictures determined how the cratered lunar surface was seen for many decades. In their book, *Nasmyth and Carpenter* (1874) showed two representations of the lunar nearside: a *picture map* and a *sketch map* (Fig. 9).

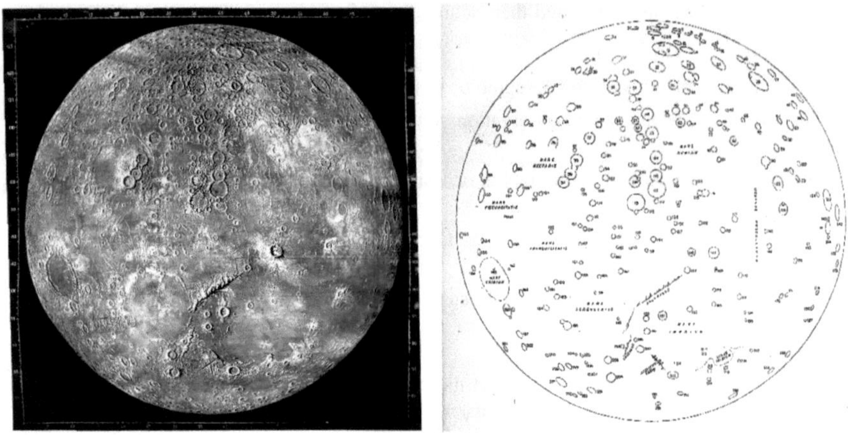

Fig. 9. Pictorial (left) and sketch map (right) of the Moon (Nasmyth and Carpenter 1874)

The picture map was copied from *Beer and Mädler's* detailed map (1841), and background tones and shadows were *manually added.* The hillshading in this map has uniform northwesterly illumination to convey a "fair impression" of the lunar landforms. USGS shaded relief maps continued this tradition by applying uniformly westerly (Batson 1991) and later northwesterly (USGS 2017) illumination.

The simple "*skeleton map*" was created using a micrometer, and showed the outlines of craters and mountains, with additional labels. This map showed the need for an interpreted, generalized (more abstract, less realistic) but accurate, *cartographic* representation of surface landforms.

This technique developed into a double-layer representation where the sketch map with nomenclature was printed on a transparent paper that was placed over a realistic drawing (e.g., Flammarion 1900).

In the early space age, similar, but more detailed, *line-drawing* outline maps served as reference to identify the crater inventory of the nearside of the Moon (44 sheets, Arthur et al. 1963, also used as nomenclature reference maps, Fig. 10), the far side of the Moon (32 sheets, Lipskiy 1967), and Mars (30 sheets, Mutch et al. 1976, this being also one of the first line-drawing maps drawn by computer). These maps using a technical drawing style were now the authoritative cartographic products.

With a similar approach, 1:50 M line-drawing "*blank maps*" were produced initially as preparation for the production of more complex maps in the Atlas of Terrestrial Planets and Their Moons (Bugaevsky et al. 1992), but they were eventually included in the atlas.

Cartographic line-drawing techniques moved to a next level in geologic maps that now showed contact lines in addition to structural feature outlines.

Image maps of airless bodies where low-sun shadows are sharp and dark frequently show the same region in multiple illumination angles. The *Rectified Lunar Atlas* displayed the same regions in five complementary views from near-terminator

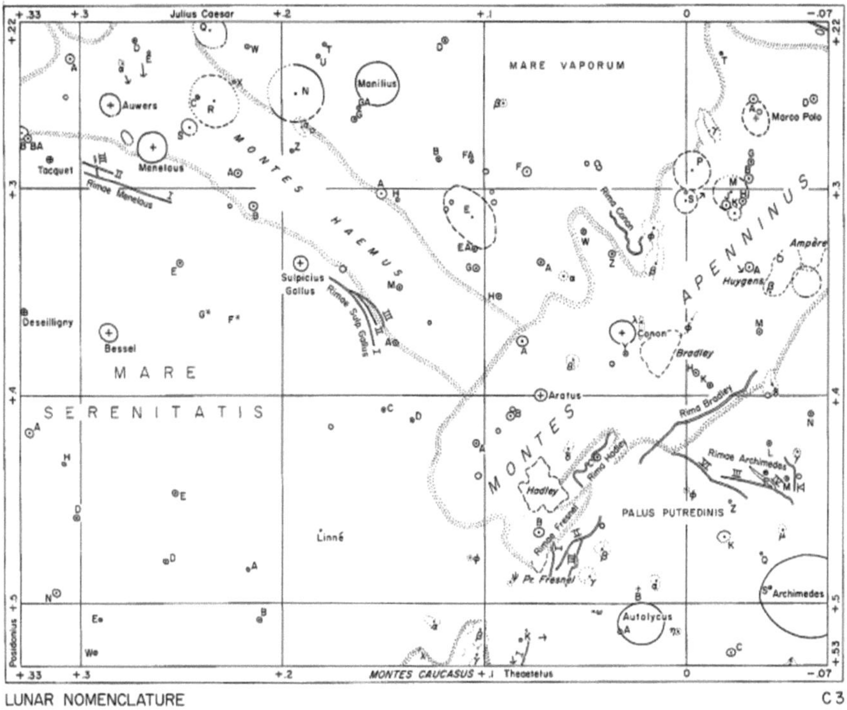

Fig. 10. Technical-drawing-style map of the Montes Apenninus region, Moon (Arthur et al. 1963, Sheet C3). South is up. Cf Fig. 11 that shows the same area

morning and evening images that accentuate low relief morphology, to high-sun mosaics that accentuate albedo variations; and a view with a nomenclature and grid overlay. *MESSENGER Mercury* images were acquired for building four different mosaics: a grazing light eastern and western mosaic, a moderate incidence angle mosaic, and a high-sun mosaic. These image mosaics are also synthetic as the mosaics combine individual images showing these regions at similar local times (Murchie et al. 2017).

W. G. Lohrman, a selenographer and professional cartographer, introduced *hachures* in 1824 (Fig. 11) to indicate length and steepness of slopes, a new technique in terrestrial cartography at that time (Whitaker 1999:116). His maps showed albedo variations using several shades of grey.

A new style of *hillshading* was developed in the 1970s where some maps were drawn by pencil and charcoal in the USA by the British cartographer C. Cross (Morton 2002:50) and in the Soviet Union at MIIGAiK. However, airbrush superseded this technique.

The realistic *airbrush hillshading* technique was introduced into planetary cartography by *P. M. Bridges* in the US Air Force ACIC lunar LAC maps in 1959 (1:1 M, 44 sheets out of the planned 144, 1959–1967; Carder 1962), which

Fig. 11. Montes Apenninus region, Moon (Lohrman 1878, Sheet IV). South is up. Cf. Fig. 10

revolutionized the representation of planetary surfaces (Inge and Bridges 1976; Batson 1991; Schaber 2005). Air Force mappers moved to the Lowell Observatory to draw maps of the Moon and Mars from telescopic views (IPPP 1971), where *J. L. Inge* augmented the airbrush technique. Bridges and Inge were hired by the USGS Astrogeology Branch also in Flagstaff (Morton 2002:51) to perform "topographic interpretation," i.e., shaded relief map production from images and visual observations.

These manually produced shaded relief maps incorporate details that never appear in a *single* photograph and *exclude* inherent albedo information ("surface markings") (Batson 1973). However, if needed, a synthetic albedo drawing can be added to hillshading (e.g., USGS Moon map i-2276, plates 1 and 2 showing manual hillshading with and without albedo markings).

Airbrush was utilized in more than a hundred sheets of lunar maps, mostly based on Lunar Orbiter and Apollo photographs. This 1960s technique reappeared with the computer-processed Mariner 9 Mars images that contained numerous artifacts and were taken by a variety of atmospheric, illumination, and surface conditions at different resolutions and therefore had to be synthesized manually (USGS 1973; Inge and Bridges 1976). Airbrush technique was also used throughout the Voyager mission.

The Mariner 9 orbiter mission, along with Earth-based radar observations, presented new challenges in data integration. It generated a diverse set of data that were used to produce the first topographic map of Mars (Wu 1978). The final map showed the relief using shaded relief airbrush drawing combined with 1-km contour line, or albedo ("surface marking") overlays (Batson 1973).

Manual airbrush technique could not be easily employed for interpreting radar mosaics which lead to the development of *computer-based* hillshade production (Kirk 1993), although radar-derived airbrush maps were also produced by Bridges (USGS 1989). Airbrush maps were gradually replaced by shaded relief maps produced from gridded digital elevation data (DEMs) (Batson et al. 1975; USGS 1984).

Another element in visualization is the use of signature *color schemes*. The application of saturated, vivid colors in the "*Geologic Atlas of the Moon*" series, also published in nearside and farside thematic views (Wilhelms 1987), became a signature of post-Apollo era lunar geologic mapping (e.g., Wilhelms and El-Baz 1977) and were coincided with and perhaps subconsciously influenced by the color schemes of the contemporary pop culture originating from the San Francisco-based Psychedelic Movement. Most other lunar reference/planning maps were published in bluish to yellowish colors (for maria and highlands, respectively), for example, in the multisheet LAC series, and Chinese maps (AMS 1963; Carder 1962 Compiling Committee 2013) modified to blueish to brownish in the LOC and LMP series (Schimeran 1973) and deep blue to gray to yellow tints in the most recent one-sheet topographic map (Hare et al. 2015a). In contrast, Soviet lunar maps had a brown-orange to white color scheme (Lipsky 1968) (Fig. 12). Modern *topographic* mapping is symbolized with the rainbow colors of the MOLA Mars topographic map (Smith et al. 1999).

3.3 Map Sections and Schemata

It was Grimaldi who first divided a planetary map into segments. His one-sheet Moon map consisted of eight circle sections or *"octants"* instead of a coordinate grid. Multisheet maps were introduced by Lohrman who constructed a lunar map in 25 square sheets, published between 1822 and 1878. The most time-consuming part of the map production was to engrave the drawings into copper plates (Whitaker 1999). Beer (a banker who provided his rooftop observatory) and Mädler produced their lunar map (1834) in four large sheets, combined from 104 manuscript sections.

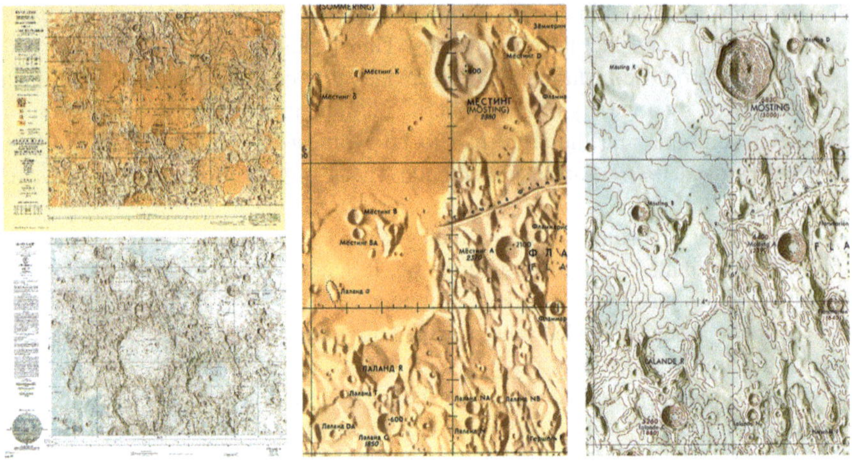

Fig. 12. Soviet (orange) and American (blue) 1:1 M general (planning) map sheets of an overlapping central nearside region of the Moon. Detail images show the same region. (A) Karta Luny, List 4, Zaliv Tsentralniy. Shaded relief with height points. Sternberg State Astronomical Institute, Nauka, Moscow, 1968. (B) Lunar Chart LAC-77 Ptolemaeus. Aeronautical Chart Information Center, United States Air Force, 1963. Shaded relief with contour lines

For systematic, long-term geologic mapping, USGS divided the surfaces of planets to schemes of named and numbered mapping *quadrangles*. These include the 44-quadrangle scheme for the Moon at 1:1,000,000 scale (Wilhelms 1972) and 30 quadrangles (Mars Charts – MCs) for mapping Mars at 1:5,000,000 scale (Batson 1973). The 62-quadrangle scheme for Venus at 1:5,000,000 scale was first used in the Soviet Venera mapping (GUGK 1988), where 27 northern quadrangles were mapped. Projections of quadrangles are also standardized, first defined for the LAC series (Carder 1962): Mercator is used for the equatorial quadrangles, Lambert Conformal Conic for intermediate latitudes, and polar stereographic for the two polar quadrangles. All these projections are conformal. Quadrangle schemes for different-sized bodies at different scales are presented in Batson (1990). Nomenclature reference maps also use the quadrangle schemata (Hunter et al., this volume).

Since the systematic mapping guided by the USGS is conducted over the timescale of decades, and quadrangle maps are updated from time to time, the ongoing standardization of the mapping and visualization methodology (Naß et al. 2017) is a basic requirement as much as possible to make adjacent quadrangles and quads in a series comparable and compatible (Galluzzi 2019).

The Dawn Mission used a 15-quadrangle scheme, based on the recommendations of Greeley and Batson (1990). Individual quadrangles were produced by different groups within the mission team during the systematic mapping of Vesta (Roatsch et al. 2012; Yingst et al. 2014; Williams et al. 2014) and Ceres (Williams et al. 2017; Roatsch et al. 2016).

3.4 *The Development of Planetary Nomenclature*

The core concepts and names in the planetary nomenclature are based on the personal choices of early planetary mappers. Since the language of science, and cartography, was Latin at that time, they uniformly used Latin for naming features. With the birth of national sentiments over Europe, German (1821, Gruithuisen), French (1860, Lecouturier and Chapuis), and English (1869 M. Ward), equivalents of all or some of the lunar names were introduced, and place-names of Mars were also given in the native tongue of the observing astronomers until Schiaparelli (1878) reintroduced Latin—not very different from his native Italian. Planetary place-names were uniformly (re-)latinized by IAU in 1961 (Sadler 1962), this time to ensure international political neutrality.

For the early mappers, it was apparently tempting to name surface features after members of royal families: The very first map of the Moon (Langrenus 1645) and the first map of Venus (Bianchini 1728) equally include such names for the most prominent features. None of these names survived (the Venus features were in fact the telescope's artifacts).

Specifics of place-names show a transition from ancient European names (nineteenth century) to American/Soviet names (1950s) to a "worldwide representation" (IAU 1974) and multiethnic naming system initially suggested by Carl Sagan for Martian valleys (de Vaucouleurs et al. 1975).

The neutrality of the nomenclature was a goal from the beginnings, when Hevelius discussed that he has chosen classical geographic names over personal names for naming lunar features in order to be as impartial as possible (Whitaker 1999:55). Although Hevelius's system did not survive, this theme of archaic geographies reappeared when Schiaparelli's new Martian nomenclature conflicted the existing one. Before Schiaparelli, features of Mars bore the names of contemporary astronomers (Lane 2005), similar to the lunar map of van Langern that preceded that of Hevelius.

Albedo features of Mars were given ancient "Old World" names (Schiaparelli 1878; Burba 1981). Spacecraft observations of topographic features that were uncorrelated with albedo features made that scheme obsolete for geologic use. The former names were "saved," however, and many albedo names were transferred to topographic features (de Vaucouleurs et al. 1975). For some time, Earth-based visual albedo mapping activities were continued at Lowell Observatory (Inge et al. 1976) and are still being conducted by amateur astronomers who actively use the "old," or parallel, albedo-based nomenclature.

Almost all selenographers added a few names for their newly discovered features here and there, even though the basis of all works was Riccioli's toponym corpus. Blagg and Müller (1935) correlated the existing maps of Mädler, Schmidth, Neison, and others, creating the "collated list," the first edition of the standardized IAU "selenographic" nomenclature. Standardization of the nomenclature also meant that IAU became the name authority where names were approved by a commission instead of individuals (Blagg and Müller 1935). The *IAU Working Group for*

Planetary System Nomenclature (WGPSN, e.g., IAU 1977) is responsible for approving name themes, descriptor terms, quad names, and astronaut-given names. The new lists of standardized planetary place-names (gazetteers) were published informally in Arthur et al. (1963), and Masursky et al. (1986), formally since Batson et al. (1995), and now online.[2]

Today IAU WGPSN, based at USGS in Flagstaff, supervises planetary nomenclature. The Gazetteer of Planetary Nomenclature is maintained by USGS, and names are approved by IAU WGPSN (USGS 2016, Hunter et al., this volume). G. A. Burba published the Russian localization of the Gazetteer in the 1980s (Gazetteer 2016), and lunar names now also have Chinese standardized equivalents (Hargitai et al. 2014).

4 The Current Practice of Geologic Mapping

Geologic maps are essential to any planetary missions. Spacecraft-based geologic mapping has served as reconnaissance for landing site selection, where maps produced from one mission justify the need and choice for landing the next mission. At USGS, planetary geologic maps are published within its *Scientific Investigations Map* (SIM) series that includes terrestrial geologic maps. In this section, we document the current practice (Williams 2016, J. Ziegler, email communication, 11/24/2017, Skinner et al. 2018) to demonstrate the need for a strictly enforced review and production procedure that follows scientific analysis.

Tasks during the production of a planetary geologic map include delineating structural units and contact lines, crater counting, GIS database management, and geologic and geomorphic analysis to answer science questions. The maps then go through technical peer review by two mappers. This could take up to 40 hours of work per reviewer, and review and revision can spread over several years. After this stage, map production continues at USGS. Following technical review, a geologic map editor edits for clarity and consistency of scientific data, as well as for grammar, spelling, and USGS style, in the printed publication. This editor inspects that the information in different parts of the printed product and the database is consistent. In addition to symbols and colors, descriptions and discussions in the pamphlet must match the information shown on the map and in the *Description of Map Units* and *Correlation of Map Units*.

After the edit is complete, the author contact reconciles the edits in collaboration with the editor. When all issues are resolved, the materials are submitted to three individuals for approval for publication. The approving officials check that the scientific content is thorough, complete, and focused on the main subject; they also make sure the content is unbiased.

[2]http://planetarynames.wr.usgs.gov.

After the publication is approved, the materials are submitted to a cartographer for production—to be formatted and laid out according to USGS policy and style. When the publication production is complete, the author contact and the editor both review the final materials for accuracy, clarity, and consistency. Part of the editor's tasks is to make sure data are not changed during the production process.

Today, USGS planetary maps are published on the Web with database, metadata, and readme files, but USGS produces a printed publication too. Generally, USGS releases publications on the Web as soon as they are final and the printed copies become available soon after. PDFs (soon XMLs) of the printed product (interpretive pamphlet and the map sheets) are also available on the publication Web page. *The database is the main product,* but it is also important to provide the compiled map, explanations, and interpretations in printable and plottable formats. Printed materials represent the content of the database in a form that can be displayed for visual comprehension, study, discussion, comparison, and understanding of the author's interpretations. Without an accompanying visual compilation of the map, a database user may not be able to quickly and easily reproduce the map that the authors are discussing and interpreting.

Statuses of ongoing works are reported at the annual Planetary Geologic Mappers Meetings and other conferences.

Calculating with two work-years over a four-year project period and an average annual salary of $125,000, the total cost of the production of a planetary geologic map is about $250,000, before editing. The total duration of a project from start to printing is typically 5–7 years.

In Europe, universities and research organizations may produce their own planetary maps, typically but not always as part of a mission where their instruments fly. In the USA, upcoming mapping projects are selected via a proposal process open for all US-based scientists. In Europe, however, mission teams or research institutions decide on the mapping projects in an internal process. Map products are typically available online through various journals and mission Web sites, but some are not published formally and only exist in printed manuscript form, or as figures in conference proceedings, research papers, or theses. Mapping is also carried out in international collaborative projects such as the cartographic mapping of Vesta. China's planetary map-editing procedures are described in this volume (Mu et al. 2019).

5 Conclusion

We distinguish three stages of planetary mapping regarding its goals. Between 1610 and ca.1960, planetary mapping was driven by the scientific interests of professional and amateur astronomers, selenographers, and areographers who predominantly worked in Europe (Germany, Italy, France, and Great Britain) and were financed from private resources. Planetary maps also were produced for outreach: Reproductions of planetary maps were engraved for encyclopedias, popular science books since the eighteenth century. Many lunar maps were produced for amateur observer astronomers when this hobby became widespread.

From 1960, as a result of Space Race, planetary mapping moved from Western Europe to the USA and Moscow. Lunar mapping became one of the top national priorities in both the USA and the Soviet Union. The Soviet Union had no planetary mapping activities previously. In 1960, both countries started planetary mapping that supported their space programs. The USA, preparing for human landing, produced more than 300 lunar maps until 1969, while the Soviet Union did about half a dozen. During the next ca. 50 years, the USA produced approx. 1400 planetary maps, while the Soviet Union and Russia produced about 50 (Hargitai and Pitura 2018). These mapping activities were financed from government funds and involved astrogeologists and professional cartographers. The turn of 1950s/1960s is perhaps the most significant turning point in the history of planetary cartography. There emerged new concepts (geologic planetary mapping, crater counting), new technologies (photogrammetry, radar), new visualization methods (airbrush hill-shading), new projections (rectified images), new motivations (landing site selection), new financing plans (public money), and planetary mapping moved almost exclusively to the USA.

This era of modern planetary cartography is ending now. Analog technology is replaced by digital, with machine learning based systems emerging. Mapping became international, now including not only Europe but Asia too. Europe restarted planetary mapping in the 2000s with German maps related to the Mars Express, and later to ExoMars 2020 missions, and an increasing number of planetary maps are also produced in other European countries in the 2010s. China produced its first lunar maps in 2008 from Chang'E data and continues to expand planetary map production. Although financing is still almost solely based on public resources such as NASA, Europlanet, Russian and Chinese central resources, private companies may soon emerge for mapping the geologic resources of asteroids or the potential landing sites of their own missions. In the 2010s, citizen mappers and private companies emerged online and produce outreach-type digital maps. Geologic mapping now does not only support earth and planetary sciences but also astrobiology where geologic data can serve as the basis for the identification of potential past, present—and perhaps, future—habitats on other planets.

6 Further Reading

The most comprehensive review on planetary cartography in general is that of Snyder (1982, 1987), and Greeley and Batson (1990). The history of planetary mapping is discussed in, e.g., Wilhelms (1972), Schimeran (1973), Moore (1984), Martin et al. (1992), Whitaker (1999), Kopal and Carder (1974), and Morton (2002), Markley (2005), Lane (2011), Stooke (2012). Historic lunar maps are available at the LPI Web site (LPI 2015). Radar mapping techniques are discussed in Ford et al. (1993). For detailed summaries on the development and evolution of planetary cartography, the reader is referred to Shevchenko et al. (2016) for the history of Soviet and Russian planetary cartography, and to Jin (2014) for Chinese lunar mapping results. Recent planetary cartographic techniques and tools are reviewed in Beyer (2015) and Hare et al. (2017, in prep). The international catalog of planetary maps is available at the Web site of the International Cartographic Association's Commission on Planetary Cartography (Hargitai and Pitura 2018).

Acknowledgements The authors are grateful for D. Portree's comments that greatly improved the manuscript.

References

Albert G, Ilyés V, Szigeti CS, Kis D., Várkonyi D. (2017) How hard is it to design maps for beginners, intermediates and experts? In: International Cartographic Conference. https://doi.org/10.1007/978-3-319-57336-6_17

AMS (1963) Topographic lunar map. Army Map Service, Corps of Engineers, U.S. Army, Washington, D.C

Arthur DWG, Agnieray AP, Horvath RA, Wood CA, Chapman CR (1963) The system of lunar craters, quadrant I. Communications of the Lunar and Planetary Laboratory, vol 2, no 30

Baldwin RB (1964) Lunar crater counts. Astron J 69:377–392

Barabashov NP, Mikhailov AA, Lipskiy YN (1960) Atlas obratnoy storony Luny. 1960. Pod redaktsiey. Izdatelstvo Akademii nauk SSSR, Moskva

Basilevsky AT, Batson RM, Burba GA (1990) Pre-magellan mapping of northern venus: completion of a Joint U.S./U.S.S.R. project. In: Lunar and Planetary Science Conference, vol 21, pp 50–51

Basilevsky AT, Burba GA, Batson RM (1989) Maps of part of the Venus Northern Hemisphere: a joint US/USSR mapping project. In: Abstracts of the Lunar and Planetary Science Conference, vol 20, pp 46–47

Batson RM (1973) Mars cartography: the 1:5,000,000 map series. Interagency Report: Astrogeology 56 Prepared under JPL Contract W0-8122

Batson RM (1990) Map formats and projections used in planetary cartography. In: Greeley R, Barson RM (eds) Planetary mapping. Cambridge University Press, Cambridge

Batson RM (1991) Mapping the planets. Endeavour New Ser 15(3):107–114

Batson RM et al (1984) Voyager 1 & 2 Atlas of the Six Saturnian Satellites, NASA-SP-474

Batson RM, Edwards K, Eliason EM (1975) Computer-generated shaded relief images. J Research US Geol Surv 3(4):401–408

Batson RM, Russell JF (eds) (1995) Gazetteer of planetary nomenclature 1994. U.S. Geological Survey Bulletin 2129, Washington

Beer W, Mädler JH (1841) Beiträge zur physischen Kenntniss der himmlischen Körper im Sonnensysteme mit 7 Kupfertaf. Weimar

Beyer RA (2015) An introduction to the data and tools of planetary geomorphology. Geomorphology 240:137–145

Bianchini F (1728) Hesperi et phosphori nova phaenomena, sive Observationes circa planetam Veneris, Roma

Blagg MA, Müller K (1935) Named lunar formations. International Astronomical Union, Percy Lund, Humphries and Co Ltd, London, p 196

Bowker DE, Hughes JK (1971) Lunar orbiter photographic atlas of the Moon. NASA SP-206, Washington, 41 p, 675 plates

Bugaevsky LM, Shingareva KB, Krasnopevtseva BV et al (1992) Atlas of terrestrial planets and their Moons, Atlas Planet Zemnoi Gruppy i ih Sputnikov (« Атлас планет земной группы и их спутников » .). MIIGAiK, Moscow (in Russian)

Buie MW, Tholen DJ, Horne K (1992) Albedo maps of Pluto and Charon: initial mutual events. Icarus 97:211–227

Buie MW, Young EF, Binzel RP (1997) Surface appearance of Pluto and Charon. In: Stern SA, Tholen JD (eds) Pluto and Charon. The University of Arizona Press Tucson, pp 269–293

Burba GA (1981) The nomenclature of Martian topographic features. Izdatel'stvo Nauka, Moscow, p 88

Bussey B, Spudis P (2004) The clementine atlas of the Moon. Cambridge University Press, Cambridge

Carder RW (1962) Lunar charting on a scale of 1:1000000. In: Kopal Z, Mikhailov ZK (eds) The Moon, IAU Symposium 14. Academic Press, pp 117–129

Carr MH, Masursky H, Saunders RS (1973) A generalized geologic map of Mars. J Geophys Res 78(20):4031–4036

Compiling Committee of the Chang'E-1 Topographic Atlas of the Moon (2013) The Chang'E-1 topographic atlas of the Moon. NAOC and SinoMap Press

Davies ME, Berg RA (1971). A preliminary control net of Mars. J Geophys Res 76:373

Davies ME, Dwornik SE, Gault DE, Strom RG (1978) Atlas of Mercury. National Aeronautics and Space Administration Scientific and Technical Information Office. https://history.nasa.gov/SP-423/sp423.htm

Davies ME, Robinson JC et al (1970) Mars chart. US Army Topographic Command, Washington, D.C.

de Vaucouleurs G, et al (1975) The new Martian nomenclature of the international astronomical union. Icarus 26:85–98

Dobinson ER, Curkendall DW, Plesea L, Hare TM (2005) Adaptation & use of OpenGIS web technologies for multi-disciplinary access to planetary data. In: Ensuring long-term preservation and adding value to scientific and technical data (PV 2005) The Royal Society, Edinburgh

Dollfus A (1961) Visual and photographic studies off planets at the Pic du Midi. In: Kuiper GP, Middlehurst B (eds) Planets and Satellites. Chicago Univ Press, pp 543–571

Dopplemayr JG (1742) Atlas Coelestis. Nurenberge

Evans JE, Maunder EW (1903) Experiments as to the actuality of the "Canals" observed on Mars. Mon Not R Astron Soc 63:488–499

Flammarion C (1900) Népszerű csillagászat (Popular Astronomy), Az égbolt egyetemes leírása. Budapest, Pallas. (in Hungarian, translated by P. Hoitsy)

Ford JP, Plaut J et al (1993) Guide to Magellan image interpretation. NASA-JPL, Pasadena

Franz J (1899) Die Figur des Mondes, Astron. Beobachtungen, Königsberg, Bd. 38

Frigeri A, Hare T, et al (2011) A working environment for digital planetary data processing and mapping using ISIS and GRASS GIS. In: Planetary and space science (PSS), vol 59, Special issue: Planetary Mapping, Elsevier Ltd., pp 1265–1272. https://doi.org/10.1016/j.pss.2010.12.008

Galluzzi V (2019) Multi-mapper projects: Collaborative mercury mapping

Garov AS, Karachevtseva IP, Matveev EV, Zubarev AE, Patratiy VD (2015) 3D planetary geoportal as online laboratory for geo-collaboraition of scientific community. In: ISPRS international workshop working group IV/8. Planetary Mapping and Spatial Databases, Berlin, 24–25 September

Gazetteer (2016). http://www.planetmaps.ru/gazetteer/

Greeley R (1994) Planetary landscapes. Springer, New York

Greeley R, Batson RM (1990) Planetary mapping. Cambridge University Press, Cambridge

Greeley R, Batson R (1997) The NASA Atlas Of The Solar System. Cambridge University Press, Cambridge

Green NE (1879) Chart of Mars. Mem R Astron Soc 44:138

Grishakina EA, Shevchenko VV, Lazarev EN, Rodionova ZhF (2014) Hypsometric map of the Moon. Moscow State University of Geodesy and Cartography, Sternberg Astronomical Institute
GUGK (1988) 27 photomap sheets of Venus. Map series of Venus based on Venera 15 and Venera 16 observations. 1:5 M. GUGK SSSR (Main Administration of Geodesy and Cartography), Moscow
Gulick VC, DearDorff DG (2003) Mars data visualization and E/PO with Marsoweb. Lunar and Planetary Science XXXIV abs. 2081
Hare TM, Tanaka KL (2001) Planetary interactive GIS-on-the-web analyzable database (PIGWAD). In: 20th ICC, Beijing, China
Hare TM, Kirk RL, et al (2009) Chapter 60: Extraterrestrial GIS. In: Madden M (ed) Manual of geographic information systems. The American Society for Photogrammetry and Remote Sensing, Bethesda, pp 1199–1219
Hare TM, Hayward RK, Blue JS, Archinal BA, Robinson MS, Speyerer EJ, Wagner RV, Smith DE, Zuber MT, Neumann GA, Mazarico E (2015a) Image mosaic and topographic map of the Moon: U.S. Geological survey scientific investigations map 3316, 2 sheets. https://doi.org/10.3133/sim3316
Hare TM, Skinner JA, et al (2015b) Planetary GIS at the U.S. Geological Survey Astrogeology Science Center. In: 2nd planetary data workshop, 8–11 June 2015, Flagstaff, Arizona. LPI No. 1846, id.7005
Hare TM, Skinner JA, Kirk RL (2017) Cartography tools. In: Rossi A-P, van Gasselt, S (eds) Planetary geology. Springer, Berlin, 540 p
Hargitai H, Li C, Zhang Z, Zuo W, Mu L, Li H, Shingareva KB, Shevchenko VV (2014) Chinese and Russian language equivalents of the IAU Gazetteer of planetary nomenclature: an overview of planetary toponym localization methods. Cartogr J. https://doi.org/10.1179/1743277413Y.0000000051
Hargitai H (2018) Catalog of planetary maps. Available at https://planetarymapping.org
Hargitai H, Pitura M (2018) International catalog of planetary maps 1600–2017. In: 49th Lunar and planetary science conference (LPI Contrib. No. 2083)
Hargitai H, Willner K, Hare T (2019) Fundamental frameworks in planetary mapping: A review
Hartmann WK (1965) Terrestrial and lunar flux of large meteorites in the last two billion years. Icarus 4:157–165
Hartmann WK (1972) Interplanet variations in scale of crater morphology earth, Mars, Moon. Icarus 17:707
Herschel W (1784) On the remarkable appearances at the polar regions of the planet Mars, the inclination of its axis, the position of its poles, and its spheroidal figure, with a few hints relating to its real diameter and atmosphere. Philos Trans LXXIV Article XIX, 233–237
Holden ES (1896) Observatory Atlas of the Moon. Lick Observatory, Mount Hamilton, Calif
IAU (1960) Transactions of the International Astronomical Union. In: Tenth General Assembly, Moscow, vol X
IAU (1974) Resolutions from the first meeting of the I.A.U. working group for planetary system nomenclature, Ottawa, Ontario, 27 and 28 June 1974. In: Transactions of the International Astronomical Union, vol Ume Xvi B - Proceedings d. Reidel Publishing Company, Dordrecht-Holland/Boston-USA, pp 322–325
IAU (1977) Transactions of the international astronomical union, vol Ume Xvi B - Proceedings d. Reidel Publishing Company Dordrecht-Holland/Boston-USA, pp 321–369
Inge JL, Martin LJ, Capen CF (1976) Mars 1976-76. Planetary Research Center, Lowell Observatory, Flagstaff
Inge JL, Bridges PM (1976) Applied photointerpretation for airbrush cartography. Photogram Eng Remote Sens 42(6):749–760
IPPP – International Planetary Patrol Program (1971) Mars 1969. In: Inge JL, Capen CF, Martin LJ, Faure BQ (eds) Cartographers, Lowell Observatory. Northland Press, Flagstaff
ISRO (2015) Mars orbiter mission (MOM) Mars atlas. Space Applications Centre, Indian Research Organisation (ISRO), Ahmedabad. https://www.issdc.gov.in/docs/mr1/Mars-atlas-MOM.pdf

Jin S (2014) Planetary geodesy and remote sensing. CRC Press, Boca Raton
JPL (1967) Mariner 69 Mars chart. 1:25 M. Jet Propulsion Laboratory
Kaiser F (1864) [Map of Mars] In: Valentiner KH (1881) Astronomische Bilder, Leipzig
Karachevtseva IP, Kokhanov AA, Konopikhin AA, Rodionova ZhF, Kozlova NA (2012) Atlas of Phobos. MIIGAiK, Moscow
Karachevtseva IP, Kokhanov AA, Rodionova ZhF (2019) Atlas planetary mapping: Phobos case
Kirk RL (1993) Separation of topographic and intrinsic backscatter variations in biscopic radar images: a "magic airbrush". In: 24th Lunar and planetary science conference, pp 803–804
Koeberl C (2001) Craters on the Moon from Galileo to Wegener: a short history of the impact hypothesis, and implications for the study of terrestrial impact craters. Earth Moon Planet 85–86:209–224
Kopal Z, Carder RW (1974) Mapping of the Moon. Springer Science + Business Media, Dordrecht
Kozlov VV, Sulidi-Kondriatiev ED, Kuznetsov YY (1969) Tectonicheskaya Karta Luny. Tectonic map of the Moon, Moscow Institute Academy of Science of Geology, Moscow
Krogt, P, van der Ormeling F (n.d.) Michiel Florent van Langren and Lunar Naming. https://doi.org/10.2436/15.8040.01.190
Kuiper GP (1959) Photographic Lunar atlas. Introduction and summary. In: Kuiper GP, Arthur DWG, Moore E, Tapscott JW, Whitaker EA (eds) Photographic lunar atlas based on photographs at the mount wilson, Lick, Pic du Midi, McDonald and Yerkes observatories. The University of Chicago Press, Chicago, pp 3–9
Lane KMD (2011) Geographies of Mars. Seeing and Knowing the Red Planet. The University of Chicago Press Chicago and London, Chicago/London
Lazarev E.N., Rodionova J.F., Shevchenko V.V. (2012) Venus Relief Map. Sternberg State Astronomical Institute, Space Research Institute IKI RAS
Ledger E (1882) The sun, its planets and their satellites, London
Lipskiy YuN (1967) Interpretation of images obtained by Zond-3 probe, compilation of a sketch maps and a catalog of detected formations. Atlas of the reverse side of the Moon, vol 2. Izdatelstvo Nauka, Moscow, pp 28–64 (in Russian)
Lipsky YuN (1968) Karta Luny, Yekvatorialnaya Zona Vidimogo Polushariya 1:1 M Sternberg Astronomical Institute Nauka Moscow
Loewy MM, Puiseux MP (1896) Atlas Photographique de la Lune. Observatoire de Paris
Lohrman WG, (1878) Mondcharte. Julius Schmidt (ed) Verlag von Johann Ambrosius Barth
Lowell P (1906) Mars and its canals. Macmillan, New York
LPI (2015) Lunar map catalog. Lunar and Planetary Institute. http://www.lpi.usra.edu/resources/mapcatalog/
MacDonald A (2017) The long space age. Yale Univesity Press, New Haven and London
Makarova NV, et al (1978) Geomorphological map of Mars. 1:20 M. Ministry of higher and secondary special education of the USSR, Ministry of Geology of the USSR
Markley R (2005) Dying planet: Mars in science and the imagination. Duke University Press, Durham
Martin LJ, James PB, Dollfus A, Iwasaki K, Beish JD (1992) Telescopic observation: visual, photographic, polarimetric. In: Kieffer HH, Jakosky BM, Snyder CW, Matthews MS (eds) Mars. The University of Arizona Press, Tucson, pp 34–70
Mason AC, Hackman RJ (1961) Engineer special study of the surface of the Moon. USGS I-351, USGS
Mason AC, Hackman RJ (1962) Photogeologic study of the Moon. In: Kopal Z, Mikhailov ZK (eds) The Moon. Academic Press. London, p 281
Masursky H, Aksnes K, Hunt GE, Marov MY, Milman PM, Morrison D, Owen TC, Shevchenko VV, Smith BA, Tejfel VG (1986) Annual gazetteer of planetary nomenclature. In: International astronomical union, working group for planetary system nomenclature. Open-File Report 84-692. U.S. Geological Survey
Moore P (1977) Requiem for the canals. J Br Astron Assoc 87:589–593
Moore P (1984) The mapping of Mars: presidential address, 1983. J Br Astron Assoc 94(2):45–54

Morrison D, Jones TJ, Cruishank DP, Murphy RE (1975) The two faces of Iapetus. Icarus 24:157–171
Morton O (2002) Mapping Mars: science, imagination, and the birth of a world. Picador, New York
Mu L, Liu J, Liu L (2019) Planetary map design: The Chang'E-1 topographic Atlas of the Moon
Murchie S, Mick A, Prockter L, Rivkin A, Guinness E, Ward J (2017) MESSENGER MDIS CDR/RDR software interface specification version 1.2.23. Updated 21 March. ftp://pdsimage2.wr.usgs.gov/cdroms/MESSENGER/MSGRMDS_4001/DOCUMENT/MDIS_CDR_RDRSIS.HTM
Mutch TA, Arvidson RE, Head JW III, Jones KL, Saunders RS (1976) The geology of Mars. Appendix C: crater maps. Princeton University Press, Princeton
Myler JL (1 November 1957) Latest red rumor: they'll bomb Moon. Pittsburgh Press, p 13
Nasmyth J, Carpenter J (1874) The Moon. John Murray, London
Naß A, Fortezzo CM, Skinner Jr JA, Hunter MA, Hare TM (2017) Specified collection and GIS-based implementation for cartographic symbolization in geologic and geomorphologic maps. Lunar and Planetary Science XLVIII, #1895
Naß A, Hargitai H, Buchroithner M (2019) Cartography: its role and interdisciplinary character in planetary science
Nass, A, van Gasselt, S, et al (2011) Implementation of cartographic symbols for planetary mapping in geographic information systems. In: Planetary and Space Science (PSS), vol 59, Special issue: Planetary Mapping, Elsevier Ltd., pp 1255–1264. https://doi.org/10.1016/j.pss.2010.08.022
Nefian AV, Alexandrov O, Kim T, Moratto Z, Beyer R (2013) Albedo reconstruction of the apollo metric camera zone. In: 44th lunar and planetary science conference abstract 1649
Ooms K, De Maeyer P, Fack V (2014) Study of the attentive behavior of novice and expert map users using eye tracking. Cartogr Geograph Inf Sci 41(1):37–54
Öpik E (1960) The lunar surface as an impact counter. MNRAS 120:404–411
Pettengill GH, Counselman CC, Rainville LP, Shapiro II (1969) Radar measurements of Martian topography. Astron J 74(3):461–482
Pettengill GH, Eliason E, Ford PG, Loriot GB, Masursky H, McGill GE (1980) Pioneer venus radar results: altimetry and surface properties. J Geophys Res 85(A13):8261–8270
Philips J (1865) Chart of the planet Mars. Proc R Soc Lond 14:42–46
Pohn HA, Wildey RL, Sutton GE (1970) Photoelectric-photographic map of the normal Albedo of the Moon. USGS Professional paper 599-E Plate I. https://pubs.usgs.gov/pp/0599e/
Portree DSF (2013) 1961: USGS astrogeology's first published map. http://astrogeology.usgs.gov/rpif/exhibits/astros-first-lunar-map
Roatsch T, Kersten E, Matz KD, Preusker T, Scholten F, Jaumann R, Raymond CA, Russel CT (2012) High resolution vesta high altitude mapping orbit (HAMO) atlas derived from dawn framing camera images. Planet Space Sci 73:283–286. https://doi.org/10.1016/j.pss.2012.08.021
Roatsch T, Kersten E, Matz KD, Preusker T, Scholten F, Jaumann R, Raymond CA, Russel CT (2016) High-resolution ceres high altitude mapping orbit atlas derived from dawn framing camera images. Planet Space Sci 129:103–107. https://doi.org/10.1016/j.pss.2016.05.011
Rodionova ZhF, Lipsky YuN, Shevchenko VV (1985) Karta Luny. 1:10,000,000, Sternberg Institute of Moscow State Lomonosov University, Moscow
Roth J, de Vaucouleurs G (1971) MM '71 Mars planning chart. 1:25 M. Jet Propulsion Laboratory, Pasadena
Sadler DH (ed) (1962) Proceedings of 11th general assembly of the IAU (Berkeley, 1961). IAU Trans. XIB, Academic Press, Berkeley, p 234
Sagan C, Fox P (1975) The canals of Mars: an assessment after Mariner 9. Icarus 25(4):602–612
Schaber GG (2005) The U.S. geological survey, branch of astrogeology—a chronology of activities from conception through the end of project apollo (1960–1973). Open-File Report 2005-1190
Schenk P (2010) Atlas of the Galilean satellites. Cambridge University Press, New York

Schiaparelli GV (1878) Osservazioni astronomiche e fisiche sull'asse di rotazione e sulla topografia del pianeta Marte fatte nella Reale Specola di Brera in Milano coll'equatoriale di Merz durante l'opposizione del 1877. Atti della R. Accademia dei Lincei, Memorie della Classe di Scienze Fisiche, Matematiche e Naturali, Roma, p 308-M. 1. ser. 3. v. 2
Schimeran LA (1973) Lunar cartographic dossier. Defense Mapping Agency, St Louis
Schröter JH (1791) Selenotopographische Fragmente. vol 1. Lilienthal, J.G. Rosenbusch, Univ. Buchdr
Scott DH, Carr MH (1978) Geologic map of Mars, Misc. Invest. Ser. Map I-1083, U.S. Geol. Surv., Reston
Seller J (1700) Atlas caelestis. Ier: Seller & Cha, London
Shevchenko V, Rodionova Z, Michael G (2016) Lunar and planetary cartography in Russia. Springer, Heidelberg. https://doi.org/10.1007/978-3-319-21039-1
Shingareva KB, Krasnopevtseva BV (2011) Atlas Solnechnaya Sistema. Luna. DIK, Moscow
Shirao M, Wood CA (2011) The Kaguya Lunar atlas. Springer, New York
Shoemaker EM (1959) Impact mechanics at Meteor Crater, Arizona. U. S. Atomic Energy Commission - USGS Open-File Report 59-108
Shoemaker E, Hackman RJ (1961) Lunar photogeologic chart LPC 58, copernicus, prototype chart, USGS, unpublished. U. S. Air Force's Chart and Information Center (ACIC) in St. Louis
Shoemaker EM, Hackman RJ (1962) Stratigraphic basis for a lunar time scale. In: Kopal Z, Mikhailov ZK (eds) The Moon. Academic Press, London, p 289
Shoemaker EM, Hackmann RJ, Eggleton RE (1963) Interplanetary correlation of geologic time. Adv Astron Sci 8:70–88
Skinner JA Jr, Huff AE, Fortezzo CM, Gaither TA, Hare TM, Hunter MA (2018) Planetary geologic mapping protocol–2018. The Planetary Geologic Map Coordination Group, Astrogeology Science Center, USGS. https://planetarymapping.wr.usgs.gov/Page/view/Guidelines
Slipher EC (1962a) The photographic story of Mars. Sky Pub. Corp., Cambridge
Slipher EC (1962b) MEC-1 map of Mars. United States Air Force
Smith DE, Zuber MT, Solomon SC, Phillips RJ, Head JW, Garvin JB, Banerdt WB, Muhleman DO, Pettengill GH, Neumann GA, Lemoine FG, Abshire JB, Aharonson O, Brown CD, Hauck SA, Ivanov AB, McGovern PJ, Zwally HJ, Duxbury TC (1999) The global topography of Mars and implications for surface evolution. Science 284:1495–1503
Snyder JP (1982) Map projections used by the U.S. geological survey. U.S. Geological Survey Bulletin 1532
Snyder JP (1987) Map projections used by the U.S. geological survey. U.S. Geological Survey Professional paper 1395
Spradley LH (1962) Lunar globe photography. Commun Lunar Planet Lab 1(6):31–34
Staff (1973) Martian map makers. Eng Sci 37(1):8–9. ISSN 0013-7812
Steno N (1669) De solido intra solidum naturaliter contento dissertationis prodromus. Florentiae
Stooke PJ (2012) The international atlas of Mars exploration: volume 1, 1953 to 2003: the first five decades. Cambridge University Press
Sukhanov AL, Pronin AA, Burba GA, Nikishin AM, Kryuchkov VP, Basilevsky AT, Markov MS, Kuzmin RO, Bobina NN, Shashkina VP, Slyuta EN, Chernaya IM (1989) Geomorphic/geologic map of part of the northern hemisphere of Venus. V15 M 90/0 G. I-2059
Tanaka KL (1986) The stratigraphy of Mars. J Geophys Res 91(B13):E139–E158. https://doi.org/10.1029/jb091ib13p0e139
Tanaka K, Skinner J, Hare T (2011) Planetary geologic mapping handbook – 2011. USGS
Tanaka KL, Skinner JA Jr, Dohm JM, Irwin RP III, Kolb EJ, Fortezzo CM, Platz T, Michael GG, Hare TM (2014) Geologic map of Mars: U.S. geological survey scientific investigations map 3292, scale 1:20,000,000, pamphlet 43. https://doi.org/10.3133/sim3292
Tefimim YuS, Krestnikova ND (1977) Karta Ustashtka Marsa. TsIIGAiK GUGK, Moscow
Trumpler R (1924) Visual and photographic observations of Mars. Astron Soc Pac 36(213):263–269
Tucci P (1998) La vita sul pianeta Marte: tre scritti di Schiaparelli su Marte e i "marziani". Mimesis

U.S. Geological Survey (2006) FGDC digital cartographic standard for geologic map symbolization (postscript implementation): U.S. geological survey techniques and methods 11-A2. http://pubs.usgs.gov/tm/2006/11A02/
Ulivi P (2004) Lunar exploration: human pioneers and robotic surveyors. Springer Science & Business Media, Heidelberg
UN (1967) Treaty on principles governing the activities of states in the exploration and use of outer space, including the Moon and other celestial bodies, London, Moscow and Washington, 27 January 1967
United States Army (1959) Project horizon, vol II, Technical considerations and plans. Army Ballistic Missile Agency (ABMA). https://nsarchive2.gwu.edu/NSAEBB/NSAEBB479/docs/EBB-Moon01A_sm.pdf
US Army (1968) Lunar equatorial zone mosaics, 1:2.5 M, 4 sheets, U.S. Army Topographic Command (2nd ed. 1969)
USAF (1961) Lunar expedition plan LUNEX. headquarters, Space Systems Division, Air Force Systems Command
USGS (1973) Shaded Relief Map of Mars. October 1973. Atlas of Mars MM 25 M IR
USGS (1984) Topographic map of Venus. VRM planning chart. Atlas of Venus. I-1562. V50 M 6/60RT Scale 1:50 M
USGS (1989) Maps of part of the Northern Hemisphere of Venus. Map I-2041. Miscellaneous investigations series. 1:15 M
USGS (2016) Gazetteer of planetary nomenclature. USGS/IAU WGPSN. http://planetarynames.wr.usgs.gov/
USGS (2017) Astrogeology glossary of terms https://astrogeology.usgs.gov/site/glossary
van Gasselt S, Nass A (2011) Planetary mapping: the datamodel's perspective and GIS framework. Planetary and space science 59, Special issue: Planetary Mapping, Elsevier Ltd., pp 1231–1242. https://doi.org/10.1016/j.pss.2010.09.012
Vermilion JR, Clar RN, Greene TF, Seamans JF, Yantis WF (1974) A low resolution map of Europa from four occultation by IO. Icarus 23:89–96
Wagstaff KL, Riloff E, Lanza NL, Mattmann CA, Ramirez PM (2016) Creating a Mars target encyclopedia by extracting information from the planetary science literature. In: AAAI-16 workshop on knowledge extraction from text
Wagstaff KL, Lu Y, Stanboli A, Grimes K, Gowda T, Padams J (2018) Deep Mars: CNN classification of mars imagery for the PDS imaging atlas. In: Proceedings of the thirtieth annual conference on innovative applications of artificial intelligence. Abstract CL#17-6072
Weir GE (2009) Lunar cartography informed early space programs. Pathfinder, NGA, pp 21–22. https://www.nga.mil/About/History/NGAinHistory/Documents/Pathfinder/July/LunarMapping.pdf
Whitaker EA, Kuiper GP, et al (1963) Rectified Lunar atlas - supplement number two to the photographic Lunar atlas. Aeronautical Chart and Information Center
Whitaker EA (1999) Mapping and naming the Moon: A history of Lunar Cartography and nomenclature. Cambridge University Press, Cambridge, UK
Wilhelms DE (1972) Geologic mapping of the second planet. Interagency report: astrogeology 55, Prepared under NASA Contract W-13,204
Wilhelms D (1987) The geologic history of the Moon. USGS professional paper 1348
Wilhelms DE (1993) To a rocky Moon: a geologist's history of Lunar exploration. University of Arizona Press
Wilhelms DE, El-Baz F (1977) Geologic map of the east side of the Moon. 1:5 M I-948. USGS/NASA/ Defense Mapping Agency
Wilkins HP (1946) 300 in map of the Moon. 100 in reproduction. In: The strolling astronomer. Association of Lunar and Planetary Observers
Wilkins HP (1953) The other side of the Moon. J Br Interplanet Soc 12(1):1–5
Williams DA (2016) NASA's planetary geologic mapping program: overview. In: ISPRS meeting, Prague, 17 July 2016

Williams DA, Yingst RA, Garry WB (2014) Introduction: the geologic mapping of Vesta. Icarus 244:1–12

Williams DA, Buczkowski DL, Mest SC, Scully JEC, Platz T, Kneissl T (2017) Introduction: the geologic mapping of Ceres. Icarus, (in press), this issue (https://doi.org/10.1016/j.icarus.2017.05.004)

Wollenhaupt WR, Sjogren WL (1972) Apollo 16 preliminary science report. NASA 30-1–30-5

Wu SSC (1978) Mars synthetic topographic mapping. Icarus 22:417–440

Yingst RA, et al (2014) Geologic mapping of Vesta. Planet Space Sci. https://doi.org/10.1016/j.pss.2013.12.014

Young EF, Binzel RP, Crane K (2001) A two-color map of Pluto's Sub-Charon Hemisphere. Astron J 121:552–561

Zisk SH (1972) Lunar topography: first radar-interferometer measurements of the Alphonsus-Ptolemaeus-Arzachel region. Science 178:977–980

Planetary Nomenclature

Marc Hunter, Rose Hayward and Trent Hare

Abstract This chapter details the purpose, conventions, and implementation of planetary nomenclature in cartography and GIS.

Keywords Nomenclature · Annotation · Labels · Planetary GIS · Planetary cartography

1 Background

1.1 What Is Planetary Nomenclature?

Nomenclature is a system or set of rules used to uniquely identify a feature on the surface of a planet or satellite so that the feature can be easily located, described, and discussed. To be effective, a set of nomenclature must be stable and readily understood throughout the field of study. Most sciences establish a central authority for all official nomenclature to ensure conformance to standards as well as fair access to all members of the community. Commonly known terrestrial examples are the binomial nomenclature system for naming biological species and the chemical nomenclature system developed by the International Union of Pure and Applied Chemistry.

In planetary science, nomenclature is critical to any cartographic product, providing essential context for research and discussion. Without a common understanding of place, collaboration is hindered, particularly when sharing information in digital formats and partnering internationally. Higher resolution cameras and sensors are being flown on missions each year, dramatically increasing the number of named features. Keeping up with such growth over a period of decades has been managed by the International Astronomical Union (IAU).

M. Hunter (✉) · R. Hayward · T. Hare
U.S. Geological Survey Astrogeology Science Center, 2255 N. Gemini Dr., Flagstaff, AZ 86001, USA
e-mail: mahunter@usgs.gov

© This is a U.S. government work and not under copyright protection in the U.S.; foreign copyright protection may apply 2019

H. Hargitai (ed.), *Planetary Cartography and GIS*, Lecture Notes in Geoinformation and Cartography, https://doi.org/10.1007/978-3-319-62849-3_3

2 Governing Rules and Conventions

2.1 *History of Planetary Nomenclature*

While planetary science matured into its current form during the latter half of the twentieth century, the naming conventions of inner planets were based primarily on the naming schemes of the first widely accepted nearside maps of the Moon made from telescopic observations in the mid-seventeenth century. These maps named albedo features, and authors created their own naming schemes which led to confusing, often contradicting, nomenclature that further divided the planetary science community. To address this growing problem, Mary A. Blagg of Great Britain's Royal Astronomical Society reconciled the Lunar nomenclature catalogs of Mädler, Schmidt, and Neison, which were adopted by the International Association of Academies in 1913 (Blagg 1913). This new set of nomenclature was used as the basis for the first nomenclature scheme put forth during the IAU's inaugural session in 1919. Known as the Blagg and Müller scheme, it was formally adopted by the IAU in 1935 and became the foundation for all planetary nomenclature as regulation was later expanded to address similar naming conflicts with the inner planets (Blagg and Müller 1935).

The 1960s brought a new challenge to planetary nomenclature: the higher resolution images of the Mariner missions. In anticipation of the rapid expansion of named features on Mars, the IAU sought to compliment the existing albedo-based scheme with a convention for naming now-visible topographic features. New, previously unclassified features were to be named by a composite of albedo names adopted from early Martian maps and Greek or Latin terms describing the feature appearance. This naming scheme was also applied to Mariner-based maps of Mercury and served as the foundation for new nomenclature in the age of satellite-based planetary exploration. For a more in-depth history on the evolution of planetary nomenclature, see Mary Strobell and Harold Masursky's chapter in *Planetary Mapping* (Greeley and Batson 1990).

In 1973, the current IAU structure for managing nomenclature was established with the Working Group for Planetary System Nomenclature (WGPSN; Masursky et al. 1986). Task Groups for the Moon, Mercury, Venus, Mars, small bodies, and the outer solar system were formed to conduct the preliminary work of choosing themes and proposing names for features on each newly discriminated planet and satellite (Swings 1986). Task Groups review requests made by research scientists and mission teams following predetermined themes and guidelines for equitable representation across the international community. Official names are typically only given to features larger than 100 m unless they have special significance, such as features mapped around a landing site (Blue et al. 2013).

2.2 *Process for Approving Feature Nomenclature*

The Gazetteer of Planetary Nomenclature Web site includes an online Name Request Form that can be used by members of the professional science community to request feature names ('gazetteer' refers to a geographic index or dictionary). Images of the feature and a scientific justification for naming the feature are required with submission. A name should be requested only if a feature is scientifically significant and if naming the feature is useful to the scientific and cartographic communities at large.

A standard request will be reviewed by the appropriate Task Group and then by the Working Group for Planetary System Nomenclature. Unusual requests may require multiple reviews. Each review group will typically have at least ten days to review the request. During review, multiple aspects are considered, including the need for a name, the appropriate choice of descriptor term, and the name suggestion. Reviewers may ask for additional images and information. If necessary, the review period may be extended.

A specific name may be suggested for a feature, but the name is subject to IAU review, and there is no guarantee it will be approved. For example, the IAU strongly supports an equitable selection of names from ethnic groups, countries, and gender on each map. A name suggestion with an overrepresented ethnicity may be replaced with a name from an underrepresented ethnicity. When possible, the IAU will involve the requestor in choosing a replacement name. The IAU has a preference for simple names that are easy to spell and pronounce. From a cartographic point of view, it is especially desirable to choose a short name if the feature is small. If a name uses diacritical marks, they will become part of the official name; however, for the sake of simplicity, if the name is being chosen from a theme with many names to choose from, such as names of towns, it may be preferable to choose a name that will not require diacritical marks. For themes where features are named for people, diacritical marks must be used if they are part of the name and should not prevent the use of a name. A person must be deceased for at least three years before a proposal suggesting the name may be submitted.

To address the recent proliferation of large-scale mapping, the IAU has also approved a new method for naming smaller groups of features within a larger named feature. In these instances, names should bear a mnemonic relationship to the larger feature, followed by the descriptor term. For example, in 2017 several *cavi* features within Noctis Labyrinthus were named using the word for 'night' from different languages (Dalu Cavus, Layl Cavus, Malam Cavus, Nat Cavus, Noc Cavus, and Usiku Cavus).

3 Nomenclature Labels

3.1 Planetary Map Label Conventions

Official IAU rules and conventions for place names have changed as the planetary community has grown; however, general guidance for the use of nomenclature remains simple: 'Nomenclature is a tool and the first consideration should be to make it simple, clear, and unambiguous' (IAU 2017). What this means is that quality nomenclature aims to meet these criteria while being as unobtrusive of viewing map content as possible. Following the standards listed below, nomenclature labels are typically given different size and capitalization cases for large, medium, and small features: 18-point upper case, 14-point upper case, and 10-point mixed case, respectively (Table 1). These values are relative to the printed scale of the map, and thresholds for separating features by size must be determined by the author. Each font group has open-source font options, as not all diacritics are supported by proprietary fonts, and some characters may be corrupted during language translation. The names listed are the singular and plural of each geomorphological feature type as well as the shorthand code used to capture all instances.

3.2 Placement Strategy

The above label feature type and font guidelines are not meant to be restrictive; rather, they are intended to afford authors discretion in how to represent labels in context of other map elements while also supporting standards readers have come to expect. For example, grayscale base maps typically use white lettering for annotation, but areas with high albedo may make those fonts illegible, in which case a mix of black and white, or third color like yellow, may be used. Similarly, maps using a conical projection may use annotation that is horizontal to page or follow the curvature of the latitude grid. That said, there are conventions for label placement that will serve as the best starting point and may be adjusted to deconflict with other labels or map features.

The classic planetary example is a crater, a largely circular feature with a known center point and radius. Ideal placement is in the center, with characters kerned (spread) to fill the space, and multiple-word labels stacked. If the name does not fit cleanly inside the feature, then it is placed outside the feature with a point or leader at the feature's center. Outside the feature boundary, the priority for placement is directly east (assuming the map is oriented north), moving counterclockwise until it no longer overlaps adjacent labels or map features. In very densely labeled areas, such as landing sites, this method may not suffice, in which case a numbered key or reference to a larger scale map is recommended (Fig. 1). This should illustrate that a number of different strategies may suit a given situation, but that the end goal should be to inform the reader within the context of the map scale and theme.

Table 1 IAU WGSPN convention for nomenclature label fonts by feature type

Arial/Helvetica/Univers Regular Font	
Albedo Feature (AL)	Planum, plana (PM)
Crater, craters (AA)	Plume, plumes (PU)
Eruptive center (ER)	Satellite Feature (SF)
Landing site name (LF)	Terra, terrae (TA)
Large ringed feature (LG)	Tessera, tesserae (TE)
Planitia, planitiae (PL)	Vastitas, vastitates (VS)
Arial/Helvetica/Univers Oblique Font	
Arcus, arcūs (AR)	Linea, lineae (LI)
Catena, catenae (CA)	Lingula, lingulae (LN)
Cavus, cavi (CB)	Mare, maria (ME)
Chaos, chaoses (CH)	Oceanus, oceani (OC)
Chasma, chasmata (CM)	Palus, paludes (PA)
Farrum, farra (FR)	Promontorium, promontoria (PR)
Flexus, flexūs (FE)	Rima, rimae (RI)
Fluctus, fluctūs (FL)	Rupes, rupēs (RU)
Fossa, fossae (FO)	Scopulus, scopuli (SC)
Flumen, flumina (FM)	Serpens, serpentes (SE)
Fretum, freta (FT)	Sinus, sinūs (SI)
Labes, labēs (LA)	Sulcus, sulci (SU)
Labyrinthus, labyrinthi (LB)	Unda, undae (UN)
Lacus, lacūs (LC)	Virga, virgae (VI)
Lacuna, lacunae (LU)	Vallis, valles (VA)
Times New Roman Italic Font	
Collis, colles (CO)	Mensa, mensae (MN)
Corona, coronae (CR)	Mons, montes (MO)
Dorsum, dorsa (DO)	Patera, paterae (PE)
Facula, faculae (FA)	Regio, regiones (RE)
Insula, insulae (IN)	Tholus, tholi (TH)
Macula, maculae (MA)	

https://planetarynames.wr.usgs.gov/DescriptorTerms (2017)

Labels for vast features, such as *planitia*e, may be spread out to fill the space using increased kerning and spaces between words, but irregular features, such as winding *valles* or curved *rupes*, typically benefit from directional labels. As with crater labels, if the entire name does not fit inside the feature boundary it is best to follow along side, mirroring the shape of the feature, rather than overlap boundary (Fig. 2). Major GIS platforms have methods for implementing this logic in label placement, which automate the placement and deconfliction of labels based on feature attributes such as size and type.

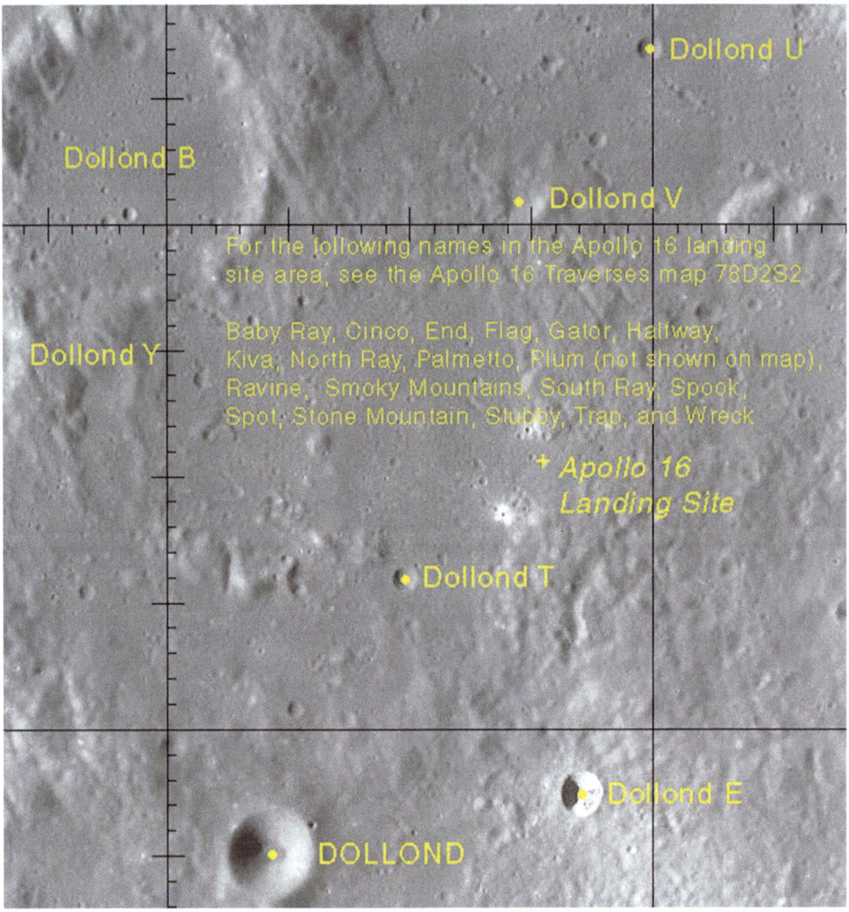

Fig. 1 Sample of a densely labeled region around the Apollo 16 landing site from the 1:1,000,000-scale Lunar Astronautical Chart (LAC) 78 map. https://planetarynames.wr.usgs.gov/images/Lunar/lac_78_wac.pdf (2013)

4 Nomenclature in a GIS

4.1 Benefits of Dynamic Nomenclature

Heavily studied bodies like the Moon and Mars have more nomenclature than can be represented on global and regional scale maps and must be filtered appropriately based on the scale and type of map being created. Though the basis for most filtering is typically feature size, it is important to consider scientific significance as well so that critical features are not omitted inadvertently. For example, at less than 800 m across, the Martian crater Airy-0 would be considered too small to be

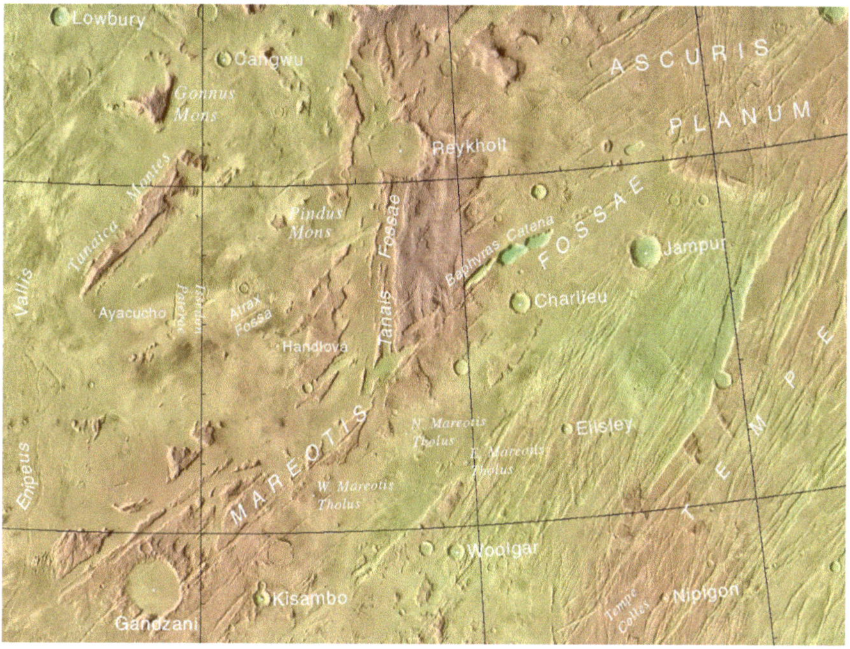

Fig. 2 Sample of directional feature labels in a complex geomorphological region from the 1:5,000,000-scale map of the Arcadia quadrangle of Mars. https://planetarynames.wr.usgs.gov/images/mc3_2014.pdf (2014)

included in a global map of Mars, but because it is used to define the planets prime meridian Airy-0 is important to label.

Due to the lengthy approval process for official IAU nomenclature and increase in non-standard maps, many authors use unofficial nomenclature to identify new features for discussion and analysis purposes. Such names should be enclosed in brackets or otherwise denoted and follow IAU conventions as closely as possible to prevent future conflicts with official nomenclature (IAU WGPSN 2017). There is no methodology that can account for all nomenclature changes over time, though; planetary science is a rapidly evolving field still, and static maps can only be as accurate as the information available at the time of their release.

The rapid pace of development in GIS has made it the ideal environment for managing planetary nomenclature. Higher resolution base maps have created a demand for discrete feature boundaries, most commonly stored as vector data, and now platforms such as Esri's ArcGIS and QGIS support nomenclature annotation as vector data as well. This allows for annotation text to be saved in a geodatabase (outside the map project) with font parameters as attributes, to be edited or shared like any other vector data (Hunter et al. 2016). Additionally, as vector data, nomenclature annotation may be managed dynamically through a relationship class (known as feature-linked annotation) that updates a nomenclature feature class when changes are made to the linked source table (Esri 2016; QGIS 2016). Much like the challenge encountered when planetary scientists transitioned from mapping albedo features to topographic features, the community has had to move from defining features by description to discrete boundaries. Resolving feature interpretations at large scales has introduced a new level of complexity, but promises to bring the entire domain into the digital era.

GIS-ready official IAU nomenclature is publicly available for download in either shapefile or KML format through the Gazetteer of Planetary Nomenclature, and as an Open Geospatial Consortium (OGC) standard Web Feature Service (WFS) 1.1 protocol. All of these Web sites are hosted by the US Geological Survey Astrogeology Science Center. All data are stored in decimal degrees on the IAU-defined spheroid or ellipsoid geographic coordinate system (Hayward et al. 2016). Table 2 describes the standard attributes recorded for each feature.

One benefit of implementing nomenclature as part of a GIS is that the author can leverage extended functionality offered by many GIS platforms. This includes map/atlas books and Web services which have previously relied on static nomenclature labels. This is most impactful with Web services because, as vector data,

Table 2 Standard attribute fields included in GIS download of IAU nomenclature

Target	Planet or Satellite where the name is found
Feature Name	As spelled by honoree or by reference shown in Sources of Planetary Names; brackets indicate that the name has been dropped or was never officially approved
Clean Feature Name	This field shows the feature name without the diacritical marks which can cause display issues in some programs
Center Latitude	Latitude of center of feature
Center Longitude	Longitude of center of feature
Northern Latitude	Northernmost latitude of feature
Southern Latitude	Southernmost latitude of feature
Eastern Longitude	Easternmost longitude of feature
Western Longitude	Westernmost longitude of feature
Diameter	Diameter or longest dimension of feature in kilometers
CT (Continent)	Continent or large geographic division that is origin of name
ET (ethnic/cultural group or country)	Country or ethnic group that is origin of name
Approval Status	Number indicates IAU level of approval
Approval Date	Date when name was adopted
Coordinate System	Coordinate system used for the latitude and longitude values
Reference	Reference from which spelling and origin were derived
Feature Type	Latin or Greek descriptor term
Quad	The specific planetary quadrant that the feature center point lies within
Origin	Short explanation of the name

https://planetarynames.wr.usgs.gov/Page/Specifics (2017)

nomenclature annotation may be served with OGC standards—like WFS—across multiple platforms. WFS nomenclature can apply filters dynamically based on scale and allow users to change fonts to better match their map layout. As OGC standards become more prevalent in the planetary science community, they will allow for greater interoperability of spatial data, including nomenclature (OGC 2017).

While current GIS platforms have greatly expanded their publication and printed map resources, many authors prefer the robust editing capabilities of vector graphics programs such as Adobe Illustrator or Inkscape for static maps. Nomenclature can be exported like any other vector layer, in many formats and options for transparency, and manipulated independently. The continued proliferation of spatial data into mediums previously restricted to tabular data or graphics promises to offer greater options for customized placement and appearance of nomenclature annotation.

4.2 *Platform-Specific Capabilities*

The number of quality GIS platforms has grown significantly in the twenty-first century, including support for planetary spatial domains, but it would not be practical to cover all of them in this text so we will focus on the two largest platforms, Esri's ArcMap at version 10.x (going forward as ArcPro) and the Open Source Geospatial Foundation's QGIS at version 3.x. It should be noted that there are planetary-specific GIS platforms such as Arizona State University's Java Mission-planning and Analysis for Remote Sensing (JMARS), but we will focus on the advanced labeling and annotation capabilities of the leading proprietary and open-source solutions.

ArcMap couples the advantages of a GIS environment (i.e., spatial and tabular data properties) with their Maplex Label Engine to apply a range of placement and deconfliction rules. This can be used with queries to apply different rules to the same dataset based on field attributes or geometric properties. Additionally, users can create dynamic relationship classes, known as feature-linked annotation, to generate labels as separate vector datasets that change as records in the source data are added or deleted. Using this method, an annotation feature class can be created to reflect all of the different font types, sizes, and placement properties used in a map at a given scale. For example, to apply the desired parameters to large, directional *dorsa* or *montes* features, a new annotation class can be created based on this SQL query:

$$(\text{code} = \text{'DO'OR code} = \text{'MO'})\text{AND diameter} \geq 85$$

Grouping annotation classes by feature type and size, each record will be accounted for and greatly reduce the time spent on custom annotation. When using this method, it is useful to create a 'default' annotation class, without a query applied, that allows for free-hand placement to handle exceptions. Users may also save these labeling properties inside of a style file for later use.

QGIS also offers customizable, data-defined label settings that are similar in appearance, but does have some limitations to creating complex labeling schemes with its native Label Tool. For simple labels, users can create multiple layers based on queries similar to those mentioned above and apply different settings; however, the greatest advantage to QGIS is the extensive network of third-party plug-ins that can provide unique solutions not available elsewhere.

Known as ‘Plugins Planet’ (http://plugins.qgis.org/plugins/), this repository of Python plug-ins is a constantly evolving list of useful tools and routines created by users to fulfill specific mapping requirements. Many of them are focused on portability to graphics editing programs and creating print-ready maps and are very useful for geologic maps. Those that are stable and most frequently used are then incorporated into future ‘master’ QGIS releases. The open, collaborative development environment of QGIS continues to make it the most popular open-source GIS platform and promises to deliver more robust labeling and annotation capabilities as the field matures.

References

Blagg MA (1913) Lunar formations named or lettered in the maps of neison, schmidt and mädler. Neill, Edinburgh

Blagg MA, Müller K (1935) Named lunar formations. Percy, Lund, Humphries, London

Blue J, et al (2013, March) Planetary nomenclature: an overview. In: Lunar and planetary science conference, March 2013, vol 45 (Abstract #2178)

Environmental Systems Research Institute (ESRI) (2016) ArcGIS Desktop 10.3 Help. About creating and editing annotation. http://desktop.arcgis.com/en/arcmap/10.3/manage-data/creating-new-features/about-creating-and-editing-annotation.htm. Accessed 27 Apr 2017

Greeley R, Batson RM (eds) (1990) Planetary mapping, vol 6. Cambridge University Press, Cambridge

Hayward RK, et al (2016) Planetary nomenclature: an update and overview. In: Lunar and planetary science conference, vol 47 (Abstract #1141)

Hunter MA, et al (2016) Feature-linked annotation of lunar and martian nomenclature. In: Lunar and planetary science conference, vol 47 (Abstract #1903)

International Astronomical Union (2017) Publications, Transactions B. http://www.iau.org/science/publications/iau/transactions_b/. Accessed 26 Apr 2017

International Astronomical Union Working Group for Planetary System Nomenclature (2017) Gazetteer of planetary nomenclature. https://planetarynames.wr.usgs.gov/. Accessed 26 Apr 2017

Masursky H, et al (1986) Annual gazetteer of planetary nomenclature: U.S. Geological Survey Open-File Report 84-692

Open Geospatial Consortium (2017) Web Feature Service. http://www.opengeospatial.org/standards/wfs. Accessed 28 Apr 2017

Quantum GIS (2016) QGIS 2.14 User Guide. General tools, annotation. http://docs.qgis.org/2.14/en/docs/user_manual/introduction/general_tools.html#annotation-tools. Accessed 27 Apr 2017

Swings JP (1986) Transactions of the international astronomical union, Volume XIXB: proceedings of the nineteenth general assembly, Delhi 1985. In: Transactions of the international astronomical union, Series B, 19

Fundamental Frameworks in Planetary Mapping: A Review

Henrik Hargitai, Konrad Willner and Trent Hare

Abstract In this chapter, we review basic concepts, measurements, and methods in mapping topographic and reflectance (image) data of planetary surfaces. This includes the definition of coordinate systems for each body, the identification of the shape of a planetary body, and the establishment of reference systems and reference bodies that are required to produce horizontally and vertically accurate representations of a planetary surface.

Keywords Reference surface · Datum · Coordinate · Projection · Geodetic control · Block adjustment

1 Introduction

1.1 Application of Planetary Maps

Planetary maps are a common instrument for many applications. Maps in general provide scientists of all disciplines with a unique tool to locate an area of surface through a set of coordinates. Horizontal positions and elevation information require a reference surface to be defined (see below). Planetary maps provide an effective way to spatially visualize surface properties. Its visual representation techniques range from symbolic to calibrated data; it can show spatial relations of features and phenomena and absolute locations that are readily comprehensible. However, the spatial accuracy, and consequently, reliability, of any planetary map depends on the methods of how data are linked to surface locations and how the framework of

H. Hargitai (✉)
Eotvos Loránd University, Budapest, Hungary
e-mail: hargitaih@caesar.elte.hu

K. Willner
German Aerospace Center (DLR), Institute of Planetary Research, Berlin, Germany

T. Hare
Astrogeology, United States Geologic Survey, Flagstaff, USA

© Springer Nature Switzerland AG 2019

H. Hargitai (ed.), *Planetary Cartography and GIS*, Lecture Notes in Geoinformation and Cartography, https://doi.org/10.1007/978-3-319-62849-3_4

surface coordinates is established. Planetary maps are one of the most important tools in investigating surface geology, evaluating the potential merits of landing sites, and planning surface operations. Although planetary mapping is based on using computers and spacecraft-transmitted data, this process is not dissimilar to the early explorers' field mappings. With planetary mapping, humanity expands its geographic knowledge to extraterrestrial territories. Maps of these faraway lands, mostly never visited by humans, can be as accurate as maps made for navigating on Earth.

Maps provide a tool to communicate not only among scientists but also with the public. The creation of maps of planetary bodies requires the application of a wide range of methods and the formulation of definitions that we discuss in the following sections.

1.2 International Astronomical Union—IAU

Mapping standards such as uniformity of coordinate systems, accurate horizontal and vertical positioning, mapping methods, scales and schemata are vital components to allow data usage across the members of the scientific community in various facilities and disciplines. In 1976, the International Astronomical Union (IAU) established the Working Group on the Cartographic Coordinates and Rotational Elements of Planets and Satellites (WCCRGE[1]) supporting the manifestation of standards. Triennially reports on the preferred rotation rate, spin axis, prime meridian, and reference surface for planets and satellites are published (Archinal et al. 2011). Thus, nearly all larger bodies in our Solar System have defined geodetic parameters, documented by the IAU, allowing studying these bodies by means of capable cartographic applications, such as geographic information systems (GISs) and remote sensing (RS) applications. Remote sensing applications are considered software systems being able to process data from remote sensing instruments in a geographic reference in various fields of application such as change detection, spectral analysis, and ortho-rectification.

2 Reference Surfaces

Reference surfaces (Table 1) are used to enable horizontal mapping of surfaces and to provide an equal level to relate height information to.

The reference surface, or geodetic datum, approximates the shape of the body and is either a geometric approximation of the rigid body, like a sphere or ellipsoid,

[1]http://astrogeology.usgs.gov/Projects/WGCCRE.

Table 1 Reference surfaces, geographic and projected coordinates

Body	IAU2000 Reference datum shape (Archinal et al. 2010; Wang et al. 2017) and radii, km	Ref datum 0 level (km) on projected maps	Surface definition of the 0° meridian (Archinal et al. 2010), in longitude	IAU coordinate system (Archinal et al. 2010) (all 0–360° except the Moon)
Mercury	Sphere 2439.7	2439.4 km MESSENGER 2440	20° = Hun Kal	ographic, +W
Venus	Sphere 6051.8 IAU2000 6051 IAU1985	6051	0° = Ariadne central peak, previously Eve (Davies et al. 1986)	ocentric, +E
Earth	Spheroid 6378 × 6356		Greenwich	
Moon	Sphere 1737.4 LOLA 2011	1737.4	0° = Sub-Earth longitude	ographic, +E (± 180)
Mars	Spheroid 3396.19 × 3376.2 IAU2000 MOLA MEGDR 3394 × 3375: IAU 1994	THEMIS, MDIM2.1: 3396.19 × 3376.2 MOLA: sphere, 3396.19	0° = Airy-0	ocentric, +E (IAU2000) ographic, +W (IAU1994)
Phobos	Irregular, approximated by a triaxial ellipsoid. Mean: 11.08	HRSC map: sphere 11.1		
Io	Ellipsoid, mean: 1821.49	1821.46 sphere	0° = sub-jovian direction	ographic, +W
Europa	Ellipsoid, mean: 1560.8	1562.0899658 sphere	182° = Cilix	ographic, +W
Ganymede	Sphere, 2631.2	2632.3449707	128° = Anat	ographic, +W
Callisto	Sphere, 2410.3	2409.3	326° = Saga	ographic, +W
Mimas	Ellipsoid		162° = Palomides	ographic, +W
Enceladus	Ellipsoid, mean: 252.1	252.1 sphere	5° = Salih	ographic, +W
Tethys	Ellipsoid, mean: 531.0	536.3 sphere	299° = Arete	ographic, +W
Dione	Ellipsoid, mean: 561.4	563.0 sphere	63° = Palinurus	ographic, +W
Rhea	Ellipsoid, mean: 763.5	764.1 sphere	340° = Tore	ographic, +W
Titan	Ellipsoid, mean: 2574.73	2575 sphere		ographic, +W
Iapetus	Ellipsoid, mean: 734.3	736 sphere	276° = Almeric	ographic, +W

(continued)

Table 1 (continued)

Body	IAU2000 Reference datum shape (Archinal et al. 2010; Wang et al. 2017) and radii, km	Ref datum 0 level (km) on projected maps	Surface definition of the 0° meridian (Archinal et al. 2010), in longitude	IAU coordinate system (Archinal et al. 2010) (all 0–360° except the Moon)
Triton	Mean: 1352.6	1350 sphere		
Ceres	Spheroid, mean: 476.2	470 sphere	0° = unnamed spot	
Itokawa	Irregular		0° = W_0 = 0°	
Pluto	Sphere, 1195	1188.3	0° = sub-Charon meridian	
Charon	Sphere, 605	606	0° = sub-Pluto meridian	

Explanations: ographic: planetographic, ocentric: planetocentric. +W: positive (increasing values) toward the western direction. For compatibility with mapping applications, many data portals may support mosaics in positive East only. Note when using a sphere (or the IAU calculated mean radius for a spheroid), there is no difference between ographic and ocentric coordinate systems

or an approximation of its gravitational potential. Geoids are surfaces of equilibrium, where the gravitational potential energy is constant.

On Earth, the geoid is related to present-day mean sea level. On extraterrestrial bodies, a potential "sea level" where the potential liquid water surface would be influenced by gravity and rotation but not affected by currents and tides would model the body's geoid. A common measure of the best-fitting shape is a reference *ellipsoid* which serves as zero-elevation surface or datum level. The actual value of the topographic reference datum is typically chosen to be at the mean planetary radius (e.g., Aeschliman 1998).

A planetary body in hydrostatic equilibrium has a round or nearly round shape. The simplest form of a planetary shape is a *sphere*, which is the equipotential of a non-rotating planetary body with a homogeneous mass distribution. Rotating bodies form oblate *spheroids* (rotational ellipsoid, also called ellipsoid of revolution) that are aligned with the spin axis of the body. Oblate spheroid bodies have larger equatorial radii (radius A) than polar radii (radius C). This phenomenon is also termed polar flattening or the equatorial bulge. Tidally deformed bodies (e.g., most synchronously rotating satellites) form triaxial *ellipsoids* (Melosh 2011) with different radii in the subplanetary equatorial, along-orbit equatorial (radius B), and polar axes.

The definition of reference surfaces varies from body to body. Ellipsoids provide *geometric height*, and a gravitational equipotential provides *geopotential height* values.

The reference surfaces for Mercury, Venus, the Moon, and several satellites are spheres; it is a spheroid for Mars, a spheroid at 1-bar pressure level for the gas giant planets; ellipsoids constitute the reference datum for most of the moons (Table 1). For small bodies, elevations may be measured as radii from the center of mass.

The *cartographic representation* of a planetary body's surface may be sphere or spheroid based. In map projections, usually spheres are used rather than spheroids or triaxial ellipsoids, because of computational costs and software capabilities (Wang et al. 2017).

Planetary shape and radii data can be obtained from several methods including limb profiles (Dermott and Thomas 1987; Thomas 1987; White et al. 2014; Oberst et al. 2011), occultation (Perry et al. 2011), laser altimetry (Zuber and Smith 1996; Smith et al. 2010; Smith et al. 1999), and stereophotogrammetry (e.g., Preusker et al. 2017; Willner et al. 2014).

3 Coordinate Systems and Coordinate Frames

Coordinate systems are a set of conventions that define the general properties of a structure that aids users in establishing a spatial orientation. The realization of the coordinate systems is called *coordinate frames*. Specific coordinates provide the means for locating points within a reference frame (NAIF 2017).

A number of coordinate systems are specified and their usage depends on the application.

Next to *inertial coordinate systems* that have *a fixed orientation* with respect to reference stars over time, *local coordinate systems* are defined as needed.

The purpose of planetary mapping is to provide a spatial relationship between different points of interest that are linked to the surface of one body. This, for instance, might be surface features, footprints of orbital remote sensing data, or the location of a space probe on the surface of a body.

For example, when an instrument observes a planetary body, data from the sensors show the spatial relations of the observed features within one observation—e.g., one image. However, the spatial relation is missing when observing several image datasets not necessarily overlapping. This is overcome by linking the data to a reference surface. Exact locations with respect to the global frame of the features represented in the data points are only then known. During the process of geodetic control, data points are connected to surface locations, using a surface coordinate system as a reference framework that is realized with a coordinate frame, by means of well-determined 3D coordinates of points on the body's surface. These points are often referred to as *ground control points* which are part of the *control point network* (e.g., Archinal et al. 2004).

As a consequence, *body-fixed coordinate systems* are defined for each planetary body. These coordinate systems are non-inertial, meaning that their orientation changes with respect to the reference stars over time. Coordinate-system-defining parameters include its origin, the surface location of the prime meridian, the spin axis direction, and a fundamental plane. *The body-fixed coordinate system rotates with the body.*

A relation between the body-fixed coordinate frame and the inertial frame can be established by (1) modeling the motion of the body through the Solar System (Jacobson and Lainey 2014) and describing this in form of ephemerides and (2) by modeling the rotation of the body with respect to the inertial reference frame (Stark et al. 2017; Burmeister et al. 2018). The rotation model includes next to the orientation of the rotation axis the model of the prime meridian. All formulations of the rotation model are time-dependent functions where the phase of the rotation—as only time-independent value in the equation—is stated with respect to the standard epoch J2000.0. It is to be noted that the rotation model is dependent on the ephemerides model (Jacobson et al. 2018; Stark et al. 2017).

In general, one can distinguish between *Cartesian* and *spherical* coordinate systems. A triple of metric coordinates, *XYZ*, specifies a location in a Cartesian coordinate system while in a spherical coordinate system a pair of angular values, *latitude and longitude*, in addition to the distance to the point of interest, the *radius*, describes the location of that point.

3.1 Cartesian Coordinates

Right-handed, orthogonal coordinate systems are commonly applied for planetary bodies. The origin is defined to be the center of mass—as opposed to the center of the figure, i.e., the geometric center—of that body.

Poles: Planets and satellites have their "north pole" above the invariable plane of the Solar System. The direction of the North Pole is specified by the value of its right ascension (α_0) and declination (δ_0) (Archinal et al. 2010).

Small bodies can have large changes in the polar axis orientation known as precession. This can cause the IAU-defined North Pole to become the South Pole—i.e., moving below the invariable plane (Archinal et al. 2010). For this reason, dwarf planets (including Pluto), minor planets (asteroids), and comets have "positive" and "negative" poles and they spin about this pole in the right-hand sense instead of defining directions relative to the ecliptic (~invariable plane). The "positive pole" may point above or below the invariable plane of the Solar System. Longitudes increase 0° to 360° using the right-hand rule (in eastern direction) (Archinal et al. 2010). To avoid confusion, Zangari (2015) recommended introducing the term spinward direction instead of eastern direction, and right-hand pole instead of "north" or "positive" pole.

Coordinate axis orientation: The Z-axis of a planetary body-fixed coordinate frame is defined to point along the mean rotational axis of the body. From the center of mass toward the North Pole, values are increasing—toward the South Pole, values are decreasing.

As a second axis, the X-axis orientation is usually defined. The X-axis is perpendicular to the Z-axis and its direction coincides with the direction of the prime meridian (see below for details).

The Y-axis completes the right-handed coordinate frame by being orthogonal to both X- and Z-axis to establish a three-dimensional Cartesian coordinate frame.

3.2 Spherical Coordinates

Spherical coordinates can directly be related to Cartesian coordinates through a functional expression as these make use of the definition of the Cartesian coordinate frame. The prime meridian and the direction of the X-axis of the Cartesian coordinate frame coincide.

The *longitude* angles are measured in the XY plane of the Cartesian coordinate system. This plane is also referred to as the equatorial plane. The longitude is measured between the prime meridian and the projection of the vector to the point into the XY plane. Whether the longitudes increase to the east or to the west depends on the conventions established for the body in question.

The prime meridian (0° longitude) of a solid surface body is defined by specifying the coordinates of an observable surface feature on the body (for example the center of a small crater). In the absence of permanent features such as craters, the 0° meridian may be defined by the mean direction relative to the parent body for synchronously rotating bodies (with 0° longitude at the mean sub-planet point, for example on Io) (Archinal et al. 2010). For Pallas, the direction of the longest axis defines the prime meridian.

The prime meridian may also be defined based on practical reasons or convenience. Historically, the prime meridian of Mercury, before orbital images became available, was defined as the subsolar point at the first perihelion of 1950 (IAU 1971). For Venus, the sub-Earth longitude on 20.06.1964 was defined as the 320° longitude (IAU 1971). This choice is also convenient for representation of Aphrodite Terra, the largest terra on Venus, because it then falls into a single hemisphere (Shevchenko et al. 2016). Several Russian maps show Venus with two hemispheres centered at 320° and 140° longitudes, which were thought to be the returning central meridians for the near and far sides of closest approach, respectively (Pettengill et al. 1980; Burba 1996). The longitude of the sub-Earth point at inferior conjunction is *approximately* repeated or cycle due to synodic resonance with Earth (Bills 2005). For aesthetic reasons, some maps show Mars' hemispheres centered at 90° and 270° longitudes.

The *latitudes* range from +90° at the North Pole to −90° at the South Pole and are the angle between the equatorial plane and the vector to the point. Points in the equatorial plane have latitude of zero. Latitudes increase from the equator toward the positive (north) pole and decrease toward the negative (south) pole.

The *radius* provides the length of the vector in the longitude and latitude direction to the point of interest.

3.3 Longitude and Latitude Systems—Discussion

In 1970, the IAU (1971) approved the use of two types of coordinates for extraterrestrial mapping (Duxbury et al. 2002) (Fig. 1).

Planetocentric ("-ocentric"/East) coordinates use geocentric latitudes and positive East longitudes resulting in a right-handed spherical coordinate frame. For planetocentric coordinate systems, latitudes are defined to be the angle between the equatorial plane, the center of mass of the body, and a vector pointing from the center of mass to the point of interest. The planetocentric longitude is the angle between the prime meridian and the projection of the vector onto the equatorial plane. Longitudes increase toward the east from the prime meridian, from 0° to 360°.

Planetographic ("-ographic"/West) coordinates consist of geographic latitudes (equivalent to geographic or geodetic latitude on the Earth). Planetographic coordinates are defined by vectors perpendicular to a reference surface, e.g., ellipsoid (Fig. 1). In a spheroid-based system, the planetographic vector does not pass through the origin (the center of mass). The planetographic longitude increases with time (left-hand rule) with respect to an observer fixed in space above the object of interest (e.g., above the sub-Earth point) from 0 to 360°. This is in the direction opposite to the rotation. For practical reasons, planetographic (geographic) coordinates were useful for surveyors on Earth (because it is perpendicular to the surface and thus easily measurable with a theodolite), and for Earth-based Mars observers, too, because longitudes increased with time (Duxbury et al. 2002).

Planetocentric and planetographic systems differ in the positive direction of longitudes and the definition of the vectors to the point of interest.

A body is in *prograde* (direct) rotation when it rotates in the same direction as the Sun rotates. One could also say the body rotates counterclockwise when viewed from above the ecliptic/invariable plane. A body is in *retrograde* rotation when the rotation is opposition to the Sun's rotation—clockwise when viewed from above the ecliptic.

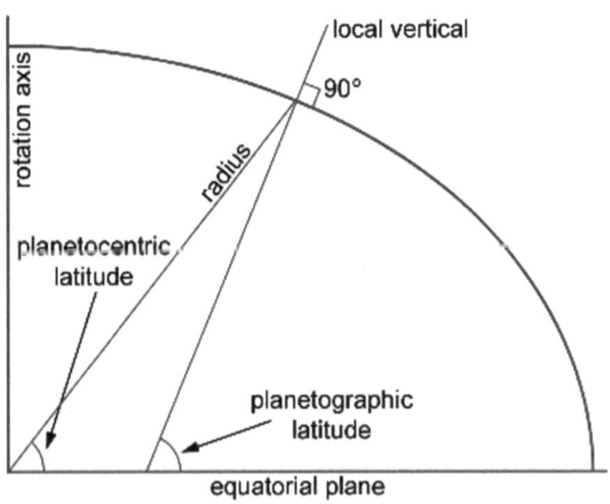

Fig. 1 Geometric distinction between planetocentric and planetographic latitudes. The degree of polar flattening in this cross section is greatly exaggerated. There is no difference between the two coordinate systems for spheres (Figure from Hargitai et al. 2017)

As a consequence, the planetographic longitude increases toward the west for prograde rotators. In case of retrograde rotation, longitudes increase toward the east. For example, Venus has a retrograde rotation while the other planets are prograde rotators (PDS 2008) (see also Table 1).

3.4 Latitudes and Longitudes in Applications

The usage of different coordinate systems is a challenge for off-the-shelf GIS and mapping applications, as these are normally designed for Earth-related data. Such applications usually expect geographic latitude and positive east longitudes.

The two kinds of latitude, planetographic and planetocentric, differ at mid-latitudes, but are identical at the equator and poles. An acceptable method to resolve this issue is to define a spherical reference system. When the body is defined as a sphere, the two *latitude* systems are identical. This technique forces the use of a planetocentric latitude system but has the potential to cause slight errors if the body is defined as an ellipse. Fortunately, only a few major bodies besides Earth are currently defined as an ellipse (e.g., Mars). This is mainly due to the lack of information about the body, the fact that the body has a shape that is very close to a sphere (e.g., the Moon), or when cartographically the use of the best-fit sphere is recommended for triaxial or irregularly shaped bodies (Archinal et al. 2010).

Changing the positive longitude direction is not available in most GIS platforms. The longitude direction does not affect the registration of the dataset when in the Cartesian plane (X, Y) in a defined map projection, but it does change the meaning if the map projection defines a longitude of central meridian other than 0 or 180 degrees. For the longitude range, many GIS applications only support values between −180 and 180. However, it is more common for planetary data to be represented in a 0 to 360 range. As a consequence, it is generally recommended to create digitally release global maps using a −180 to 180 range to help with greater interoperability across different applications.

3.5 Body-Specific Planetary Coordinate Systems

MESSENGER products for Mercury use planetocentric coordinates while previous (Mariner 10) mission used the planetographic IAU system.

For the Moon, Earth, and the Sun, a longitude range of −180° to +180° has been traditionally used; however, it has been recently recommended that in the future, only the 0° to 360° range can be used for the Moon. It is helpful to show both on maps.

For Mars, products prior to 2002 used planetographic coordinates (West longitudes), including MGS MOC, while more recent products (since the release of the 100-m-accuracy MOLA digital terrain model) use planetocentric coordinates (East longitudes). Maps may display both systems (Seidelman et al. 2002).

The direction of the North Pole of Pluto and Charon was inverted in 2009, applying the right-hand rule of dwarf planets (Archinal 2010; Zangari 2015).

For irregular bodies, where the vector from the center of the body may intersect the surface in more than one place, surface coordinates can only be identified by latitude, longitude, and radius values (PDS 2008; NAIF 2017). For irregular bodies, the longest *diameter* may replace the role of the regular bodies' longest axis.

We note that the coordinate systems used in research papers and maps may and often do differ from those recommended by IAU (Zangari 2015). Different planetary datasets of the same body may use different coordinate systems, depending on the given mission's standards or the time of publication.

4 Map Projections

Map projections are mathematical equations for mapping a three-dimensional body onto a two-dimensional plane or Cartesian coordinate system (Fig. 2). Conversion between the three-dimensional coordinates requires the choice of a map center, e.g., center latitude and center longitude, at which no geometric distortions will occur. As stated by Snyder (1987), there is no one "best" map projection for mapping and care must be taken choosing a projection that is suitable for the area of study or use given that every projection incurs some type of spatial distortion when leaving the center of the map projection. The majority of printed maps of planets and satellites have been based on conformal projections: *Mercator* for low latitudes (~0–22°) (Fig. 2/b), *Polar Stereographic* for high latitudes (~65–90°) (Fig. 2/d), *Lambert Conformal Conic* for intermediate latitudes (~21–66°) at small scales (Fig. 2/e), and *Transverse Mercator* for large-scale (small-area) maps (Fig. 2/c). Conformal projections preserve the local angles (shapes), and thus, craters will remain circular at any location on the map (although their size can be greatly distorted the further you move from the center).

Digital map products, on the other hand, have been based on a different set of projections to help facilitate digital archival. It is useful to think of these as *"database" projections*, because their most important properties have been their suitability for holding global datasets. The most significant of these qualities are (1) global applicability (i.e., there no "unmappable" areas, like polar regions in the Mercator projection); (2) simple formulation; and (3) at least roughly equal area (to avoid oversampling some areas and inflating the volume of the digital dataset). Starting in the 1990s, the preferred database projection for digital map products was the *Sinusoidal* projection (Fig. 2/f). The spherical version of this projection was used in place of the much more complicated and iterative ellipsoidal version. However, the Sinusoidal projection was never considered entirely satisfactory for several reasons. First, meridians in Sinusoidal projection curve relative to the pixel grid (apart from the central one), which limits neighboring maps from being easily combined. Second, before neighboring maps can be combined, one of the maps must be resampled even at their original scale. Finally, extra pixels must be

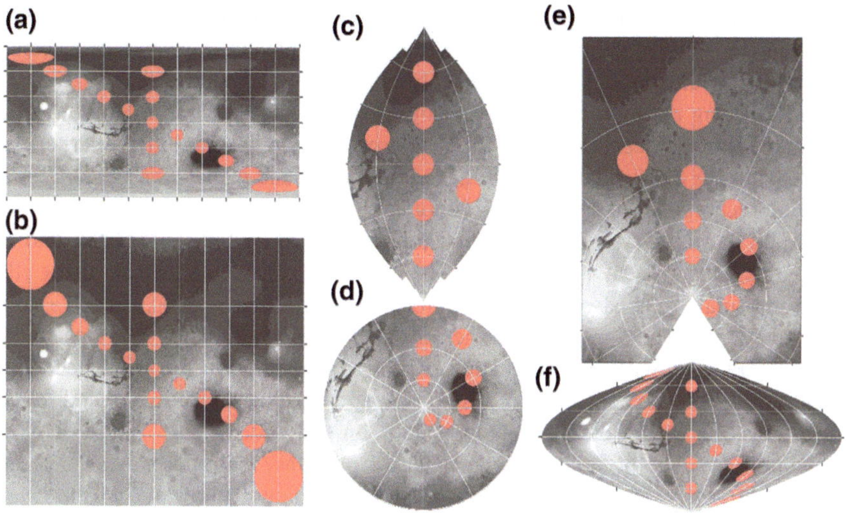

Fig. 2 Examples showing different map projections with Tissot's indicatrix distortion ellipses. The reference circle is at the origin of the projection. Projections: (a) Simple Cylindrical/ Equirectangular; (b) Mercator (truncated at ± 82° latitudes); (c) Transverse Mercator (truncated at ± 45° longitudes); (d) South Polar Stereographic (truncated at 0° latitude); (e) Lambert conformal conic (detail); (f) Sinusoidal (Mercator-Sanson). Background: Mars MOLA data

provided in order to prevent the formation of gaps along the curved edges of the archived map files when resampling is performed.

An alternative to Sinusoidal is the *Simple Cylindrical* projection, wherein the grid is simply an equally spaced raster in latitude–longitude coordinates (Fig. 2/a). Though this approach eliminates the problems caused by curved meridians, it greatly oversamples the polar regions, inflating the size of a global dataset by as much as 57%.

A useful compromise that has been adopted for an increasing number of digital databases in recent years is the *Equirectangular (Equidistant Cylindrical)* projection, where the standard parallel (where the line of latitude touches the globe) $\varphi_1 = 0°$ (the equator). In Equirectangular projection, the latitude and longitude form the grid, rather than having equal latitude and longitude dimensions, the grid cells have dimensions that give equal kilometer lengths at some specified latitude. This is called the "center latitude"—though "reference latitude" might be more precise, because it does not need to be located at the center of a given map. The Equirectangular projection is identical to the Simple Cylindrical projection when the reference latitude is set to 0°. For the Sinusoidal projection, the east–west grid spacing in degrees is adjusted on every row to keep the kilometer spacing nearly constant, whereas in the Equirectangular projection this spacing is constant for a given map file but can be adjusted for files in different latitude zones. The result is a compromise between Sinusoidal and Simple Cylindrical in the complexity and areal distortion of the dataset. Reference latitudes may be assigned freely to the latitude zones of the map series, or they may be chosen to give simple integer relations between the sample intervals in latitude and longitude. In the latter case, center latitude 0° could be used everywhere from 0° to 60° latitude, then 60° center latitude (giving a longitudinal grid spacing twice as large as that in latitude) from 60° to roughly 70°, center latitude 70.5° (giving 3:1 grid spacing) thereafter, and so on. As with the Sinusoidal projection, planetographic and planetocentric variants of the Equirectangular and Simple Cylindrical projections must be distinguished, according to which type of latitude is equally sampled by the rows of the grid.

A relatively recent development has been the acceptance of using the non-conformal, "database" projections for some large-scale maps. This was first done for Venus: The 5° Sinusoidal tiles of the USGS FMAP dataset were printed at 1:1,500,000 scale without reprojection on the grounds that (a) the effort needed to transform the files to conformal projections would have been prohibitive and (b) the distortions incurred were acceptably small, on the order of 4% at most. This decision highlights the pragmatic character of planetary cartographers. The precedent set by the Venus 1:1,500,000 series was later followed by the Mars Express HRSC team, which adopted the Sinusoidal projection for both digital products and printed maps at scales of 1:200,000 and larger (2° size, < 1.7% distortion) for single strips. On a regional scale, e.g., multi-orbit data products, an Equidistant Cylindrical projection is chosen (Gwinner et al. 2016).

One of the criteria for selecting a "database" projection is that it can be used for an entire global dataset. The polar sections of such datasets are entirely adequate as sources of data for resampling to other projections, but are too severely distorted to be directly useful as maps. Separate files are therefore usually provided, showing the polar regions in a more appropriate projection such as Polar Stereographic.

In practice, the most common aesthetic goal of selecting a projection for maps is to display craters as close to their true (generally) circular shape. For measurements, instead of relying on planar distances as defined by the map projection, distances, geodesic measures should be used when possible.

This discussion demonstrates Snyder's premise that there is no one "best" map projection, and when splitting up a planetary body into a series of discrete regions (or quadrangles), it is truly more of an art than a science and involves reconciling several interlocking sets of constraints. The resolution of available data and the desired density of pixels in the output image are the main factors dictating the choice of a hardcopy map scale (expressed as a ratio of sizes 1:x). Hardcopy scale and the size limits for printing dictate the *physical size* of map quadrangles for a given body or group of bodies of similar radius. These considerations have led to the definition of a rather large but finite number of *quadrangle schemes* (Fig. 3) for mapping the planets and satellites (Batson 1990). In some cases, similar but slightly different schemes have arisen over time, such as the 144-quadrangle scheme for lunar maps at 1:1,000,000 scale (abbreviated as 1:1M) and the 140-quadrangle series of Mars maps at 1:2,000,000 scale. In many cases, schemes for larger-scale maps have been developed by subdividing the quadrangles of a smaller-scale map series. Division of

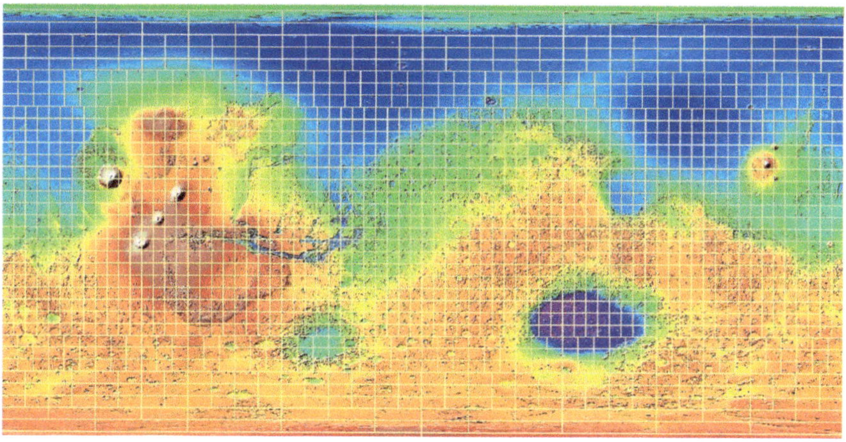

Fig. 3 1:500k Mars MTM (Mars Transverse Mercator) quadrangle scheme. Quads are 5 × 5°. Names are generated from center latitude (xx) and longitude (yyy), e.g., 40292

quadrangles into quarters is most common, but other schemes have been used, and the division of the near-polar quadrangles is often more complex. Because the set of "round" scales is based on powers of 5 as well as powers of 2, the process of cutting quadrangles into quarters cannot be continued indefinitely. For example, the 1964-quadrangle scheme for Mars maps at 1:500,000 scale is unrelated to the quadrangles used at larger scales, and it is not further subdivided (Fig. 3).

5 Image Mosaics

Planetary maps are derived from image data that are obtained from Earth-based or space-based remote sensing sensors, or historically from visual observations. *Passive remote sensing sensors* record the amount of the *reflected* solar irradiation (e.g., in visual wavelengths); the thermal (infrared) or radio-thermal (radar) *emission* of the surface; while *active remote sensing instruments* illuminate the surface and receive the return signal from the same surface. These may provide single or complementary image or topographic datasets.

An image basemap is produced in the form of a Digital Image Model (DIM) (Table 2) where body brightness is defined as a function of cartographic latitude and longitude in a specific spectral band or bands (PDS 2008). These global image mosaic files can be imported into any GIS and combined with elevation or other data.

Table 2 DIMs (global digital image mosaics) available at USGS Astrogeology Science Center https://astrogeology.usgs.gov/, except those marked with an asterisk: "Icy Moons" that is available at https://www.lpi.usra.edu/icy_moons/ and Chang'E images that are available at http://159.226.88.61/CLEPWebMaps/CEAtlas/CE1_2/Atlas.html#page/1

Body	Source	Resolution (meter/pixel)	Reference
Mercury	MESSENGER MDIS NAC or WAC 750-nm high-incidence (78–86°) east illumination (HIE), high-incidence (78–86°) west illumination (HIW), low-incidence (LOI ~45°), morphology base map (BDR, 74° moderate incidence)	166	Hawkins et al. (2007)
Mercury	MESSENGER MDIS 8-band color	665	Hawkins et al. (2007)
Venus −80° – +84° lat	Magellan SAR FMAP left look, right look, and stereo	75	Saunders et al. (1990)
Venus	Magellan C3-MDIR	2025	Saunders et al. (1990)
Venus	Global Fresnel Reflectivity GREDR	4641	Ford (1992)
Moon	Lunar Orbiter	59	Gaddis et al. (2001)
Moon	Lunar Orbiter + Clementine	59	Gaddis et al. (2001); Lee et al. (2009)
Moon	Clementine UVVIS 5 band	200	Eliason (1999); Hare (2008)
Moon	Kaguya Terrain Camera (TC) Ortho (for all TC, 7.4 m/p tiles are available)	474	Gaddis et al. (2015); Isbell et al. (2014), JAXA/SELENE, http://jda.jaxa.jp
Moon	Kaguya Reflectance 750 nm	237	Ohtake et al. (2013)
Moon	Kaguya TC Morning low-angle solar illumination	474	Haruyama et al. (2008); Isbell et al. (2014)
Moon	Kaguya TC Evening low-angle solar illumination	474	Haruyama et al. (2008); Isbell et al. (2014)
Moon	LRO LROC-WAC	100	Sato et al. (2014); Wagner et al. (2015); Speyerer et al. (2011)
Moon*	Chang'E	120	Li et al. (2010)
Mars	Viking MDIM2.1, *controlled* to MOLA	232	Archinal (2004)
Mars	Mars Odyssey THEMIS daytime infrared, controlled	100	Edwards et al. 2011; Fergason et al. (2013)
Mars	Mars Global Surveyor Thermal Emission Spectrometer (TES) Bolometric Albedo	7410	Christensen et al. (2001)
Phobos	Viking	4.8	Stooke (2012); Simonelli et al. (1993)
Phobos	Mars Express SRC	12	Willner et al. (2008)
Ceres	Dawn Framing Camera (FC)	400	Russell and Raymond (2011)
Vesta	Dawn Framing Camera (FC)	140	Russell and Raymond (2011)

(continued)

Table 2 (continued)

Body	Source	Resolution (meter/pixel)	Reference
Io	Galileo Solid-State Imaging (SSI) + Voyager Color	1000 (1.3–21)	Becker and Geissler (2005); Belton et al. (1992)
Io	Galileo SSI Color	1000	Becker and Geissler (2005); Belton et al. (1992)
Io	Galileo False Color	1000	Becker and Geissler (2005); Belton et al. (1992)
Europa	Voyager + Galileo SSI	500 (0.2–20)	Belton et al. (1992); USGS (2002)
Ganymede	Galileo/Voyager Color	1400 (0.4–20) (Greyscale 1000)	Becker et al. (2001)
Callisto	Galileo/Voyager	1000 (0.4–60)	Becker et al. (2001)
Enceladus	Cassini Imaging Science Subsystem (ISS)	100	Becker et al. (2016)
Tethys	Cassini Imaging Science Subsystem (ISS)	292.5	Roatsch et al. (2009)
Dione	Cassini + Voyager	154	Roatsch et al. (2006)
Rhea	Cassini + Voyager	417	Roatsch et al. (2012)
Titan −65° – +45° lat	Cassini ISS, controlled	450	Archinal et al. (2013)
Titan	Cassini ISS	4004	Archinal et al. (2013)
Titan	Cassini Synthetic Aperture Radar (SAR) and High Altitude Synthetic Aperture Radar (HiSAR)	351	Elachi et al. (2005); Stephan et al. (2009)
Iapetus	Cassini + Voyager	783	Roatsch et al. (2009)
Triton	Voyager 2 Color	600	Schenk (2008)
Pluto	New Horizons Long-Range Reconnaissance Imager (LORRI) and the Multispectral Visible Imaging Camera (MVIC)	300	Moore et al. (2016); Cheng et al. (2008)
Charon	New Horizons Long-Range Reconnaissance Imager (LORRI) and the Multispectral Visible Imaging Camera (MVIC)	300	Moore et al. (2016); Cheng et al. (2008)
Icy Moons*	Galileo, Cassini and New Horizons, color maps of Mimas, Enceladus, Tethys, Dione, Rhea, Iapetus, and Triton	100–600	Schenk et al. (https://www.lpi.usra.edu/icy_moons/)

5.1 Geometric Control

The assembly of geometrically correct image mosaics requires precise knowledge of the camera position and orientation at the time of the image acquisition. Though good preliminary knowledge is available from spacecraft tracking and attitude prediction, this knowledge is often insufficient for the cartographic application leading to seams and discontinuities between neighboring images in an image mosaic if left uncorrected. Correcting these geometric inaccuracies by, e.g., stereo-photogrammetric processes is called *geometric control*. If the data is additionally correctly placed within the global reference coordinate frame, the data is referred to as *geo-referenced*.

In the last decade, there has been a push to educate users on the idea of a *Spatial Data Infrastructures* (SDIs), which include the users, data, data access, policies, and standards for a research team or a community (Laura 2017). One of the major tenants for any SDI is a well-established control network to tie or register all data products to. As stated above, the IAU helps define the parameters for establishing the body's prime meridian and reference surface, but the generation of a control network goes beyond this initial definition. A *control network* consists of a set of well-defined topographic points whose latitudes, longitudes, and radii have been computed precisely (Batson 1990).

The *construction* of planetary control networks has been either (1) derived from photogrammetrically tying images together as a group—a technique called *bundle block adjustment* (Batson 1973; Edwards et al. 2011, also see details below) or (2) constructed from spacecraft-mounted laser altimeter instrument (Light Detection and Ranging, LIDAR) such as MOLA, MLA, BELA, or GALA. When photographic image data are combined to a mosaic by bundle block adjustment, it is called a *controlled mosaic*. An example is the Mars Digital Image Mosaic (MDIM v2.1, Archinal et al. 2004). When the control information is derived from a LIDAR instrument, large numbers of distance measurements are adjusted as a group and generally converted to a digital elevation model (DEM).

DEM-derived control networks have the added benefit to support the production of *orthographically rectified* images where *each pixel* in the image is modeled as to be observed from a zenith position (straight down looking), rather than with an oblique viewing geometry. In this process, the geometric (parallax-related) distortions are removed as caused by the variations in surface topography. Relief displacement (See Methods in Planetary Mapping, this volume) should be eliminated in *ortho-rectified images*, including global orthomosaics. In this process, the DEM provides elevation points for the corrections in which pixel shift is calculated form.

This is a required process for producing, e.g., controlled photomaps. These images will be both horizontally and vertically controlled at each pixel.

Batch-mode registration and ortho-rectification of high-resolution planetary images require fully automatic algorithms that could process large volumes of data with a single set of parameters (i.e., without requiring parameter tuning in each image). Such a pipeline has been recently introduced and validated (Sidiropoulos and Muller 2018).

Controlled mosaics and controlled DEMs are the foundation for any SDI and allow other datasets to be co-aligned or co-registered to them. Full geometric control helps accurately determine the ground position of the image and makes them ready for location measurements. The community should always strive to support controlled mosaics (cf. Batson 1990), as foundation products, as they facilitate science results and interoperability of data as gathered by different missions.

For data with too large-resolution difference, control may be obtained from the *hierarchical (pyramidal) co-registration* of one dataset to another, for example, registering any high-resolution new dataset to a controlled, lower-resolution dataset (e.g., Kim and Muller 2008). For the Moon, control is provided by LOLA DEM data. For Mars, MOLA provides a global high precision, medium resolution control for registering images (Shan et al. 2005), while HRSC has covered approximately 50% of the planet with high-resolution ortho-rectified products registered to MOLA (Sidiropoulos and Muller 2015). A HiRISE image may be controlled by the following registration sequence: MOLA ← HRSC ← CTX ← HiRISE. The resulting final horizontal errors should be less than 1 pixel in the scale of the original controlled dataset.

There is a lower tier of provisional products that are also created but should not be considered foundational. These are called semi-controlled or uncontrolled mosaics.

Semi-controlled mosaics are made from orthographically rectified, spatially filtered images produced during the systematic image processing procedure. Images are processed applying spacecraft tracking and predicted attitude data, which give the expected location and orientation of spacecraft and camera. Mosaics are locally adjusted by matching features within adjacent, overlapping images but *not* to an overall global control network (Batson 1990). Hence, single images of the mosaic have a geometrically correct orientation relative to each other but not to the body-fixed reference frame. Today, for some missions, the ability to more precisely track the spacecraft's location and attitude (orientation) has allowed for more accurate semi-controlled mosaics.

Uncontrolled mosaics are generally quick-look mosaics which also are geometrically corrected using predict tracking data. No attempt is made to adjust the images to each other or a control network. This mosaic type will contain geometric and radiometric distortions and overlapping images may have large errors and discontinuities.

5.1.1 Bundle Block Adjustment (Aerotriangulation)

In *bundle block adjustment* (`jigsaw` in ISIS3), target surface (object space or ground space) points are connected to the corresponding image pixel coordinates (image space) on two overlapping images. This is achieved by aerotriangulation: finding the position where the perspective center, the image point and the corresponding point on the ground form a straight line (i.e., meets the collinearity condition). All lines connecting the ground and one image plane are bundled in the perspective center. Object space coordinates can be calculated from at least two images, observing one, and the same surface point from different positions, and with a sufficient large angular separation. The measured image coordinates as well as the camera positions and orientations need to be known (USGS 2013). For the calculation, the sensor model also must be known that are different for each camera and contain different parameters such as the focal length and the pixel size. The latter provide the scale between object and image information.

Several camera types operate on spacecraft platforms. *Frame cameras* obtain all pixel values for a rectangular area at once while *line scanning*—push broom—*sensors* scan each image line separately while moving with respect to the object, having a different center of perspective per line. Aerotriangulation can be used for both data obtaining techniques.

In aerotriangulation, one selects prominent pixels (e.g., intersections of linear features, center points of small craters, and high-contrast features) on at least two images, ideally evenly distributed on the overlapping area of the images. A point that connects two images (pixel to pixel) is called *tie point.*

A point where an image pixel is linked to known object space, e.g., body-fixed coordinates (USGS 2013) containing latitude, longitude and height with respect to the reference surface, is called *ground control point* (GCP).

GCPs form planetary *control point network* (USGS 2017b). These control nets are continuously refined for each body through the inclusion of new data and application of revised techniques in the bundle block adjustment. GCPs are fundamental to establish controlled image mosaics (Archinal et al. 2004) or topographic maps (Shan et al. 2005).

For planetary bodies where controlled datasets are available (e.g., MOLA for Mars), those can be used as reference surface ("ground") in the production of controlled maps. Where control net is not available, new control net can be produced using bundle adjustment using globally distributed tie points.

Aerotriangulation results in *object point coordinates* (latitude, longitude, and radius) and *revised orientation information* (e.g., camera pointing) for the applied sensor. In addition to producing image mosaics, these solutions can also provide fundamental data on the *size and shape of the body*, its *rotation period*, and the *direction of its polar axis* in space (Burmeister et al. 2018; Willner et al. 2010). The object point coordinates also usually serve as a *coordinate reference frame* (USGS 2017a).

Vertical control is established by linking horizontal (2D) control points to reference elevation data (Kirk et al. 2003).

There are a number of software applications to perform a bundle adjustment of planetary image data. Examples include the proprietary *SOCET SET* (BAE Systems) and *ArcGIS* (ESRI) software. Open-source solutions include the *Ames Stereo Pipeline* (ASP) (Moratto et al. 2010) developed by NASA, and the *ISIS3* with its jigsaw component created by USGS. In Europe, there are several active research groups that provide high-quality controlled mosaics and reference DEMs based on their in-house processing pipelines. The German Aerospace Center (DLR) utilizes a self-developed pipeline grounded on the JPL's *VICAR* software and library, while University College London (UCL) applies additional techniques together with the ASP to process large numbers of image data (Kim and Muller 2008).

5.1.2 Ancillary Data Sources

The camera's position and orientation in space define the surface location of the image information. A priori spacecraft navigation and pointing data (before the data collection) have uncertainties, which lead to registration errors that can, however, be augmented by photogrammetric control (bundle adjustment). Information needed to perform bundle adjustment includes the position and orientation of the spacecraft, onboard sensors, Solar System body orientations to each other, as well as information on frame relations and time conversions. These data are stored in *Spacecraft, Planet, Instrument, Camera-matrix, Events* (SPICE) kernel data files (Acton 1996). This information is also termed "ancillary data."

Next to the ancillary data, SPICE provides functions and tools to derive all necessary input information for bundle adjustment. For example, SPICE is providing capabilities to convert between terrestrial UTC, Ephemeris (Solar System Barycenter) Time and the spacecraft onboard clock to establish a single reference time for computation. To reconstruct geometry, position vectors of *ephemeris objects* are determined relative to each other along with their reference frame orientations. In other words, orientations are reconstructed by referencing one object to another. The instruments' state is referenced to spacecraft and the spacecraft is referenced to the mass center of the body-fixed frame. For NASA and ESA space exploration missions, SPICE kernels are maintained and archived, e.g., in NASA's PDS and at ESA's SPICE Service (ESS) (Costa 2017).

5.2 Radiometric Calibration and Photometric Normalization

Images are obtained successively at various epochs and under various illumination conditions that have to be made nearly uniform for a mosaic.

Next to geometric control, image data require *radiometric calibration*, resulting in images suitable for reflectance-based measurements, by restoring the brightness values actually *received* at the camera, creating a "flat-field image." During this process, pixel values are converted to reflectance or radiance values, by correcting radiometric distortions caused by external factors, and those generated during data acquisition and transmission. Sub-processes include the removal of camera artifacts such as marginal camera shading, the removal of systematic noise including striping, random white noise (salt-and-pepper noise, i.e., bright and dark speckles), reseau points, and the filling of data drop-outs (null values).

Finally, *photometric normalization* results in mosaics that are aesthetically acceptable. It corrects the effects of different illumination conditions (Solar phase, emission, incidence angles at each pixel), which have resulted from using observations at different local times in the mosaic (Michael et al. 2016). Photometric (brightness) normalization also equalizes the contrast of surface albedo. This process includes planet-specific surface photometric parameters and atmospheric parameters such as scattering, or limb darkening, normalization of column and lines. Photometric correction results in brightness values at synthetically produced uniform illumination conditions without atmospheric effects over the scene.

5.3 Historic and Recent Examples

During the 1960s in the Soviet Union, the first technique to provide a control of the lunar coordinates was to use star-calibrated lunar photographs where background stars were used as reference points (Rizvanov et al. 2007). Star trackers along with Sun and Earth sensors also are used to determine spacecraft attitude and camera pointing (Wong and Lai 1980).

In early planetary reconnaissance missions, such as Mariner 9, preliminary ("real time") mosaics were made of radiometrically corrected images; uncontrolled mosaics were produced by scaling the images. These two were used to support mission operations. Semi-controlled mosaics were produced with further geometric correction, using tracking and orientation data and control points for internal reference. This product served as the basis for preliminary geologic mapping. Final, controlled products were made by a geodetic control net and mutual fitting of the images (Batson 1973).

More recent missions have various approaches depending on data resolution, surface coverage, and computation complexity. The European *Mars Express* mission (MEX) has the goal to map Mars globally based on the image data of the onboard *High-resolution Stereo Camera* (HRSC). This is a line scanning camera (Jaumann et al. 2007) that obtains image information suitable for stereo-photogrammetric reduction during one pass over the surface. The HRSC team provides controlled data products of different levels (Gwinner et al. 2010) (see next section).

5.4 A Standardized Sequence of Image Processing

Calibration, projection, and mosaicking of planetary image data are routinely performed by the mission teams using a customized or adapted processing chain. The various steps of the data processing are indicated by the *processing levels* of the given data. Each processing level defines the processes the data have been processed so far. The definitions of the single levels vary between processing chains.

Some processing steps do not require any human interaction and are performed fully automated and systematic. High-level processing, however, requires quality control and thus human interaction. We describe two different processing lines here as examples.

The USGS developed the unix-based *Integrated Software for Imagers and Spectrometers* (ISIS3), a collection of tools and software to display, process, and manipulate data of planetary missions. Images of differing ISIS3 levels can be used for different cartographic purposes. Although the process scheme described below is ISIS3-specific, similar workflow can be produced by other tools that could be significantly easier to use. One such web-based tool is the *Map Projection On the Web* (POW) (Hare et al. 2013) that uses the recommended image processing pipeline and outputs radiometrically calibrated and map projected images.

In ISIS3, the following image types can be produced (USGS 2004, 2013).

Level 0 (Experiment Data Record, EDR) images are raw uncalibrated images in the native geometry of the camera. Source images are taken from the main archive of all NASA planetary mission produced data, the Planetary Data System (PDS) where images are stored in a standardized format called EDR. Level 0 multiband *image cubes* are produced by ingesting PDS-formatted raw images [`xxx2isis`]. This process also imports SPICE kernels into the ISIS3 label for further processing [spiceinit].

Level 1 (Calibrated Data Record, CDR) images are radiometrically calibrated [`xxxcal`].

Level 2 (Reduced Data Record, RDR) images are geometrically calibrated, rectified, and map projected (to Sinusoidal projection by default) [`cam2map`], using SPICE kernels, a planetary body shape model, or a control network produced by the user. Camera distortion correction is applied using camera distortion models that improve camera pointing information.

Level 3 images are photometrically normalized [`photomet`].

Level 4 images are mosaicked together [`automos`], also applying cosmetic seam removal. To perform this step, input images must be in the same resolution and projection with the same center coordinates.

The HRSC team uses a processing chain that is based on the *Video Image Communication And Retrieval* (VICAR) tool developed by JPL and used also at various other institutes. Additional software was developed in-house to obtain all necessary functionality. For this pipeline, the following processing level definition applies (Scholten et al. 2005; Gwinner et al. 2016).

Level 1 data is de-compressed from download stream transmitted to Earth.

Level 2 includes the radiometric correction of the image data based on calibration information.

Level 3 data have been ortho-rectified onto the MOLA DEM with initial SPICE orbit and pointing data leading to an semi-controlled data product. A semi-controlled, single-strip, ortho-rectified data product is produced with the MOLA DEM as a reference surface (Gwinner et al. 2009). This data product is here considered semi-controlled since the resolution of the MOLA reference surface is by far lower than the HRSC image data and changing topography is not fully modeled at the scale needed.

Level 4 includes the determination of local digital elevation models based on the obtained stereo image data now tied to and thus controlled with the MOLA DEM reference. Ortho-rectification of the image data is performed using the determined DEM (Gwinner et al. 2010). The final results are ortho-rectified single-strip images.

Level 5 mosaics image data covering one quadrangle to one large mosaic with *full control*. Here, a bundle adjustment including all image data of one quadrangle is performed to achieve correct relative orientation of the participating images to each other and to tie the mosaic to the global reference (Gwinner et al. 2016).

6 Summary

Planetary mapping requires a framework of fundamental definitions and conventions. This includes coordinate systems that are realized by ground control point networks. Point networks define also the general shape of a body leading to a mean approximation of the body's shape that can be used as a reference body for cartographic purposes. Such a reference body can be used for lateral mapping but also serves as height reference. Once reference systems are fully established, a geo-reference for all other data is present. Image data can be correctly placed with respect to the reference systems providing the opportunity to derive fully controlled planetary image mosaics.

Acknowledgements The authors are grateful to R. Kirk for the helpful discussions during the planning and reviewing of the manuscript and to P. Sidiropoulos who provided useful additions to the manuscript.

References

Acton C (1996) Ancillary data services of NASA's navigation and ancillary information facility. Planet Space Sci 44:65–70

Aeschliman R (1998) Topographic map of the Guinevere Planitia of Venus. V10M 30/240 RTK, USGS

Archinal B, Becker T, Lee E, Edmundson K (2013) Initial global control network and mosaicking of ISS Images of titan. In: 44th lunar and planetary science conference, p 2957

Archinal BA, Lee EM, Kirk RL, Duxbury TC, Sucharski RM, Cook DA, Barrett JM (2004) A new mars digital image model (MDIM 2.1) control network. ISPRS Working Group IV/p Workshop

Archinal BA et al (2010) Report of the IAU working group on cartographic coordinates and rotational elements: 2009. Celest Mech Dyn Astr. https://doi.org/10.1007/s10569-010-9320-4

Archinal BA, A'Hearn MF, Bowell E, Conrad A, Consolmagno GJ, Courtin R, Fukushima T, Hestroffer D, Hilton JL, Krasinsky GA, Neumann G, Oberst J, Seidelmann PK, Stooke P, Tholen DJ, Thomas PC, Williams IP (2011) Report of the IAU/IAG working group on cartographic coordinates and rotational elements of the planets and satellites: 2009. Celest Mech Dyn Astr 109(2):101–135

Batson RM (1990) Map formats and projections used in planetary cartography. In: Greeley R, Barson RM (eds) Planetary mapping. Cambridge University Press, Cambridge

Batson RM (1973) Cartographic products from the mariner 9 mission. J Geophys Res 78 (20):4424–4435

Becker TL, Geissler PE (2005) Galileo global color mosaics of Io. In: Lunar and planetary institute science conference abstracts, vol 36. http://www.lpi.usra.edu/meetings/lpsc2005/pdf/1862.pdf

Becker TL, Archinal B, Colvin TR, Davies ME, Gitlin A, Kirk RL, Weller L (2001) Final digital global maps of Ganymede, Europa, and Callisto, in Lunar and Planetary Science Conference XXXII: Houston, Lunar and Planetary Institute, abs. no. 2009

Becker TL et al (2016) Completed global control network and Basemap of Enceladus. In: Lunar and Planetary Science Conference XLVII, Abs. #2342. http://www.hou.usra.edu/meetings/lpsc2016/pdf/2342.pdf

Belton MJS, Klaasen KP, Clary MC, Anderson JL, Anger CD, Carr MH, Chapman CR, Davies ME, Greeley R, Anderson D (1992) The Galileo solid-state imaging experiment. Space Sci Rev 60. https://doi.org/10.1007/bf00216864

Bills BG (2005) Variations in the rotation rate of Venus due to orbital eccentricity modulation of solar tidal torques. J Geophys Res 110, E11007. https://doi.org/10.1029/2003je002190

Burba GA (1996) Cartographic aspects of Venus global geologic mapping at 1:10,000,000 scale. Vernadskiy-Brown Micro 24 abstracts 11

Burmeister S, Willner K, Schmidt V, Oberst J (2018) Determination of Phobos' rotational parameters by an inertial frame bundle block adjustment. J Geodesy 1–11. https://doi.org/10.1007/s00190-018-1112-8

Cheng AF et al (2008) Long-range reconnaissance imager on new horizons. Space Sci Rev 140:189–215

Christensen PR et al (2001) Mars global surveyor thermal emission spectrometer experiment: investigation description and surface science results. J Geophys Res 106:23823–23872

Costa M (2017) SPICE for ESA planetary missions. EPSC Abstracts, vol 11, EPSC2017-654-1

Dermott SF, Thomas PC (1987) The shape and internal structure of mimas. Icarus 73:25–65

Duxbury TC, Kirk RL, Archinal BA, Neumann GA (2002) Mars geodesy/cartography working group recommendations on mars cartographic constants and coordinate systems. ISPRS, vol 34, part 4, "Geospatial Theory, Processing and Applications," Ottawa

Edwards CS, Nowicki, KJ, Christensen, PR, Hill, J, Gorelick, N, Murray, K (2011) Mosaicking of global planetary image datasets: 1. Techniques and data processing for Thermal Emission Imaging System (THEMIS) multi-spectral data. J Geophys Res 116:E10008. https://doi.org/10.1029/2010je003755

Elachi C et al (2005) Cassini radar views the surface of Titan. Science 308:970–974

Eliason E, Isbell C, Lee E, Becker T, Gaddis L, McEwen A, Robinson, M (1999) Mission to the Moon: the clementine UVVIS global lunar mosaic, PDS Volumes USA_NASA_PDS_CL_4001 through 4078, produced by the U.S. Geological Survey and distributed on CD media by the Planetary Data System

Fergason RL, Lee EM, Weller L (2013) THEMIS geodetically controlled Mosaics of Mars, 44th Lunar and Planetary Science Conference, The Woodlands, TX, Abstract #1642

Ford PG (1992) MGN V RDRS 5 global data record reflectivity V1.0, MGN-V-RDRS-5-GDR-REFLECTIVITY-V1.0, NASA planetary data system from cassini-ISS images. Planet Space Sci 57:83–92

Gaddis L, Barrett J, Laura J, Milazzo M (2015) USGS ISIS tools supporting lunar selene "Kaguya" data from terrain camera, multiband imager and spectral profiler instruments. In: Second Planetary Data Workshop, 7040
Gaddis LR, Sucharski, T, Becker, T, Gitlin, A (2001) Cartographic processing of digital lunar orbiter data, LPS XXXII, abs. #1892. http://www.lpi.usra.edu/meetings/lpsc2001/pdf/1892.pdf
Gwinner K, Jaumann R, Hauber E, Hoffmann H, Heipke C, Oberst J, Neukum G, Ansan V, Bostelmann J, Dumke A, Elgner S, Erkeling G, Fueten F, Hiesinger H, Hoekzema NM, Kersten E, Loizeau D, Matz KD, McGuire PC, Mertens V, Michael G, Pasewaldt A, Pinet P, Preusker F, Reiss D, Roatsch T, Schmidt R, Scholten F, Spiegel M, Stesky R, Tirsch D, van Gasselt S, Walter S, Wählisch M, Willner K (2016) The High Resolution Stereo Camera (HRSC) of Mars Express and its approach to science analysis and mapping for Mars and its satellites. Planet Space Sci 126:93–138. https://doi.org/10.1016/j.pss.2016.02.014
Gwinner K, Scholten F, Preusker F, Elgner S, Roatsch T, Spiegel M, Schmidt R, Oberst J, Jaumann R, Heipke C (2010) Topography of Mars from global mapping by HRSC high-resolution digital terrain models and orthoimages: characteristics and performance. Earth Planet Sci Lett 294:506–519. https://doi.org/10.1016/j.epsl.2009.11.007
Gwinner K, Scholten F, Spiegel M, Schmidt R, Giese B, Oberst J, Heipke C, Jaumann R, Neukum G (2009) Derivation and validation of high-resolution digital terrain models from Mars Express HRSC-Data. PE&RS 75:1127–1142
Hare TM, Archinal BA, Becker TL, Lee EM, Gaddis LR, Redding BL, Rosiek MR (2008) Clementine mosaics warped to ULCN 2005 network, LPSC XXXIX, abstract#2337
Hare TM et al (2013) Map projection web service for PDS images. LPSC XLIV, abstract 2068
Haruyama J, Matsunaga T, Ohtake M, Morota T, Honda C, Yokota Y, Torii M, Ogawa Y (2008) Global lunar-surface mapping experiment using the Lunar Imager/Spectrometer on SELENE. Earth Planets Space 60:243–255
Hawkins SE III et al (2007) The mercury dual imaging system on the MESSENGER spacecraft. Space Sci Rev 131:247–338. https://doi.org/10.1007/s11214-007-9266-3
IAU (1971) Commission 16: physical study of planets and satellites. In: Proceedings of the 14th General Assembly, Brighton 1970. Trans Int Astron Union 14B:128–137
Isbell C, Gaddis L, Garcia P, Hare T, Bailen M (2014) Kaguya terrain camera mosaics. In: 45th lunar and planetary science conference 2268
Jacobson RA, Konopliv AS, Park RS, Folkner WM (2018) The rotational elements of Mars and its satellites. Planet Space Sci 152:107–115. https://doi.org/10.1016/j.pss.2017.12.020
Jacobson RA, Lainey V (2014) Martian satellite orbits and ephemerides. Planet Space Sci 102:35–44. https://doi.org/10.1016/j.pss.2013.06.003
Jaumann R, Neukum G, Behnke T, Duxbury TC, Eichentopf K, Flohrer J, Gasselt SV, Giese B, Gwinner K, Hauber E, Hoffmann H, Hoffmeister A, Köhler U, Matz K-D, McCord TB, Mertens V, Oberst J, Pischel R, Reiss D, Ress E, Roatsch T, Saiger P, Scholten F, Schwarz G, Stephan K, Wählisch M (2007) The high-resolution stereo camera (HRSC) experiment on Mars Express: instrument aspects and experiment conduct from interplanetary cruise through the nominal mission. Planet Space Sci 55:928–952
Kim JR, Muller J-P (2008) Very high resolution stereo DTM extraction and its application to surace roughness estimation over Martian surface. Int Arch Photogram Remote Sens Spatial Inf Sci. XXXVII(B4):993–998
Kirk RL, Howington-Kraus E, Redding B, Galuszka D, Hare TM, Archinal BA, Soderblom LA, Barrett JM (2003) High-resolution topomapping of candidate MER landing sites with Mars Orbiter Camera narrow-angle images. J Geophys Res 108(E12):8088. https://doi.org/10.1029/2003JE002131
Laura JR, Hare TM, Gaddis LR, Fergason RL, Skinner JA, Hagerty JJ, Archinal BA (2017) Towards a planetary spatial data infrastructure. ISPRS Int J Geo-Inf 6:181
Lee EM, Gaddis LR, Weller L, Richie JO, Becker T, Shinaman J, Rosiek MR, Archinal BA, USG (2009) A new clementine basemap of the Moon. In: Lunar and planetary science conference XL, Houston, TX. http://www.lpi.usra.edu/meetings/lpsc2009/pdf/2445.pdf

Li C, Ren X et al (2010) Laser altimetry data of Chang'E-1 and the global lunar DEM model. Sci China Earth Sci 53(11):1582–1593

Melosh JH (2011) Planetary surface processes. Cambridge University Press, New York

Michael GG, Walter SHG, Kneissl T, Zuschneid W, Gross C, McGuire PC, Dumke A, Schreiner B, van Gasselt S, Gwinner K, Jaumann R (2016) Systematic processing of Mars Express HRSC panchromatic and colour image mosaics: image equalisation using an external brightness reference. Planet Space Sci 121:18–26. https://doi.org/10.1016/j.pss.2015.12.002

Moore JM et al (2016) The geology of Pluto and Charon through the eyes of New Horizons. Science 351(6279):1284–1293. https://doi.org/10.1126/science.aad7055 https://arxiv.org/abs/1604.05702

Moratto ZM, Broxton MJ, Beyer RA, Lundy M, Husmann K (2010) Ames stereo pipeline, NASA's open source automated stereogrammetry software. In: LPSC, vol 41, p 2364

NAIF (2017) An overview of reference frames and coordinate systems in the SPICE context. Navigation and Ancillary Information Facility. https://naif.jpl.nasa.gov/pub/naif/toolkit_docs/Tutorials/pdf/individual_docs/17_frames_and_coordinate_systems.pdf

Oberst J, Elgner S, Turner FS, Perry ME, Gaskell RW, Zuber MT, Robinson MS, Solomon SC (2011) Radius and limb topography of Mercury obtained from images acquired during the MESSENGER flybys. Planet Space Sci 59:1918–1924. https://doi.org/10.1016/j.pss.2011.07.003

Ohtake M, Pieters CM, Isaacson P, Besse S, Yokota Y, Matsunaga T, Boardman J, Yamomoto S, Haruyama J, Staid M, Mall U, Green RO (2013) One Moon, many measurements 3: spectral reflectance. Icarus 226(1):364–374

PDS (2008) PDS standards reference, chapter 2. Cartographic standards. Draft: v. 4.3, 12.10.08. https://pds.jpl.nasa.gov/documents/sr/stdref3.7/Chapter2_20081210_v4_3_final_rev.pdf

Perry ME et al (2011) Measurement of the radius of Mercury by radio occultation during the MESSENGER flybys. Planet Space Sci. https://doi.org/10.1016/j.pss.2011.07.022

Pettengill GH, Eliason E, Ford PG, Loriot GB, Masursky H, McGill GE (1980) Pioneer venus radar results: altimetry and surface properties. J Geophys Res 85(A13):8261–8270

Preusker F et al (2017) The global meter-level shape model of comet 67P/Churyumov-Gerasimenko. Astron Astrophys 607. https://doi.org/10.1051/0004-6361/201731798

Rizvanov NG, Nefed'ev YA, Kibardina, MI (2007) Research on selenodesy and dynamics of the Moon in Kazan. Solar Syst Res 41(2):140–149

Roatsch T, Wählisch M, Hoffmeister A, Kersten E, Matz K-D, Scholten F, Wagner R, Denk T, Neukum F, Helfenstein P, Porco C (2009) High-resolution Atlases of Mimas, Tethys, and iapetus derived from Cassini-ISS images. Planet Space Sci 57(1):83–92

Roatsch T, Wählisch M, Scholten F, Hoffmeister A, Neukum F, Porco C (2006) Mapping of the icy saturnian satellites. ISPRS XXXVI Commission IV, WG IV/7

Roatsch T, Kersten E, Wählisch M, Hoffmeister A, Matz K-D, Scholten F, Wagner R, Denk T, Neukum G, Porco CC (2012) High-resolution atlas of Rhea derived from Cassini-ISS images. Planet Space Sci 61(1):135–141

Russell CT, Raymond CA (2011) The dawn mission to vesta and ceres. Space Sci Rev 163(1–4): 3–23. https://doi.org/10.1007/s11214-011-9836-2

Sato H, Robinson MS, Hapke B, Denevi BW, Boyd AK (2014) Resolved Hapke parameter maps of the Moon. J Geophys Res: Planets 119:1775–1805. https://doi.org/10.1002/2013je004580

Saunders RS, Pettengill GH, Arvidson RE, Sjogren WL, Johnson WTK, Pieri L (1990) The magellan venus radar mapping mission. J Geophys Res 95(B6):8339–8355. https://doi.org/10.1029/JB095iB06p08339

Schenk PM (2008) Cartographic and topographic mapping of the icy satellites of the outer solar system. ISPRS XXXVII Commission IV, WG IV/7

Scholten F, Gwinner K, Roatsch T et al (2005) Mars Express HRSC data processing – methods and operational aspects. Photogram Eng Remote Sens 71(10):1143–1152

Seidelmann PK, Abalakin VK, Bursa M, Davies ME, De Bergh C, Lieske JH, Oberst J, Simon JL, Standish EM, Stooke P, Thomas PC (2002) Report of the IAU/IAG working group on cartographic coordinates and rotational elements of the planets, and satellites: 2000. Celest Mech Dyn Astr 82:83–110

Shan J, Yoon J, Lee DS, Kirk RL, Neumann GA, Acton CH (2005) Photogrammetric analysis of the mars global surveyor mapping data. Photogram Eng Remote Sens 71:97–108

Shevchenko V, Rodionova Z, Michael G (2016) Lunar and planetary cartography in Russia. Springer, Heidelberg. https://doi.org/10.1007/978-3-319-21039-1

Sidiropoulos P, Muller J-P (2015) On the status of orbital high-resolution repeat imaging of Mars for the observation of dynamic surface processes. Planet Space Sci 117:207–222

Sidiropoulos P, Muller J-P (2018) A systematic solution to multi-instrument co-registration of high-resolution planetary images to an orthorectified baseline. IEEE Trans Geosci Remote Sens 56(1):78–92

Simonelli DP, Thomas PC, Carcich BT, Veverka J (1993) The generation and use of numerical shape models for irregular solar system objects. Icarus 103:49–61

Smith DE et al (2010) The lunar orbiter laser altimeter investigation on the lunar reconnaissance orbiter mission. Space Sci Rev. https://doi.org/10.1007/s11214-009-9512-y

Smith DE, Zuber MT, Solomon SC, Phillips RJ, Head JW, Garvin JB, Banerdt WB, Muhleman DO, Pettengill GH, Neumann GA, Lemoine FG, Abshire JB, Aharonson O, Brown CD, Hauck SA, Ivanov AB, McGovern PJ, Zwally HJ, Duxbury TC (1999) The global topography of Mars and implications for surface evolution. Science 284:1495–1503

Speyerer EJ, Robinson MS, Denevi BW, The LROC Science Team (2011) Lunar reconnaissance orbiter camera global morphological map of the Moon. In: Lunar planetary science conference, abstract #2387. https://www.lpi.usra.edu/meetings/lpsc2011/pdf/2387.pdf

Snyder JP (1987) Map projections–a working manual, U.S. Government Printing Office, U.S. Geological Survey professional paper, no 1395, vol 1395

Stark A, Willner K, Burmeister S, Oberst J (2017) Geodetic framework for martian satellite exploration i: reference rotation models. In: European Planetary Science Congress 11

Stephan K et al (2009) Mapping products of Titan's surface. In: Brown RH, Dougherty M (eds) Titan From Cassini-Huygens. Springer, New York, pp 489–510

Stooke P (2012) Stooke small bodies maps V2.0. MULTI-SA-MULTI-6-STOOKEMAPS-V2.0. NASA Planetary Data System

Thomas P (1987) Limb topography of Uranian satellites. LPSC XVIII 1010-1011

USGS (2002) Controlled photomosaic map of Europa, Je 15M CMN: U.S. Geological Survey Geologic Investigations Series I–2757. http://pubs.usgs.gov/imap/i2757/

USGS (2004) Production of digital image models with ISIS. ISIS 2 documentation. https://isis.astrogeology.usgs.gov/Isis2/isis-bin/intro_digi_mosaic.cgi

USGS (2013) Stereo processing of planetary stereo imagery using ISIS3 and SOCET SET® a primer. Astrogeology Science Center, USGS

USGS (2017a) Mimas Voyager Image Control Network (RAND)

USGS (2017b) Control Networks. https://astrogeology.usgs.gov/maps/control-networks

Wagner RV, Speyerer EJ, Robinson MS, LROC Team (2015) New mosaicked data products from the LROC team. In: 46th lunar and planetary science conference, abstract #1473. https://www.hou.usra.edu/meetings/lpsc2015/pdf/1473.pdf. Eposter: http://www.lpi.usra.edu/meetings/lpsc2015/eposter/1473.pdf

Wang J, Scholes D, Zhou F, Bennette K (2017) COORDINATE SYSTEM? In: Planetary data system (PDS) geosciences node orbital data explorer version 3.0 user's manual. http://ode.rsl.wustl.edu/moon/pagehelp/quickstartguide/index.html?introduction.htm

White OL, Schenk PM, Nimmo F, Hoogenboom T (2014) A new stereo topographic map of Io: Implications for geology from global to local scales. J Geophys Res Planets 119:1276–1301. https://doi.org/10.1002/2013JE004591

Willner K, Oberst J, Hussmann H, Giese B, Hoffmann H, Matz K-D, Roatsch T, Duxbury T (2010) Phobos control point network, rotation, and shape. Earth Planet Sci Lett 294:541–546. https://doi.org/10.1016/j.epsl.2009.07.033

Willner K, Oberst J, Wählisch M, Matz K, Hoffmann H, Roatsch T, Jaumann R, Mertens V (2008) New astrometric observations of Phobos with the SRC on Mars Express. Astron Astrophys 488:361–364
Willner K, Shi X, Oberst J (2014) Phobos' shape and topography models. Planet Space Sci (PSS) 102:51–59
Wong EC, Lai JY (1980) Attitude determination of Galileo spacecraft from star data. In: Guidance and Control Conference, Danvers, MA, 11–13 August 1980. AIAA PAPER, pp 80–1732
Zangari A (2015) A meta-analysis of coordinate systems and bibliography of their use on Pluto from Charon's discovery to the present day. Icarus 246:93–145
Zuber M, Smith DE (1996) Topographic mapping of the Moon. Int Arch Photogram Remote Sens 31(B4):1011–1015

Part II
Specialized Planetary Mapping

Planetary Geologic Mapping

Ernst Hauber, Andrea Naß, James A. Skinner and Alexandra Huff

Abstract Geologic mapping is a key to understanding the geology of extraterrestrial bodies and is based on the same principles as mapping in terrestrial geology. However, there are some important differences, with respect to map scale and data availability. Whereas the terrestrial geologic mapper typically has access to the study area and can use data of all spatial scales, planetary geologists have to rely on remote sensing data or robotic in situ data that only cover a certain range of scales. Careful selection of mapping scale, data, and purpose are therefore essential. The advance of modern GIS techniques has recently enabled efficient mapping approaches and digital dissemination of mapping results for further usage.

Keywords Geology · Mapping · Geologic maps · Stratigraphy · Chronology · GIS

1 Introduction

Geologic maps are two-dimensional products that represent the three-dimensional distribution of rocks, sediments, and soils at and near the solid surface, including key geologic features or characteristics that occur on, within, or across these distributed materials. A map and its associated components afford a spatial view of how geologic materials are distributed and a temporal view of how these materials were originally emplaced and then modified. Geologic maps are fundamental tools for interrogating a planet's surface and uppermost crust in that they simultaneously result from scientific investigation and promote additional research at a variety of map scales.

E. Hauber (✉) · A. Naß
Institute of Planetary Research, German Aerospace Center (DLR), Berlin, Germany
e-mail: Ernst.Hauber@dlr.de

J. A. Skinner · A. Huff
U.S. Geological Survey, Flagstaff, AZ, USA

© Springer Nature Switzerland AG 2019

H. Hargitai (ed.), *Planetary Cartography and GIS*, Lecture Notes in Geoinformation and Cartography, https://doi.org/10.1007/978-3-319-62849-3_5

The process of geologic mapping is a learned endeavor, and using Geographical Information Systems (GIS), to create and enhance geologic map products is a significant modern challenge. As the planetary geologic mapping process has evolved with the development of new technology, the discipline-specific expectations of a planetary geologic mapper have increased to the point where the investigating scientist (or team) is often expected to be (or include) a geologist, image processor/analyst, GIS specialist, and cartographer. Therefore, the production of peer-reviewed geologic maps is a complex process involving a wide range of data, software tools, and technical procedures, which requires planetary geologic mapping to be a learned discipline.

In this chapter, we provide an introduction to planetary geologic mapping. After a short overview on the history of geologic mapping and current programs, we will give an overview of the mapping techniques, emphasizing the inherent problems of planetary science, where typically no ground truth is available. A section on the data sets available to planetary geologists is followed by information on the technical aspects of modern planetary geological mapping, and future developments and challenges. Finally, some useful (but not exhaustive) links provide basic and advanced information on data archives, image processing and cartography software, and guidelines for planetary geologic mapping.

1.1 History

Modern geologic mapping started seriously in the eighteenth century in the mining regions of Europe. Later, more comprehensive efforts attempted to graphically represent the spatial and temporal distribution of rocks over wide regions (e.g., Winchester 2009). Geological maps are a quintessential tool for the Earth scientist, as they visualize the three-dimensional architecture of a given rock record and describe the relationships of rock units to each other in space and time. They provide the fundamental basis for further interpretations of the geologic evolution of a region, which may ultimately reveal the spatiotemporal sequence of events and the range of processes which contributed to the geologic history (Hansen 2000; Tanaka 2014a). Geologic maps portray the surface geology in plan view and contain geometrical information that enables projecting the surface geology to the subsurface or into regions without surface control, i.e., into regions which cannot be accessed and inspected. For terrestrial and planetary geologists alike, understanding the inherent predictive capability of geologic maps and the ability to exploit a geological map's three-dimensional character is of utmost importance.

Ever since telescopes enabled studying the surface of the Moon in detail, scientists tried to depict their ideas about the geology of our celestial neighbor in the form of maps. Although the first maps used to infer the geology of the Moon were drawn on the basis of telescopic observations already in the nineteenth century (Sheehan and Dobbins 2001), an approach that persisted well into the twentieth century (Hackman 1961), the true age of planetary geology began in the early 1960s when spacecraft exploration enabled studying the surfaces of the Moon and planets in detail (Carr 2013). Naturally, caused by the closest distance to Earth, it was again the Moon that was the first extraterrestrial object to be mapped according to modern geological principles (Fig. 1; see also Wilhelms 1972). Now, after almost six decades of spaceflight, space missions with a plethora of instruments have visited every planet and most of their satellites, some dwarf planets, and several small Solar System bodies. The growing number and diversity of data sets have spurred a variety of mapping efforts, and new challenges arise due to the completely different nature and scale of mapped bodies (Skinner 2015).

Fig. 1 The first USGS map of the Moon using modern geological principles was produced on the basis of telescopic observations by two pioneers of planetary mapping, Eugene Shoemaker and Robert Hackman (Shoemaker and Hackman 1962). It shows the Copernicus crater in Lambert conformal conic projection and has a scale of 1:1,000,000. The map was printed in small quantity by the USAF Aeronautical Chart and Information Center in April 1961, but never formally published (Portree 2013). (image: US Geological Survey)

1.2 Status of Planetary Geologic Mapping

Although planetary geologic mapping has a long history in the Soviet Union and later in Russia (e.g., MIIGAiK 1992; Savinykh 2015; Fig. 2), and the Italian Space Agency (ASI) tried to establish their own map series (Pacifici 2008), formal planetary geologic maps are currently mostly produced with funding from NASA under the auspices of the US Geological Survey (USGS) (Williams 2016). USGS maps conform to strict conventions and are published in the Scientific Investigation Maps (SIM) series. Some planetary geologic maps are published with less stringent formalism, for example, in dedicated map journals (e.g., Molina et al. 2014), but most currently produced geologic or geomorphologic maps are less formal and contained in regular articles in planetary science journals such as *Icarus* or *Journal of Geophysical Research*: *Planets* (e.g., Ivanov and Head 2011; Le Deit et al. 2013) or other more general geoscientific publications (e.g., Pondrelli et al. 2015). Apart from the USGS maps, the majority of these maps does not conform to common standards and are therefore not easily comparable. Moreover, they lack adequate

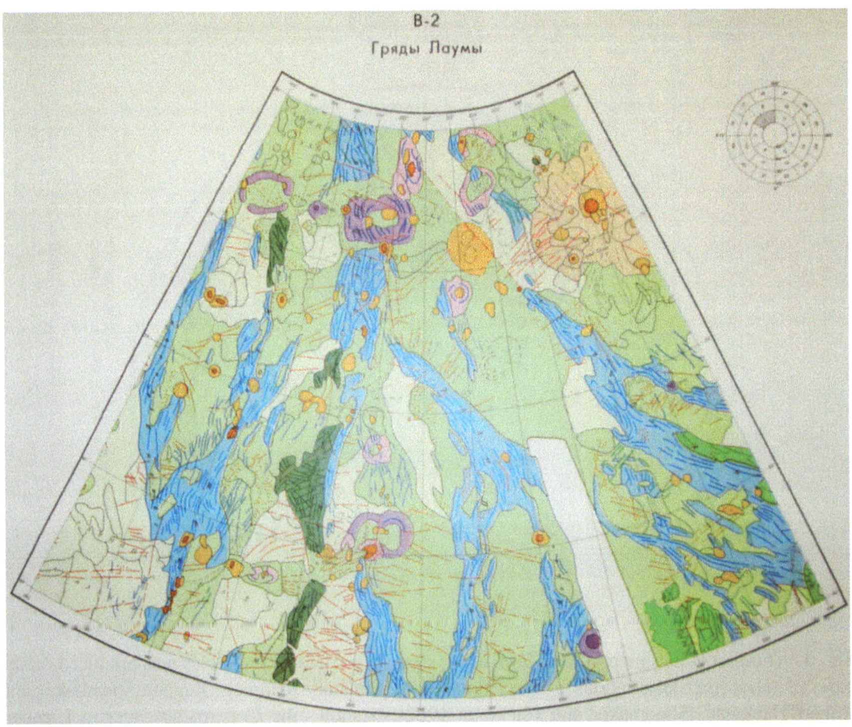

Fig. 2 Geologic map at a scale of 1:10,000,000 of a portion of Venus, Laūma Dorsa, prepared by Soviet scientists of the Vernadsky Institute of Geochemistry and Analytical Chemistry of the Soviet Academy of Sciences (GEOKHI) in the 1970s (kindly provided by A. Basilevsky). The map was prepared in a Lambert projection on the basis of Venera data (Kotelnikov 1989)

metadata information and are consequently not easy to find by simple searches and typically cannot be further used by interested users either digitally or at the proper scale.

It is important to note, however, that both standardized geologic maps (such as those reviewed and published by the USGS) and non-standardized geologic maps (such as those reviewed and published by scientific journals) both serve an equally important role in conducting scientific investigation and disseminating investigative results. There are many planetary geologic and topical (thematic) maps that are produced through scientific investigations that do not—and are not intended to—strictly adhere to community standards (Skinner 2017a, b). These maps are encouraged and should not be perceived as scientifically deficient. Likewise, standardized geologic maps should be perceived as an approved geologic framework that does not require further refinement with the given base data and techniques. In general, the strict preparation and iterative review process for standardized maps attempt to ensure that resulting maps are not only objective but also have a common look and feel. For example, standardized maps have particular map components, use common symbol and unit label schemes, and delineate and describe units as objectively as possible. As a result, standardized maps are intended to be longer-lived, benchmark products. However, the process of standardization is lengthy and tedious (wherein scientific and technical edits by authors are often required) and less responsive to rapid changes in the type and volume of existing (and often increasing) data. For example, it is often intractable to seamlessly incorporate all data into an objective, standard map product. Consequently, standardized geologic maps products must rely on a limited set of strategically selected data. It is therefore important to keep in mind that not all geologic problems should (or actually can) be solved by creating standardized geological maps, and in such cases, alternative approaches should be considered. Non-standardized maps, on the other hand, are not required to adhere to all community standards in process of product (though a minimum expectation is that some semblance of standards is implemented). As a result, non-standardized maps are able to leverage a fuller range of data, and the resulting map and map components can be produced more rapidly and be more readily beneficial to the community, often at the expense of objectivity and comparability afforded by standardization. Non-standardized maps are more adept at promoting a particular scientific hypothesis rather than an objective range of potential interpretive scenarios. Ideally, standardized and non-standardized geologic maps should work in tandem in order to establish context, promote comparability, and push the envelope advance of science and scientific communication.

2 Basic Methods

Geologic mapping is based on the premise that geologic materials such as rocks, sediments, dust, or ices distributed across the surface of a planet can be subdivided into discrete three-dimensional bodies (units) based on a common suite of characteristics (e.g., mineralogy, grain size, and color). These characteristics allude to a particular geologic process that occurred within a particular period of time. Geologic units are separated on a map by geologic contacts, which effectively denote a transition in geologic process or event. The three-dimensionality of a geologic unit alludes not only to the spatial and temporal pervasiveness of that particular geologic process when it was deposited but also how that unit was modified after emplacement. Thus, the process of geologic mapping is intended to catalog the characteristics of geologic units and the nature of their adjoining contacts as a means to infer how those rocks were originally emplaced (e.g., by intrusion) and how they have been modified (e.g., by faulting) since emplacement (Fig. 3).

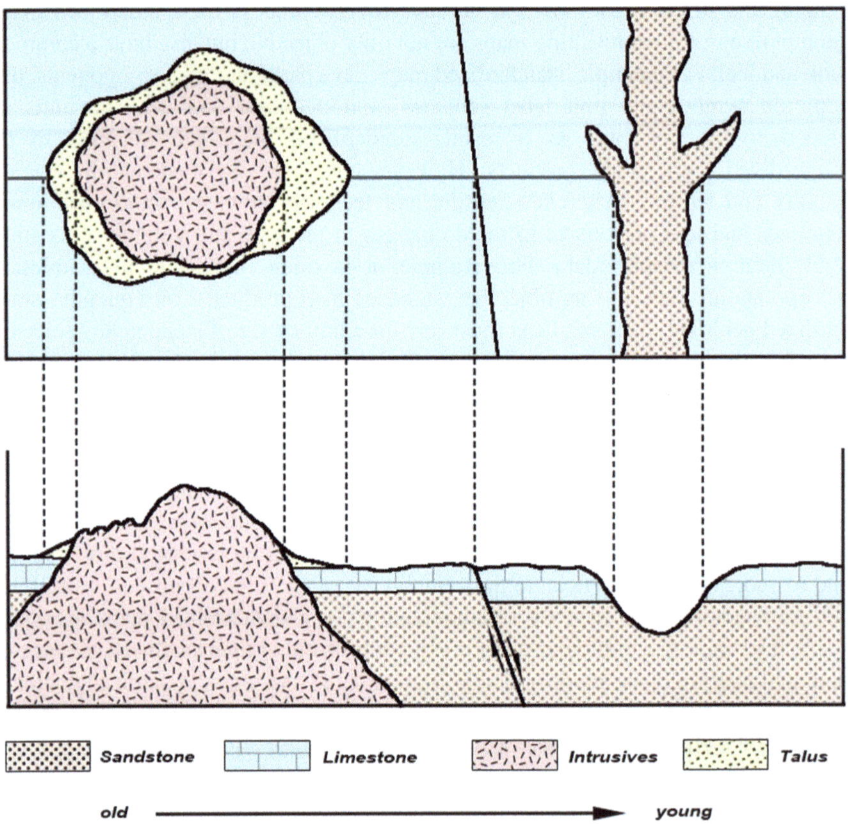

Fig. 3 A geologic sketch map (top) depicts the spatial distribution of rocks at the surface. The cross section (bottom) illustrates the three-dimensional character of geologic bodies. A key principle is the stratigraphy, i.e., the relative ages (e.g., the intrusives are younger than the sandstone) which are typically displayed in a stratigraphic column (e.g., Fig. 9) [Redrawn from Greeley (2013)]

2.1 Approach

Understanding that geologic maps are two-dimensional representations of a three-dimensional environment is advantageous. However, the act of interrogating an area of interest, discriminating discrete units, partitioning these units, and then using the distribution to determine geologic history is a tedious exercise that benefits from a regularized approach. In the simplest terms, the preparation of a geologic map can be perceived as a four-step process that encompasses (1) a clear definition of the scientific and technical scope of the map, (2) interrogation of relevant data sets to discriminate geologic details, and (3) coordinating observations into a cohesive map package, and (4) reviewing and refining map components. We outline the basic workflow that is undertaken at each step below.

The sheer volume of modern data sets and visualization and analysis software (i.e., GIS) have revolutionized planetary science generally and planetary geologic mapping specifically. GIS software easily ingests diverse spatially enabled data, allowing geologists to bring to bear the full range of available data sets on a geologic problem and to easily share geologic map files. However, despite the renaissance afforded by modern data sets and software, these advances have imposed some significant challenges to modern planetary geologic mappers. For example, GIS has essentially become the standard means by which geologic maps are created, produced, shared, and also analyzed. However, using GIS software is a learned skill, meaning that mapping geologists must either have the requisite skills set or be prepared to employ someone who does. Data set volume and diversity pose another significant hurdle that must be negotiated when preparing a planetary geologic map. It is often neither technically feasible nor scientifically desirable to use all of the available data when constructing a planetary geologic map because not all data are relevant at all scales of investigation. Therefore, a geologic mapping project must be well scoped.

2.2 Defining the Scope of a Map

Geologists must carefully consider and make important decisions regarding map boundary, scale, base data, and projection before they can begin the mapping process. Before starting a new geologic map, geologists must carefully consider the answers to a variety of questions: Has the region of interest been previously mapped at any scale and, if so, what geologic problems remain that have not been sufficiently addressed? What type, etc., of data has been acquired since the last geologic map was made that may reveal new information? Are there multiple interpretations regarding the geologic character and/or history of the region of interest that can be reasonably addressed through a geologic map? What type and spatial resolution of data exist within the region of interest and is it continuous or discontinuous across that region? What geologic relationships are known and which are unknown within the region of interest? How do these relationships compare to other regions of the body or even to other bodies that exhibit similar geologic characteristics? Are the geologic relationships observed within the region of interest scale dependent? The answers to these, and other, questions can help provide critical information about the need for a geologic map and the spatial extent and scale that most appropriately captures and relates the geologic context to enable future scientific studies.

The map boundary, or *region of interest*, is essentially the area within which a particular geologic problem occurs on a planetary surface, which will be addressed through the construction of a geologic map. In other cases, the region of interest is defined by a map quadrangle or by the extent of individual images, especially if these have a very high spatial resolution but limited extent (e.g., HiRISE data of Mars). When considering the region of interest, which can be local, regional, or global in scale, the geologist should consider not just whether the region contains interesting geologic features but whether the broader scientific community has a demonstrated need for the strict discrimination of geologic features in the form of a geologic map. It might be preferable in some instances to avoid the creation of a geologic map in favor of, for example, the creation and analysis of a landform database that might more succinctly help answer the outstanding scientific questions at hand. It is important to recognize that just because a geologic map can be created does not mean that it should be created. Similarly, just because new data exist for a particular region of interest does not mean that a new geologic map is required. The geologist should carefully consider whether there is an established need for a new geologic map of a particular region of interest and if so, what kind of geologic map is most appropriate to answer those questions.

Geologists must be cognizant of what data exist within the region of interest and which are the most relevant for establishing the geologic context and history through mapping. The *base map* is the most critical data set upon which geologic units are identified and described. When selecting a base map for a geologic mapping project, authors should typically select the highest resolution, most spatially continuous product available for the region of interest (e.g., in the case of Mars: CTX or HRSC images for local and regional maps, THEMIS IR images for global maps). When multiple data sets satisfy one or the other of these requirements, authors should select the type of data that is most apt to provide objective observations. When no data set satisfies the requirements, authors should seriously consider whether a geologic map can be reasonably and objectively constructed for the region of interest. The base map is generally either a single image, an image mosaic, or a digital elevation model that is either geodetically controlled or georeferenced (with errors documented) to a controlled product. The base map provides a known reference point for the geologist and the map user so that the data set upon which observations were made is not questioned. *Supplemental data* are those data products that support the identification and description of geologic units. They help to place these units into a spatial and temporal framework but are not compatible with the scale of the geologic map. They may also be specialty (derived, interpretive) products and/or do not have continuous coverage across the mapped region. Supplemental data augment and can help verify and expand upon observations made in the base map but should not be used to establish geologic contacts. Supplemental data sets should be strategically selected and limited in number; with increased supplemental data sets, objectivity and repeatability decrease because it becomes less clear to the map user how each data set influenced the identification and description of geologic units. Often, it is not necessary, preferred, or feasible to objectively use all available data to create a planetary geologic map within the scope of a proposed study, and the monetary and temporal limitations therein.

The map scale of a map is directly related to the geologic problem, i.e., the purpose of the mapping, and the features of interest, especially within the confines of a scientific study and a non-standardized map. The importance of problem-specific scale is only diminished when pondering a map series with a predetermined scale, e.g., of standardized USGS maps. Other factors that may also be relevant for the selection of map scale, especially for standardized maps, are the map boundary and the base map. Map scale identifies the relationship between the distance on a map (M) and the corresponding distance on the planetary surface (S), represented as a ratio (M:S). For example, a 1:1,000,000-scale map means that 1 mm on the map is equivalent to 1,000,000 mm (1 km) on the planetary surface. An increase in surface distance relative to map distance results in a smaller ratio and is thus considered a "small-scale" map (1:20,000,000). Contrarily, a decrease in surface distance relative to map distance results in a larger ratio and is thus considered a "large-scale" map (1:18,000).

Consideration of map scale is essential both prior to and during the creation of a geologic map. Map scale controls the size of features that can be represented at a particular scale as well as the vertex spacing for drafting features (e.g., contacts, linear features) within a GIS. For example, the creation of a geologic map at 1:5,000,000 scale can reasonably represent features that are >15 km (three times the minimum length; Table 1) and should employ data that are suitable at that scale. Map authors should select base data that are relevant to the map scale and vice versa. For example, it is unrealistic to use THEMIS IR data (100 m/px) to create a geologic map on Mars at 1:18,000 scale (Table 1). The size of geologic features and correlations can also be used to estimate map scale (and base material). For example, a geologic map that intends to delineate surficial units and correlate them with local outcrops of particular units is more likely to be compiled at large (local) scale using the highest-resolution data sets currently available.

Every map has limitations on the size of feature and relationships that can be reasonably represented (cartographic generalization). However, these limitations are not only dependent on the map scale but also on the region being mapped. For example, it cannot be definitively stated that a 1:1,000,000 scale geologic map should identify the rim of every impact crater >1 km (or >5 or >10 km) in diameter. Even though, according to Table 1, this is technically feasible, it might not be tractable or worthwhile for various scientific and technical reasons. Map authors should be cognizant of the interplay between map scale, base (and supplemental) data, and features and relationships that can be represented at a particular map scale.

Table 1 Example map scales for Mars geologic maps relative to ground distance, use, and potential informative data set. In general, it is advised that the minimum sized feature that can be represented on a map is 3 mm, though variations might occur depending on the map area and subject. In particular, point features can be used to show features that are ~1 mm or less on a particular map.

Relative Size	Scale	Ground distance of 1 mm on map	Modern scale relevant base data for Mars	Example terrains and map uses	Historical and modern example publications
Largest (local)	1:18,000	18 m	HiRISE (0.25 m/px), CTX (6 m/px)	Stratified deposits, local structure, strike dip, local unconformities, outcrop-scale geology, rover traverse, surficial textures	Okubo (2014) USGS SIM 3309; Okubo and Gaither (2017) USGS SIM 3359
	1:200,000	200 m	CTX (6 m/px), HRSC (12.5–25 m/px), THEMIS VIS (18 m/px)	Geomorphic facies, image and rover context and planning, contact characteristics	Mouginis-Mark (2015) USGS SIM 3297
	1:500,000	500 m	HRSC (12.5 m/px), THEMIS IR (100 m/px)	Bridge-scale map for global to regional/local	De Hon et al. (1999) USGS I-2579 (MTM quadrangle-based)
	1:5,000,000	5 km	THEMIS IR (100 m/px), MOLA (463 m/px)	General reference, bridge-scale map for global to regional/local, detailed landform geomorphology and distribution	Carr (1975) USGS I-893 (Mariner 9-based Mars Chart Series); Dohm et al. (2001) USGS I-2650 (1:5 M quadrangle-based)
Smallest (global)	1:15,000,000 1:20,000,000 1:25,000,000	15 km 20 km 20 km	THEMIS IR (100 m/px), MOLA (463 m/px)	General reference, very large landform geomorphology and distributions, geomorphic contacts and gradations	Greeley and Guest (1987) USGS 1-1802 (Viking-based); Tanaka et al. (2014b) USGS SIM 3292 (THEMIS IR-based); Scott and Carr (1978) USGS I-1083 (Mariner 9-based)

Note, image resolution is comparable across bodies, but scale will vary depending on the body's radius

The geologist should recognize the difference between the *printed map scale* and the *digital map scale* (or *digital mapping scale*). GIS software permits geologists to identify, describe, and define geologic contacts at a range of map scales, regardless of whether these scales are compatible with the printed map. Objectivity of geologic maps relies on establishment and adherence to a specified map scale. The *printed map scale* is the scale at which the map is intended to be viewed in hard copy format (regardless of whether the map is actually being printed or not). Effectively, the printed map scale is the scale at which the geologic characteristics are being represented. The *digital map scale* is the scale at which data are being viewed in a GIS and the scale at which geologic contacts and features are being mapped. The general rule of thumb is that the printed map scale is 4 times the digital map scale. That is, if a map is being printed at 1:1,000,000, then the geologist will identify geologic contacts at 1:250,000 scale. Map bases and supplemental data should be selected with these relationships in mind. Establishing a scale of observation and representation helps limit the number of features that can be feasibly mapped and represented within the region of interest so that "over-representation" of geologic units, contacts, and features is avoided. The relationship between printed and digital map scale does not mean that spatially incompatible data sets cannot be used as supplemental data or that observations cannot be made at very local scales. Indeed, local observations are critically useful for the geologist.

2.3 The Mapping Process

Terrestrial geologic maps are most commonly created by physically traversing an area of interest. For planetary surfaces, where physical access is typically impossible or, in the case of robotic exploration, extremely limited (Eppler et al. 2014), geologists rely entirely on observations made in remotely acquired data. The amount of detail that can be cataloged for a particular geologic unit is drastically reduced compared to the terrestrial equivalent (e.g., Stack et al. 2016). Where terrestrial geologists can make in situ observations regarding grain size and lithology, planetary geologists rely on remote sensing-based, macro-scale (often meter- to decameter-scale) characteristics such as thermal properties, surface roughness and texture, tone, and cross-cutting relationships. Geologists in general, and planetary geologists in particular, rely on stratigraphic concepts to help define and temporally order geologic units.

Stratigraphy is the science that deals with the bulk characteristics of rocks that occur in sheet-like bodies, termed *strata*. There are four basic principles of stratigraphy, originally defined by Nicolaus Steno in 1669 (e.g., Nichols 2009). These include the principles of superposition (in a sequence of strata, any stratum is younger than the stratum upon which it resides and is older than the stratum that resides above it), the principle of original horizontality (strata are deposited horizontally and deformed to different attitudes later), the principle of lateral continuity (strata were deposited in a continual sheet across a broad surface), and the principle

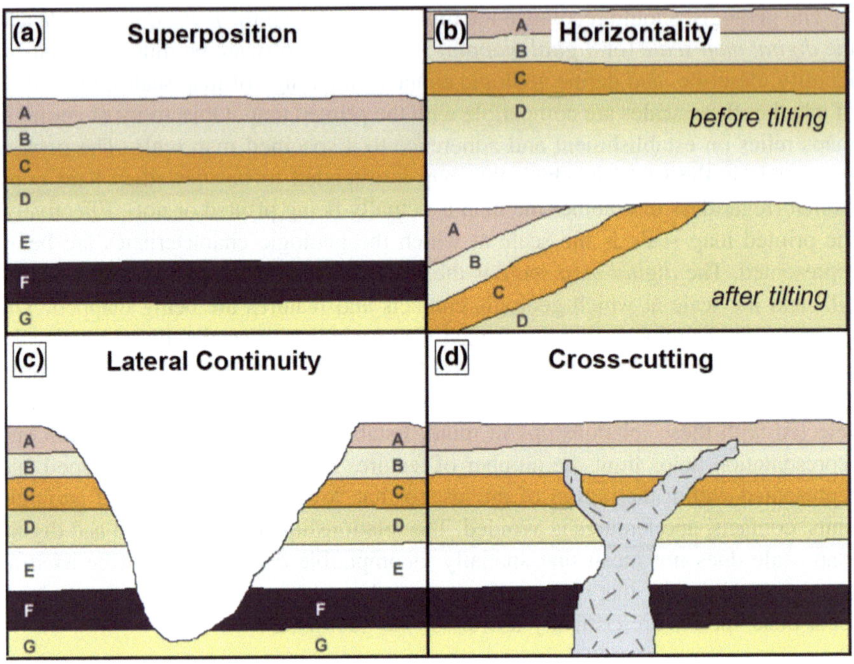

Fig. 4 Illustration of Steno's laws. **a** Older strata are superposed by younger strata (in this example, layer G is the oldest, and layer A is the youngest). **b** Layers are originally deposited horizontally and may be tilted later. Layers A–D are first horizontal (upper part) and were then tilted (lower part). **c** Layers are horizontally continuous and can be traced across wider areas (in this example, across a valley). **d** Cross-cutting relationships can be used to determine the relative emplacement times of rock units. The intrusion was emplaced after the layers. (Modified after Kurt Rosenkrantz, *Illustration of Steno's Laws*. CC-BY-SA)

of cross-cutting relationships (abrupt discontinuity across strata reflects activities that post-date original deposition) (Fig. 4). It is important to note, however, that neither orbital observations nor in situ 2D traverses by rovers may always enable an unambiguous interpretation of the local and regional stratigraphy (Fig. 5 and Stack et al. 2016).

There is no one process by which a geologic map can or should be constructed. The specifics of the mapping process should be dictated by the preferences of the geologist (or geologists if mapping as a team) and by consideration of the project scope, including the complexity of the region being mapped and the volume and type of data being used as the base and supplemental maps. The most important aspect of planetary geologic mapping is the final product, which should be an objective, concise, familiar, and repeatable product. Objective, in that the map should be prepared without promoting or favoring a particular scientific result. Concise, in that the map should be limited to only those components and discussions that are critical to conveying the observations. Familiar, in that the map should employ symbols, colors, and formats that are recognizable and adhere

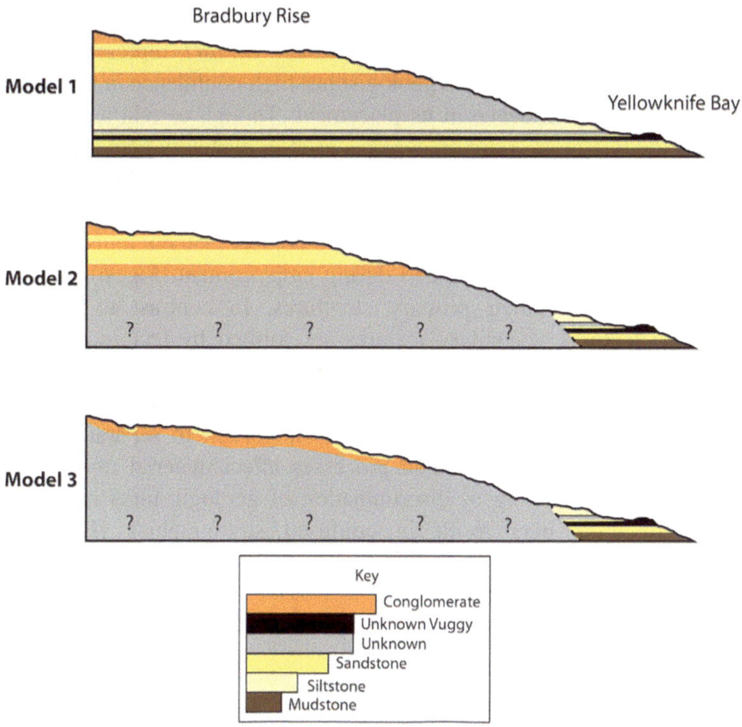

Fig. 5 Different possible stratigraphic interpretations of the local geology at the Curiosity landing site, based on 2D rover traverses. In each scenario, coarse gravel and pebbles ("conglomerates") lie on top, representing sediments deposited by ancient stream flow. Thickly bedded mudstones on the bottom reflect fine-grained sediments, perhaps deposited at the distal portions of an alluvial fan. Drill cores are essential to help reconstructing the subsurface stratigraphy on Earth, but are missing (yet) for other bodies. (Image by K. Stack: http://redplanet.asu.edu/?p=2362)

closely to published standards and, thus, promote use and comparability. And finally, maps should be repeatable, in that the observations upon which the map is constructed should be reproducible by other professional geologists. Thus, the process is not as important as the integrity of the product itself.

Once a mapping project is properly scoped, the construction of the geologic map begins with locating a single geologic contact through cross-cutting relationships and/or tonal or textural differences, describing the contact as well as the potential units that reside on either side of the contact and then attempting to extend the contact laterally for as far as possible. Similar to terrestrial field geologists, the planetary geologist should take copious notes and make sketches regarding the character and variation of the geologic units and contacts that are being identified and how those might be oriented with respect to one another in space. When tracing a geologic contact, the geologist should be cognizant of how and when the contact changes character. Different line types are used to depict different contacts, the most common being certain and approximate. A *certain contact*, represented by a thin, solid, black line, implies that the geologist has high confidence that the contact exists and its location is known. An *approximate contact*, represented by a thin, dashed, black line, implies that the geologist has high confidence in the existence of the contact but low confidence in its placement. Though certain and approximate contacts are the two most common contacts used in planetary geologic maps, there are others that may be used as long as they are rationalized and used consistently.

Contacts should be identified between different geological units, each defined by a consistent lithology. In this sense, they are geological material units, characterized by primary features that were formed during emplacement (Fig. 6). Ideally, units should be mapped on their primary structures. In contrast to primary (i.e., emplacement) features, secondary features are formed by tectonic, erosional, or in situ alteration processes. Although the mapping of secondary features (e.g., structural geologic mapping, or tectonic mapping) is an essential part of planetary geologic mapping (Tanaka et al. 2010), it is important to separate primary and secondary features (Fig. 6). Tectonic processes affect material units of different ages, and thus, the mapping of discrimination of geologic units on the basis of secondary structures may result in confused stratigraphies (Hansen 2000).

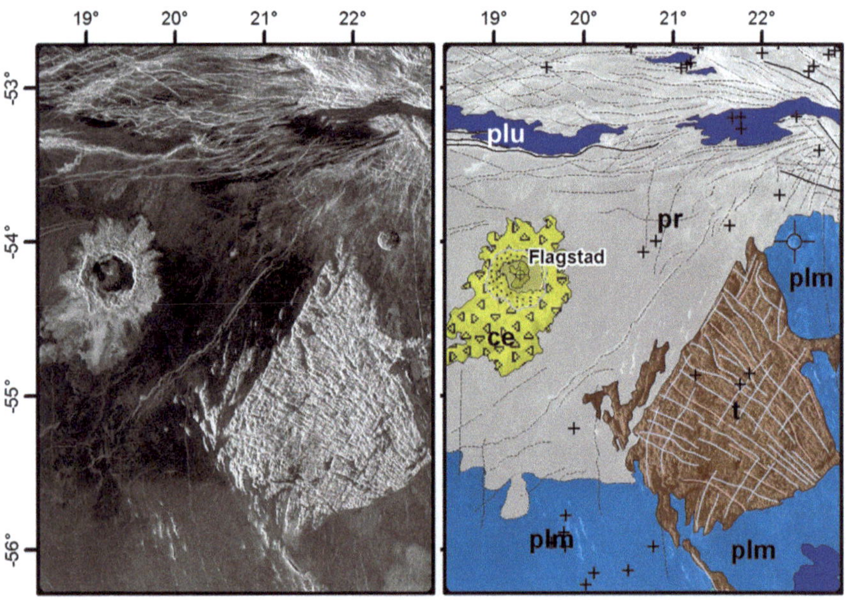

Fig. 6 Base data and photogeologic map. This example is a detail of a 1:5 M map showing the Flagstad crater region, Venus (Kumar and Head 2013), at the original print scale. *(left)* Magellan radar base image. *(right)* Map detail. Key: cross/registration mark: shield; pr—regional plains material; plu—lobate plains material, upper; plm—lobate plains material, middle; t—tessera material; ce—impact crater ejecta material; dashed lines—fracture; solid line—graben, lineament. Primary structures (colored ares) and secondary structures (line features, e.g., fractures) are separated

In general, it is important to keep in mind that—wherever possible—materials should be mapped, not topographic forms of whatever origin (structural, erosional) (Wilhelms 1972).

It is beyond the scope of this chapter to provide a comprehensive description of all the geologic principles that should guide the mapping process. More detailed explanations are given by Wilhelms (1972, 1990) and Tanaka et al. (2011). It has to be noted, however, that some of the content in Wilhelms (1990) is outdated in the sense that it states the goal to establish a global stratigraphy. On tectonically active bodies, which turns out to be the case even for bodies that were previously considered tectonically inactive (e.g., Pluto and Charon), such globally applicable stratigraphy is elusive (Hansen 2000) because it simply may not exist.

2.4 Packaging, Review, and Production

A planetary geologic map should be prepared and used as a package of *map components* rather than a single, typically large-format map sheet. The map package contains at a minimum a scaled geologic map, a description of map units (DMU), a correlation of map units (CMU), and an explanation of map symbols (EOMS) (Fig. 7). The geologic map is prepared at a scale that has been specifically determined by the map author to be appropriate for identifying the critical geologic details that can be observed in the map region using the base map and supplemental data as well as for conveying these details. The DMU is either a tabulated or prose-based description of the geologic units that are identified by the map author using base map and supplemental data at the specified map scale (Fig. 8a). In addition to the name and label of the geologic unit, the DMU specifically separates unit definitions (those geologic details identified in the base map, such as tone, texture, lateral continuity, thickness) and additional characteristics (those geologic details identified in supplemental data, such as color, thickness, mineralogy, three-dimensional orientation). For clarity, details per unit are often presented in a similar sequence and definitions may include a type locality, which is essentially the latitude and longitude of the best representation of a particular geologic unit. The DMU also provides a range of interpretations of the identified geologic units. The CMU provides a pictorial (graphical) representation of how the mapped geologic units are oriented in space and time with respect to one another (Fig. 8b). The CMU is typically a series of colored and labeled boxes wherein each box represents one geologic unit, the vertical axis represents time before present (either relative or modeled absolute), and the horizontal axis is spatial distribution or unit grouping, e.g., by process or region. The CMU is used to show, based on the identified geologic units and their contact relationships, how particular units are spatially and/or temporally associated (i.e., how geologic units were being emplaced relative to one another). These details are demonstrated by lengthening a particular box in the CMU or placing it in contact with another box, vertically or horizontally (Fig. 8b). We note that the CMU can be formatted in a variety of ways

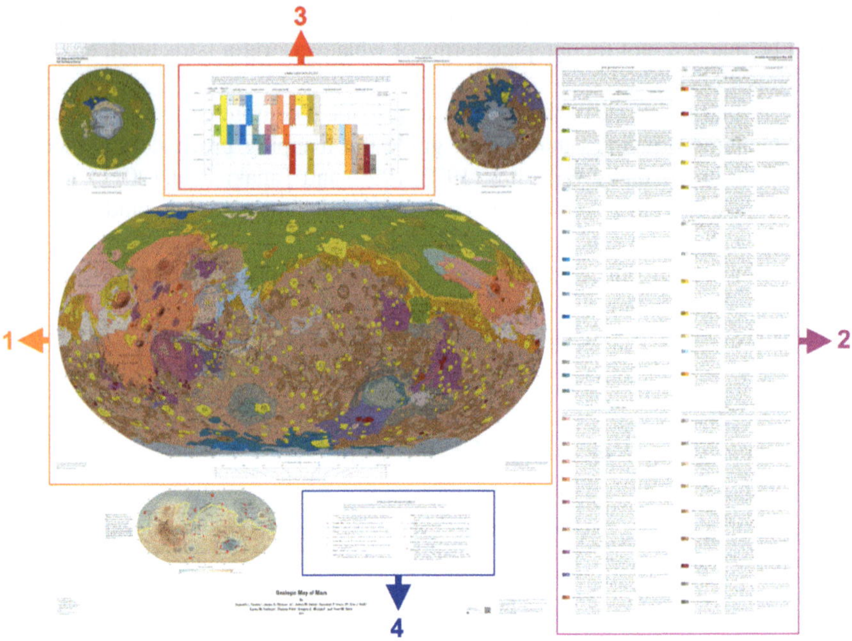

Fig. 7 Example of a formal digital planetary geologic map (the example shows the global, 1: 20,000,000 scale geologic map of Mars; Tanaka et al. 2014a, b). 1: Scaled geological map (north and south polar regions on top left and right, respectively). 2: Description of map units (DMU; see Fig. 8a). 3: Correlation of map units (CMU; see Fig. 8b). 4: Explanation of map symbols (see Fig. 8c)

and, as a result, often benefits from a descriptive legend. The EOMS is essentially the map's legend, where all point, line, and polygon symbols are identified and described (Fig. 8c). Standard cartographic symbols and definitions (e.g., FGDC 2006) should be used in order to maximize the familiarity and comparability between geologic maps, though some adaptation might be required, if explicitly defined and appropriately rationalized. Additional map elements can be added as required (Fig. 8d).

Though the map components described above typically form the basic components of the map package, they may be accompanied by other materials that are intended to support and/or clarify the observations made in the geologic map. These might include images of type localities, key geologic units, or graphical figures of relationships or features, tables of unit characteristics (e.g., cross-cutting relationships or unit statistics), geologic cross sections, crater counts (see Sect. 4.3), and the global context of the mapped region. Another element that is commonly associated with USGS maps is a pamphlet that provides explanations of mapping methods and detailed information on the geologic units. It is important to note that though the aforementioned map components are generally expected for published planetary geologic maps, they are not required unless publishing through the US Geological

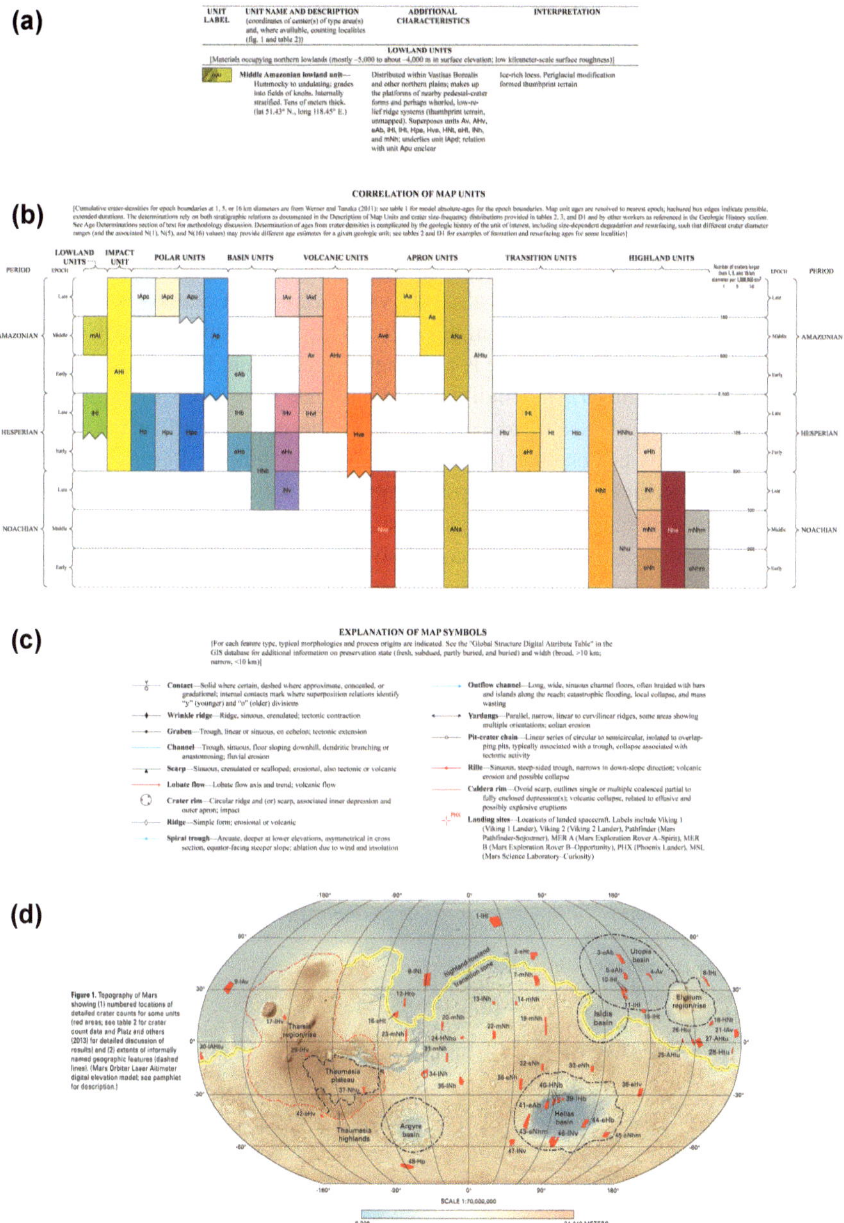

Fig. 8 Elements of a formal planetary geologic map (cf. Fig. 7; Tanaka et al. 2014a, b). **a** Description of map units (DMU). **b** Correlation of map units (CMU). **c** Explanation of map symbols. **d** Context for crater counting areas used for determining the model chronostratigraphy

Survey, which is currently the only institution that coordinates the review and publication of planetary geologic maps. However, even though certain components are not required for non-USGS publication, the community can only benefit from consistent preparation and publication of map documents.

2.5 *Additional Means of Visualization*

Geological mapping results can also be visualized by a variety of additional graphical representations, apart from geological maps. An important way to present the stratigraphy of a rock sequence is a stratigraphic column. It shows, from bottom (oldest) to top (youngest), the individual rocks that have been identified during mapping (Fig. 9). Colors should correspond to the colors used in the map, and established textures correspond to the lithology. The lithostratigraphy (also rock stratigraphy) describes the relative arrangement or *relative ages* of rock units; i.e., older units are superposed by younger ones. If the *absolute time* when the units have been formed and deposited can also be determined (in addition to the superposition relationships and *relative ages*), a chrono- or time-stratigraphy of the mapped units can be reconstructed (Wilhelms 1987, 1990).

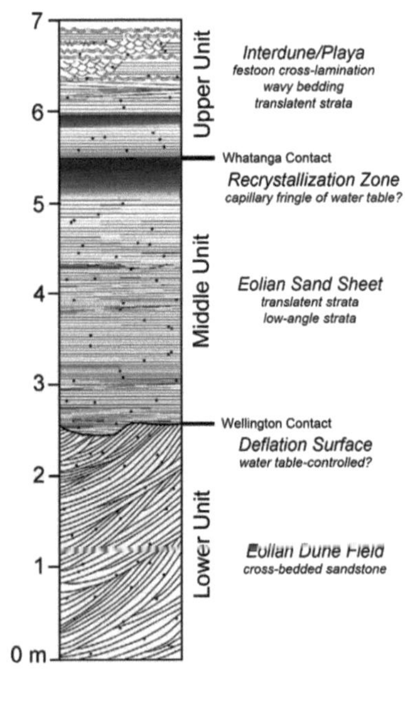

Fig. 9 Example of a stratigraphic column. A stratigraphic column is a description of the (relative) vertical location of rock units in a particular area (units on vertical axis are meters). It is typically used to illustrate sequences of sedimentary rocks (the example shows the Burns formation exposed at Burns cliff and within Eagle Crater, Meridiani Planum, Mars; from Grotzinger et al. 2005)

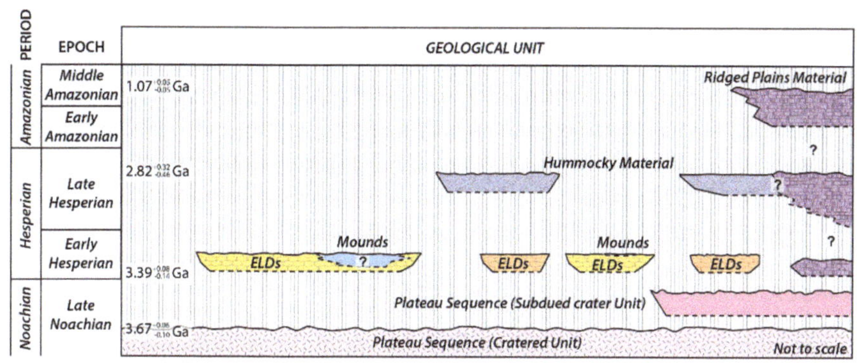

Fig. 10 Space-time diagram, highlighting the duration of sedimentary units through time as well as the time which is not represented in the geological record (gray area with vertical lines). (Courtesy M. Pondrelli)

Additional means of visualization may accompany geologic maps, such as cross sections (Fig. 3) and space-time diagrams, which show the geological history of the study region through time (Fig. 10). It will depend on the specific scope of the scientific investigation, if such visualization tools are used to present the stratigraphy of the mapped region. A more detailed description of such illustrations is given by Pondrelli et al. (2018).

3 Data

Remote sensing data are the basis for basically all planetary geologic mapping. They serve a dual purpose, i.e., as a base map and as the source or basis for interpretation. As in terrestrial geologic mapping, a base map is required as the geodetic reference to which the description of the geologic units is spatially registered. In contrast to typical terrestrial geologic mapping, no detailed field inspection of rocks and other surface materials is yet possible for other planets, except for a few very restricted locations along lander positions and rover traverses, and as a consequence, geologic mapping must rely on the interpretation of remote sensing data. In the following two sections, we first give an overview on the types of data available for different Solar System objects and then discuss the proper selection of data to be used for different geologic mapping objectives. More information on remote sensing data and methods can be found, e.g., in Sabins (2007) and Lillesand et al. (2015).

3.1 Types of Data for Planetary Geologic Mapping

Various passive and active sensor systems on orbiting spacecraft and landed missions provide spatially resolved data of planetary surfaces, and basically, all of these data sets are used as a basis for geologic mapping. The level of detail to which individual bodies were mapped is highly variable and depends on several factors such as size, the presence of a dense atmosphere, the distance to Earth, and—last but not least—the scientific interest. For example, the best-known surfaces of the terrestrial bodies in the inner Solar System are those of the Moon and Mars, because they are scientifically of high interest and relatively easy to reach. On the other hand, the dense cloud cover of Venus prohibits mapping the surface with cameras in the visible wavelengths; hence, most of our knowledge of its surface geology comes from radar measurements and, to a lesser extent, from thermal infrared spectroscopy. Fewer spacecraft have visited the more remote parts of the Solar System, and therefore, the amount and the spatial resolution of data from asteroids, comets, and the icy moons of the outer Solar System are generally poorer than in the inner Solar System. Nevertheless, the data availability for these objects is highly diverse, and some small bodies are now well covered by imaging data due to dedicated missions (e.g., the Dawn mission to asteroids Vesta and Ceres, and the Rosetta mission to comet 67P/Churyumov–Gerasimenko), while the surfaces of many others are virtually unknown.

Images obtained by cameras are by far the most widely used data for geologic mapping. The "information density" contained in an image exceeds that of most other data types, and most missions exploring the Solar System carry one or more camera experiments. In recent years, other sensors such as hyperspectral spectrometers, gamma-ray spectrometers, and radar experiments have also acquired data for geological mapping, although at resolutions that are lower than those of cameras.

3.1.1 Imaging

The basic and most often used data sets for virtually all geologic planetary mapping are images acquired by camera instruments on flyby and orbiter missions. Such images have commonly been panchromatic (i.e., sensitive in a single channel to a wide range of wavelengths, typically all wavelengths of visible light) in the early years of planetary exploration, but are increasingly complemented by color and multispectral information in the visible and near-infrared wavelengths, respectively, often as red–green–blue composites (RGB true- or false-color images; Fig. 11a). Images provide information on the surface texture, albedo, and, qualitatively, on the relief, which are all key to (photo) geologic interpretations. If stereo (i.e., overlapping) images from different viewing angles are available, they contain information on the surface topography. This is increasingly used to derive digital elevation models (DEM), which enable quantitative investigations of geologically relevant surface properties such as volumes, slopes, and derived parameters (i.e., surface curvature).

The amount of information provided by images is dependent on the radiometric and spatial resolution. The radiometric resolution can be described as the differences in brightness levels that the sensor can resolve; for example, 8- or 16-bit images can contain a maximum of 2^8 or 2^{16} gray levels or digital numbers (DN), respectively. The better the radiometric resolution of a camera system, the more sensitive it is to detecting small differences in reflected or emitted radiation (e.g., subtle brightness nuances of the surface materials). The spatial resolution depends on the instantaneous field of view, i.e., the angular cone of visibility of the sensor, which determines the area on a planetary surface which is visible, the detector itself (e.g., the number of pixels on the detector), and on the sampling speed of the detector (e.g., readout frequency). The size of the visible ground area depends on the flight altitude (the lower the latter, the smaller the former) and is often expressed in terms of meters per pixel [m/px or m px^{-1}]. As more than one pixel is required to identify an object (or *"resolve"* it), the image resolution is typically a factor of 3 or 4 less than the ground pixel size. For example, the High-Resolution Imaging Experiment on the Mars Reconnaissance Orbiter has a ground pixel size of $\sim$25 cm, which means that a table with a plan view area of 1 m $\times$ 1 m could be "resolved." As basically all planetary missions had (and will have) cameras onboard, images are available for a multitude of Solar System objects. Image availability is excellent at all scales for the Moon and Mars and at selected scales for many smaller bodies. On the other hand, no orbiter images in the visible wavelengths are available for Venus due to its dense atmosphere. Similarly, the surface of the Saturnian satellite, Titan, cannot be readily investigated on visible wavelength images.

3.1.2 Laser Altimetry

Topography or relief is a quintessential type of information in geologic mapping. The most accurate information on planetary topography comes from laser altimetry, which can yield absolute accuracies of centimeters to decimeters. However, the spatial resolution is typically much worse than that of images (e.g., the shot-to-shot distance on the Martian surface of single measurements was $\sim$330 m for the Mars Orbiter Laser Altimeter). Laser altimetry has been applied to the Moon, Mars, Mercury, and the asteroid 433 Eros, and laser altimeters will fly on future missions to Mercury (BepiColombo, ESA), Jupiter's Moon, Ganymede (JUICE, ESA), and asteroid 101955, Bennu (OSIRIS-REx, NASA). Laser altimetry data are especially useful for geologic mapping when combined with geometrically co-registered images. This is valid both for the individual laser "shots" along the spacecraft track, but also for interpolated DEM, which can also be used as base maps for small-scale maps (Fig. 11e). The relief of planetary surfaces as revealed by DEM is a result of primary (emplacement) and secondary (e.g., erosional, tectonic) processes. Different materials (geologic units) respond differently to these processes; for example, they are more easily eroded than others, and their topographic characteristics can therefore yield important clues on the compositional differences of surface materials (see also Hargitai et al. this volume).

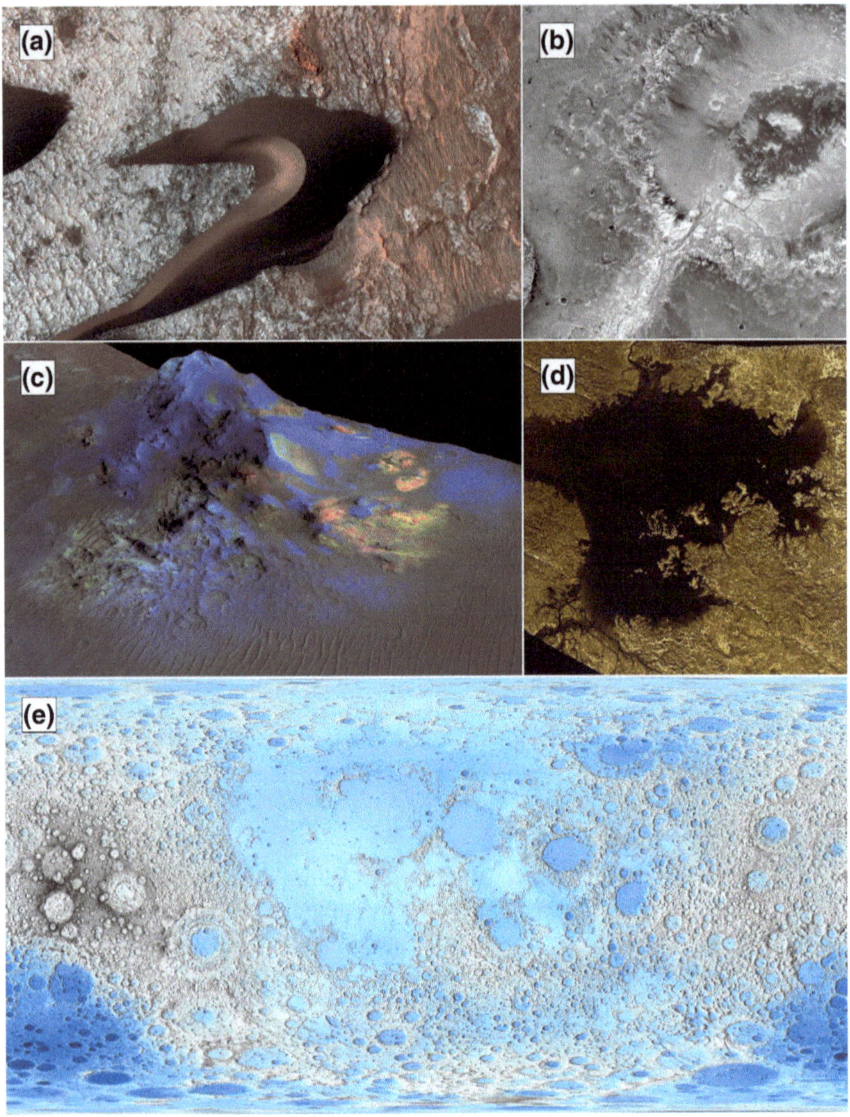
(a)
(b)
(c)
(d)
(e)

◂**Fig. 11** Examples of data sets and derived products commonly used for planetary geologic mapping. **a** Visible or (false) color images are ideal to investigate fine details of surface texture and tone. This picture is about 1 km wide and was acquired by the High-Resolution Imaging Science Experiment (HiRISE) and shows a sand dune and bright fractured bedrock on Mars (image: NASA/University of Arizona). **b** Thermal infrared images show differences in the thermophysical properties of surface materials. In this THEMIS nighttime image of the 140 km diameter Martian impact crater, Holden, bright materials have a relatively higher thermal inertia (i.e., they are still warmer in the night), whereas the dark surfaces correspond to materials which cool down faster after sunset (image: NASA/ASU/USGS). **c** Information from a hyperspectral imager, CRISM, is overlain on a digital elevation model and an image of Alga Crater's central peak (ø 5 km) on Mars, both derived from HiRISE data. This merged data set presents compositional information and geological context together (image: NASA/JPL-Caltech/JHUAPL/ University of Arizona). **d** Radar images can reveal surface details even when clouds or a dense atmosphere obscure the view for camera experiments. This example, covering an area of ~530 × 490 km, shows rugged terrain around the lake Ligeia Mare on Titan (black area in this colored representation; image: NASA/JPL-Caltech/ASI/Cornell University). **e** Topographic data are extremely useful for geologic mapping and are also good base maps at large scales. This view shows a global color-coded hillshade map of the lunar surface, based on data acquired by the Lunar Orbiter Laser Altimeter (LOLA). (image: Goddard Space Flight Center, NASA, JPL, PDS)

3.1.3 Radar

Radar data are indispensable for the mapping of objects with a non-transparent atmosphere, as radar instruments actively emit microwaves that can penetrate haze and permanent clouds. Synthetic-aperture radar (SAR) instruments return information on the radar brightness of terrain (a proxy for roughness) as well as on topography and dielectric characteristics. They were used to map Venus and Titan (Fig. 11d). Sounding or ground-penetrating radar instruments such as Mars Advanced Radar for Subsurface and Ionosphere Sounding (MARSIS) and Mars SHAllow RADar (SHARAD) provide insights into the subsurface, as they are able to detect boundary layers between materials with different dielectric properties, which can be useful in geologic mapping as the detection of such boundary layers can help establishing the local and regional stratigraphy (e.g., Brothers et al. 2015). When hazy or cloudy atmospheres prohibit the use of camera systems, radar images and radar-derived altimetry can be the only source of information for geologic mapping (e.g., in the case of Venus). However, radar images are substantially different from conventional images, and any interpretations must consider the different observation geometries and the physics behind the recording sensor systems (Tanaka 1994).

3.1.4 Spectrometers

Hyperspectral imaging spectrometers, often sensitive to visible- and near-infrared wavelengths, provide information on the mineralogical composition of surface materials, often in the form of three-dimensional (x, y, λ) image "cubes," where the third dimension (λ) is represented by individual images taken at different wavelengths.

Such hyperspectral images have enormous science potential for planetary geologic mapping, as they enable remotely discriminating surface units on the basis of composition. For example, near-global coverage by the Observatoire pour la Minéralogie, l'Eau, les Glaces et l'Activité (OMEGA) instrument on ESA's Mars Express mission has enabled generating mineral maps of the Martian surface, which are valuable resources for geological mapping (Ody et al. 2014). Traditional panchromatic or color images, on the other hand, only provide indirect hints at lithology (or compositional) variations via the examination of texture and morphology, which can vary with composition. Typically, certain spectral features that are indicative of specific mineralogies are displayed as false-color images overlaid on texture (from panchromatic images) or relief (e.g., from hillshaded DEMs), or both (Fig. 11c). Thermal spectrometers are also highly useful for geologic mapping as they can reveal differences in the thermophysical properties of surface materials (e.g., thermal inertia). Moreover, global daytime and nighttime infrared maps of Mars (Fig. 11b) are standard base maps for USGS-supported geologic mapping. Spectrometers operating at other wavelengths, such as neutron spectrometers or gamma-ray spectrometers, have been used to map the abundance of hydrogen and elemental chemistry, respectively.

3.1.5 Ground Truth

Rover and lander data may be useful as local "ground truth" to constrain the interpretation of wider regional geologic mapping based on remote sensing data (Heiken et al. 1991; Crumpler 2016). The very limited availability of such data, however, restricts this approach to the few sites where landers have been successfully placed (e.g., Mars, Venus, Titan). The Moon is an exception, as the huge amount of information collected by the manned Apollo missions proved to be important for later mapping efforts.

3.2 *Selection of Data for Planetary Geologic Mapping*

The selection of the most appropriate data set to prepare a base map for the generation of a geologic map is a fundamental decision which needs to consider several factors. Perhaps, the most important of these are data availability (Fig. 12) and scale (Fig. 13). As the mapping scale will depend on the mapping objective, it will be different for different science questions, and so will be the data for the base data set. For example, the mapping of a landing site will require very large scales, and so mosaics of the highest-resolution image data (CTX, HiRISE) were used as base maps for recent mapping of candidate landing sites for the ExoMars and Mars 2020 rovers. On the other hand, regional mapping of Mars, commonly organized in map quadrangles or subsets of these, is commonly performed on mosaics made of many individual infrared images. It is important to keep in mind that regardless of the choice of the base map (data set and scale), diverse data sets at different scales will

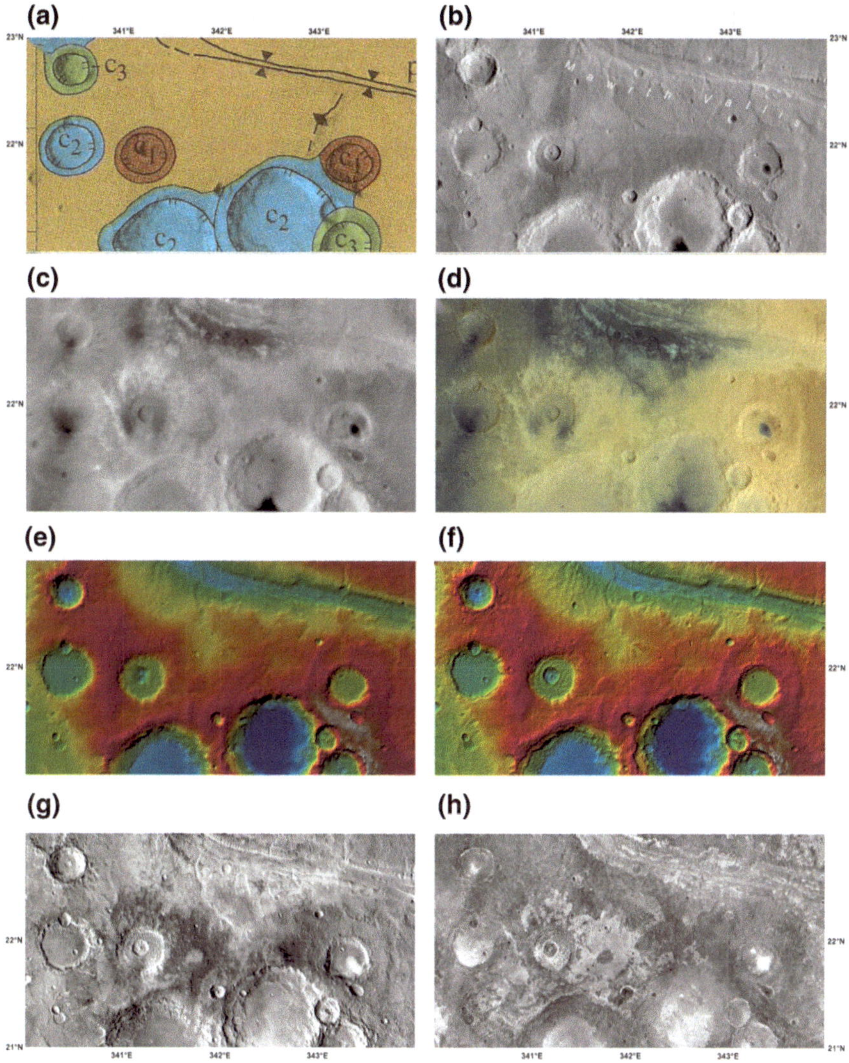
(a)
(b)
(c)
(d)
(e)
(f)
(g)
(h)
341°E
342°E
343°E
23°N
22°N
21°N
c3
c2
c1
c2
c2
c1
c3

◂**Fig. 12** Comparison of an early geologic map with various base map data sets. The various data sets were acquired at different times and with different sensor systems, covering the regional context of a former candidate landing site for ESA's ExoMars 2020 rover mission just south of Mawrth Vallis. **a** Geologic map based on Mariner 9 images (100 m/px to 1 km/px) which were assembled to a mosaic at a scale of 1:5,000,000 in Mercator projection. Note how little detail was resolved on the basis of these images (Wilhelms 1976). **b** The MDIM2.1 was produced at a digital scale of 1/256° or ~231 m/px on the basis of Viking Orbiter images that were acquired in the mid-1970s (Archinal et al. 2003). **c** The global mosaic of MOC (Mars Orbiter Camera) Wide-Angle Camera images has a scale of 250 m/px or better (Niedermaier et al. 2002). In comparison to the MDIM2.1, it has a better geometric accuracy and improved radiometric resolution (8 bits instead of 7 bits). **d** False-color image mosaic of High-Resolution Stereo Camera (HRSC) images acquired at an original ground pixel scale of 12–15 m (Gwinner et al. 2016). **e** Hillshade version of the global MOLA (Mars Orbiter Laser Altimeter) Digital Elevation Model (DEM) at a grid size of ~463 m (1/128°) (Smith et al. 2001). **f** Hillshade version of HRSC DEM (50 m grid spacing; Gwinner et al. 2016). **g** Infrared image mosaic generated from Thermal Emission Imaging System (THEMIS) daytime images at a scale of 100 m/px (Edwards et al. 2011). **h** THEMIS infrared nighttime image mosaic (100 m/px). Latitudes are planetodetic in all panels

need to be interrogated for geologic mapping. Some of them will not be available for the entire mapping area, but can nonetheless yield important clues. For instance, the analysis of local high-resolution images may reveal stratigraphic relationships that, once locally established, can then be applied over wide areas even where no high-resolution data are available. Similarly, the availability of radar or spectrometer data may be spatially limited, but may provide key insights into subsurface stratigraphy or composition, respectively.

4 Technical Aspects

As planetary geologic mapping is based on preprocessed remote sensing data, the geological analysis, interpretation, and mapping process need to be performed within general and valid environments for achieving understandable and usable results. These environments will accompany the whole mapping process and address the nomenclature, cartographic symbology, data description (metadata), and finally the usage of data models, archiving, and infrastructure of the mapping data. The description and current state of the art of these topics are summarized in the following sections.

4.1 Data Preparation, Processing, and Environments

The conduction and communication of geological maps are mainly based on the comparison and interpretation of base data. The whole process reaching from a database to derived information and knowledge is comparable to the common

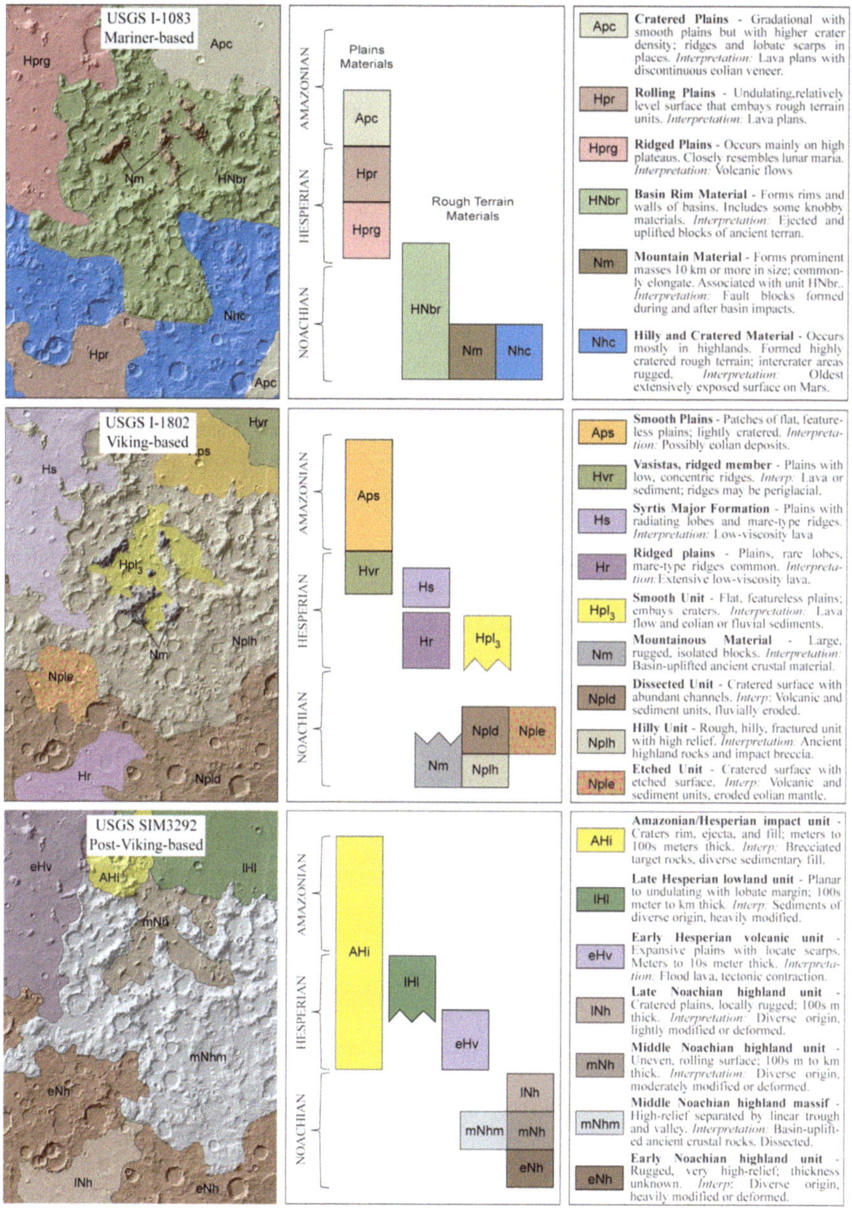

Fig. 13 Map elements and their variations as a function of base data and scale (as demonstrated at an example from the Libya Montes region, Mars). Shown are the map on the left, the Correlation of Map Units (CMU) in the middle, and the Description of Map Units (DMU) on the right. Top: Base data: Mariner 9 (Scott and Carr 1978). Middle: Viking Orbiter (Scott and Tanaka 1986). Bottom: THEMIS (Tanaka et al. 2014b)

visualization and mapping process in cartography (see Naß and Hargitai this volume a). Historically established, the major part of such data originates from orbital observations. Consequently, geological mapping in planetary science will be mainly conducted by interpretation of different remote sensing data. However, the amount of ground truth data collected by landers and rovers also increased in recent times.

"For any given instrument, an image processing system always exists that the instrument team uses and that tool will have the most accurate algorithms for that instrument" (Beyer 2015). These steps of data processing and correction used by the instrument teams can be adapted individually to the particular data set. However, the potential users are not limited to apply these procedures but can also use more general and universal environments for data processing. After necessary radiometric and photogrammetric calibration steps, the data have to be referenced to a three-dimensional reference body and finally projected onto a two-dimensional plane (see e.g., Hare et al. 2018). Final and high-level data products are archived on platforms like the *Planetary Science Archive* (PSA 2017) or the *Planetary Data System* (PDS 2017).

Using the mentioned variety of data, the geological mapping process can be started. While former mapping procedures were conducted by a sheet of paper and a pencil (based on field work in terrestrial science), the mapping process has been heavily changed in the last decades. After starting digital mapping in graphic software systems, nowadays planetary scientists make use of database-driven software systems called *Geographic Information Systems* (GIS). These systems can efficiently handle the spatial context of data; i.e., data can be managed, analyzed, compared, and visualized by extrapolating, generalizing, classifying, and merging the essential information. The users can choose between proprietary desktop systems like ArcGIS™ or open-source environments like *Quantum GIS* (2017), *Saga GIS* (2017), or *Grass GIS* (2017), or Web-based mapping systems like *geoportal*, *map-a-planet (2017)*, *jMars* (2017), and *openplanetary*. For more detailed descriptions about the application of GIS in Planetary Science, see Hare et al. (2018) and Nass and Hargitai (this volume a).

4.2 Map Components

4.2.1 Cartographic Visualization—Symbols

Just like any spoken language, a language for communicating spatial information by cartographic visualization with symbols has to follow common rules and standards. These symbols are summarized and described in various models, which are made available in recommendation documents, also known as standards. One such document, collecting a huge amount of different symbols and corresponding guidelines for the use of the symbols, is the *Digital Cartographic Standard for Geological Map Symbolization* (FGDC 2006) and was developed by the *Federal*

Geographic Data Committee (FGDC) and the *United States Geological Survey* (USGS). In chapter 25 (Planetary Geology) of this standard, a wide range of symbol requirements are dedicated to planetary geological mapping. However, as also mentioned in the document guidelines, this standard is not intended as a static repository; i.e., it is recommended to use it as a modifiable and expandable basis. Especially, when interpreting planetary surfaces by remote sensing data alone, an individual and flexible use and creation of symbols are needed, because there are a variety of spatial objects that cannot be addressed and interpreted appropriately during the time of mapping (Nass et al. 2011).

The most prominent and important information in geological maps are the different geological units, represented by different colors. For the visualization of surface ages and stratigraphic relationships between different surface areas, time-stratigraphic color recommendations have been developed (e.g., *International Stratigraphic Guide* for terrestrial mapping; Salvador 1994). In these charts, various chronological and chronostratigraphic subdivisions are defined based on a color scheme. In the context of planetary chronostratigraphy and associated geological mapping, this subject has been addressed by Tanaka (1986) and Tanaka and Skinner (2003), who mention the *International Stratigraphic Guide* (Salvador 1994) as a reference basis for coloring lithologic units (Nass et al. 2011).

Currently, there are no formal stratigraphic standards in planetary science and cartography. However, there is a clear informal color scheme based on former geological maps produced by the USGS, which visualize, e.g., Noachian units on Mars in brownish, volcanic units in reddish, and crater ejecta in yellowish colors.

4.2.2 Nomenclature and Unit Labels

Besides colors, representing geological units, and symbols for point, linear, and areal objects, feature and region names or *toponyms* (i.e., nomenclature) are essential parts of geological maps. Together with the labeling for the geological units, they represent a major map component. Planetary nomenclature, like terrestrial nomenclature, is used to uniquely identify a feature on the surface of a planet or satellite so that the feature can be easily located, described, and discussed (IAU 2017). Thus, the *International Astronomical Union* (IAU 2017) has been founded in 1919 with the aim to regularize the chaotic Lunar and Martian nomenclatures existing until that time. The definition of the nomenclature follows rules set up by the IAU, and within the *Working Group for Planetary System Nomenclature* (WGPSN), to homogenize various systems used in Lunar and Martian nomenclatures across different countries (Blagg and Mueller 1935). In 1957, requirements for extraterrestrial nomenclature were dramatically changed when the age of space exploration began—starting by the successful flight of Sputnik and by the US commitment to land a man on the Moon in the 1960s (IAU 2017). The amount of detail visible in the new high-resolution images of extraterrestrial surfaces required large amounts of names; hence, the official naming of features became a more complex task (Strobell and Masursky 1990).

One very important requirement is that the descriptor terms within the maps intentionally should "avoid interpretation and stick to descriptions" (Rothery 2010, p. 59). Informal names, often used in the mapping of previously unexplored regions, should follow the same guideline. This topic is essential, because during the interpretation process and geological mapping, a large numbers of objects may become apparent that do not have a formal name yet. Other questions related to nomenclature were recently discussed by Bruno and Ruban (2017). As for the nomenclature, there are also conventions for labeling, which in geological maps especially concern the unit names. Commonly, a combination of the unit age (e.g., for Mars Amazonian, Hesperian, or Noachian) and the initials of the complete unit name is used. Problems that can possibly arise in unit naming as well as some recommendations are discussed in Tanaka et al. (2011). The best placing for the labels varies from map to map and depends on the unit size and other overlaying mapped objects shown on the map. Some rules on label placement and font and font size selection are described in FGDC (2006).

4.2.3 Map Projections

The choice of the most appropriate map projection is an important task. The projection of a spheroidal surface, i.e., the surface of a planetary body, onto a plane, i.e., the visualization on a map, creates distortions in one or the other way (area, length, or/and angle). Which projection should be used for a map in general, and a geologic map in particular, depends on the mapping objective. For instance, is it necessary to measure and compare size- and length-relationships? Is the mapped area local, regional, or global in scale? As there are many different projections in the field of cartography, we will only mention those which are the most important and commonly used in planetary science (Fig. 14). The *Mollweide* projection is ideal for global maps showing the mapped geologic units in their true size; thus, the units can easily be compared to each other. However, the distortion in length (except along

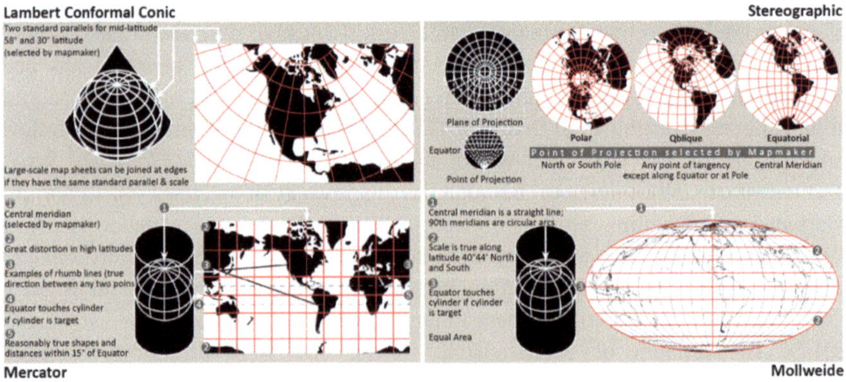

Fig. 14 Overview of commonly used map projections for geological mapping in planetary science (figure modified by the authors, based on USGS (1995) and Snyder (1987))

central meridian and along equator) and angle increase steadily with growing distance from the equator. For small-scale regional maps, the *Stereographic* projection is used for polar regions, the *Mercator* projection for equatorial regions, and the *Lambert Conformal Conic* projection for mid-latitude regions. For larger-scale local maps, the accurate visualization of the length or area of the mapped objects is typically the most important objective. It is therefore necessary to place the projection center at the middle of the mapping area and then to choose the required projection. Apart from the extent of the objects, their location, distribution, and orientation within the map area are other decisive factors for the choice of the proper projection. As there are many factors that are important for large-scale local maps, it is necessary to decide on a case-by-case basis which the optimal projection is.

For further information about map projections in general, see Snyder (1987). A historical overview of common map projections used for geological maps and map series in planetary science is given by Greeley and Batson (1990). More recent reviews are provided by Hare et al. (2018), Hargitai et al. (2017), and Hare (this volume).

4.3 *Analyzing and Measurement Tools*

As defined by Wilhelms (1990), "a geologic map is a two-dimensional representation of the three-dimensional spatial relations and chronologic sequences of the materials and structures of a planetary crust" (Wilhelms 1990, p. 208). As mentioned before, planetary geologists typically map on the basis of the interpretation of various remote sensing data sets. As there is no possibility to verify their mapping by fieldwork (e.g., field measurements, sampling), they have to rely on photogeologic measurements and tools using the different base data (see Sects. 3 and 4.1). The main tools in planetary geology are *calculations* of volumes, lengths, areas, and slopes of the mapped objects and surface units, which enable assessing specific objects in relation to other objects and finding similarities and differences. Because GIS environments primarily operate with the geometry and raster values of the spatial data, these calculations can be easily performed within such environments.

A key aspect of geological maps is the chronology of mapped units, i.e., the age of mapped materials. The determination of an absolute age for a given unit is problematic, and apart from radiometric ages for meteorites and lunar samples returned by the Apollo and Luna programs, a detailed assessment of ages is typically missing (van Gasselt and Nass 2014). Instead, a well-established method in planetary science enables deriving model ages on the basis of remotely sensed images. This method is called *crater counting*, and it is based on the statistical analysis of crater size–frequency distributions (CSFDs) of impact craters on planetary surfaces (Michael 2013). A software for the GIS-based usage of this method was recently made available by Kneissl et al. (2011, 2015).

Another fundamental property of (layered) rocks commonly depicted in geologic maps is their *strike* and *dip*, i.e., the orientation of a planar layer in space. The first GIS-based implementation of strike-and-dip measurements is described by Kneissl et al. (2010). It was used, e.g., in Okubo and Gaithner (2017) and was updated by the USGS (see below). A current overview of the available tools is given within Rossi and van Gasselt (2018).

A comprehensive list of tools which are necessary for geological mapping is available on the Web page of the Astrogeology Science Center (MRCTR GIS Lab) of the USGS. Within the *PGM toolbox*, generated for using directly in GIS environments, the user can perform *topology checks* of the borders of the geological units or *build polygons* automatically (as geological units) after drawing the contacts as lines. The tool for *graphics and shapes* focuses on geodetically correct calculations of areas, lengths, and angles. In addition to the list of tools, general documentation, tutorials, and resources (regarding data and software) are also provided here.

4.4 Data Description and Structure

In order to allow for an efficient collaboration within the geological mapping community, in terrestrial as well as planetary science, one of the most important tasks is to uniformly prepare, describe, manage, and archive the mapped data. Only if the comparability of the mapping projects is ensured, the results can be used as starting points for further investigations. On printed maps (or "static" PDF versions thereof), the descriptive information such as projection, author, legend, title, scale, etc., was mainly provided by the facts included in the frame of the map sheet (as so-called marginalia). For digital maps, this context is handled by detailed map descriptions and an efficient management of mapping results by using metadata information. Such data descriptions, e.g., of the geometry, extent, quality, contents, and conditions of source data, enable detailed queries for context information (Nass et al. 2010). As mentioned before (see Sect. 4.2.3), such detailed descriptions are especially necessary for geological interpretations in order to understand and being able to reproduce the results for further work.

In addition to the descriptive data information, the digital structure of the mapped objects and units is essential. The digital structure should reflect, as far as possible, the fundamental geological principles and the mutual relations of the objects as observed in nature.

For example, the stratigraphic relations of objects should also be documented in the digital structure as attributes of the mapped objects. Furthermore, additional information like the description of object origin is derived within the interpretational mapping process. In order to handle all these object attributes in a most realistic (regarding their nature) and efficient way (regarding their relations to each other), one can benefit from spatial data models. These data models are common in

GIS environments and allow a logical data structure for arranging all the different objects in predefined data tables, linked by different key attributes and values.

Several data models have been discussed for terrestrial geological mapping in GIS environments. The most recent and available ones are the *Data Specification on Geology* (INSPIRE 2013), and the *North American Geologic Map Data Model* (NADM Steering Committee 2015). A more detailed overview is given by van Gasselt and Nass (2011). The implementation of a geological data model in a proprietary system and its application to geologic map publications within the *National Geologic Map Database* (NGMD) is shown in the example of the geologic map of the Mount Baker quadrangle.[1]

One of the first implementations of a planetary geological data model is presented by van Gasselt and Nass (2011). Within this data model, all the geological attributes and relations, comparable to terrestrial geology, can be handled, but the model takes also care about the special information needed in planetary geology mapping (see Fig. 15). Here, the different reference systems, as well as the chronology systems of the individual planetary bodies, play the most important role.

The first implementation of such a general data model adaptable for geological mapping projects is presented by Nass (2017). In this example, the geological mapping content of NASA's DAWN mission is organized in a predefined data structure. As mentioned before, the stratigraphic arrangement of the units is one of the most important issues in geological sciences. To handle all these different systems (i.e., the absolute and relative ages, and the chronology and chronostratigraphy), a more detailed geological data model is needed which focuses on the ontological attributes of all mapped units. This topic is discussed in van Gasselt and Nass (2014).

5 Future Developments and Challenges

Geologic mapping will remain one of the key methods to investigate the geologic history of Solar System bodies with solid surfaces. It is important to note that there is no such thing as a single and uniform approach to planetary geologic mapping that can be applied to any area. The diversity of bodies that can now be examined by imaging data, many of them at high spatial resolution, results in a corresponding diversity of scientific objectives that are addressed by mapping studies (e.g., McSween 2015). For example, recent spacecraft missions have returned excellent images from previously unknown types of surfaces, such as those from the comet, 67P/Churyumov–Gerasimenko, the asteroid Itokawa, or from the Pluto–Charon system. These surfaces are unlike the "standard" rocky surfaces in the inner Solar System, and they are also different from those of the icy satellites of the gas and ice

[1]http://www.arcgis.com/home/esriNcgmpOnlineGeologicMappingTemplate.

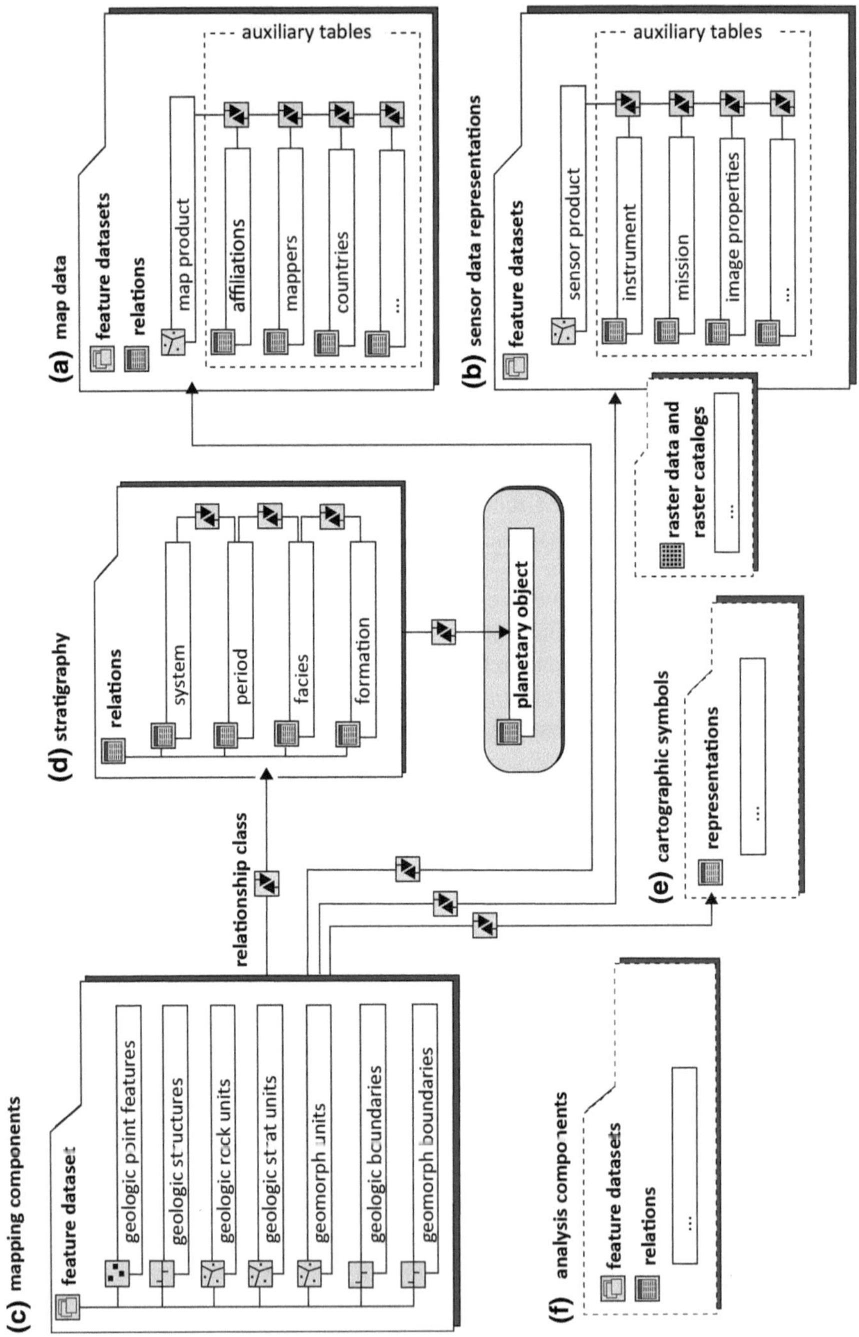
(a) map data
feature datasets
relations
map product
auxiliary tables
affiliations
mappers
countries
...
(b) sensor data representations
feature datasets
sensor product
auxiliary tables
instrument
mission
image properties
...
raster data and raster catalogs
...
(d) stratigraphy
relations
system
period
facies
formation
planetary object
relationship class
(c) mapping components
feature dataset
geologic point features
geologic structures
geologic rock units
geologic strat units
geomorph units
geologic boundaries
geomorph boundaries
(e) cartographic symbols
representations
...
(f) analysis components
feature datasets
relations
...

◂**Fig. 15** Low-level entity relationship (ER) diagram depicting major key components of the data model by focusing on data types and major relations rather than attribute level. However, the listed attributes and relations are examples and raise no claim of being complete. **a** and **b** cover the data search component, **c** and **d** refer to the main mapping components, **e** marks additional elements discussed in the work by Nass et al. (2011), and **f** stands for the nomenclature and annotation (**e** + **f** linked by dotted relation frame). The arrows indicate the possible relations among each other (figure slightly modified from van Gasselt and Nass 2011)

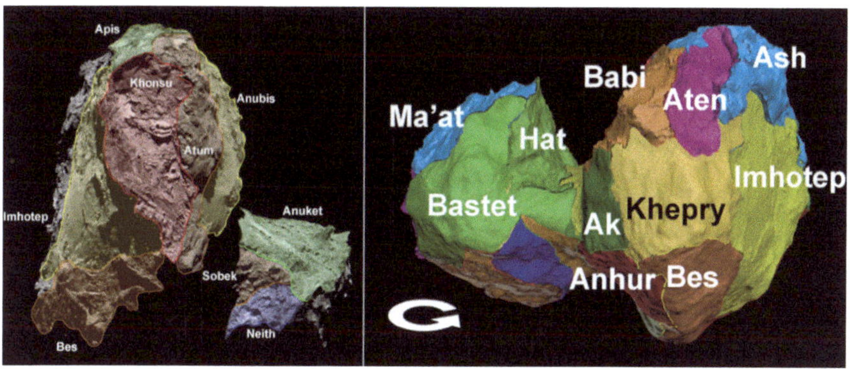

Fig. 16 Example of mapping of irregular bodies. The images show different views of regional surface units (not yet formally named by the IAU) of comet 67P/Churyumov–Gerasimenko (from El-Maarry et al. 2016)

giants. Consequently, specific mapping approaches, at many different scales, and based on different sensor systems, may be required to reveal the surface and subsurface characteristics of such bodies. Another factor that will demand variable approaches is the increasing amount of data, and the different types of sensor systems that are applied in planetary exploration (Milazzo et al. 2017). Standardized geologic mapping programs, although indispensable for generating geologic maps that are comparable and adhere to certain quality standards (e.g., Skinner 2015), cannot cover the need for multiple, often local studies of increasingly complicated terrain, both in terms of geometry and in composition. For example, the highly complex geometry of comet 67P/Churyumov–Gerasimenko implies that there are no standard map projections; hence, any mapping has to be performed on irregular geometries (see Stooke and Pajola, this volume) (Fig. 16). Another new field for planetary geologic mapping has recently been opened by long-term rover operations on the surface of Mars. Similar to what has been done on the basis of the Apollo landings, detailed maps can now be constructed on the basis of rover images and spectrometers (Crumpler 2016), and the combination of field observations (e.g., detailed stratigraphic context) made by rovers and remote sensing data holds even more potential for geologic mapping (Crumpler et al. 2011). In the more distant future, the geological mapping may be a prerequisite for the in situ utilization of resources (ISRU; e.g., Anand et al. 2012).

In many such future studies, interpretation will outweigh the observations, so there is a reduced need for formal and standardized documentation, from which new discoveries could be made (such as it is possible from standardized maps). Both standardized and non-standardized mapping approaches will need to be used, depending on scientific questions, data availability, and the nature of the study object (Skinner 2017a, b).

Acknowledgements We thank the editor, Henrik Hargitai, for preparing Fig. 6. Detailed and constructive comments from Corey Fortezzo and two anonymous reviewers as well as the editor are highly appreciated and helped to improve the manuscript. Special thanks go to Sasha Basilevsky who kindly provided maps that were produced in the context of Soviet planetary exploration. Marita Wählisch helped with translations from publications in Russian.

Web References

Data Archives

Planetary Data System (PDS) https://pds.nasa.gov/
Planetary Science Archive (PSA) http://psa-tools.cosmos.esa.int/

Data Processing

ISIS (Integrated Software for Imagers and Spectrometers) https://isis.astrogeology.usgs.gov/
GDAL (Geospatial Data Abstraction Library) http://www.gdal.org/
GMT (Generic Mapping Tools) https://www.soest.hawaii.edu/gmt/
QGIS http://qgis.org/de/site/
USGS MRCTR GIS Lab https://astrogeology.usgs.gov/facilities/mrctr-gis-lab

Mapping Guidelines and Tools

USGS Planetary Geologic Mapping: https://planetarymapping.wr.usgs.gov/
FGDC (Federal Geographic Data Committee) Digital Cartographic Standard for Geologic Map Symbolization: http://ngmdb.usgs.gov/fgdc_gds/geolsymstd.php
IAU Gazetteer of Planetary Nomenclature home page: http://planetarynames.wr.usgs.gov/
IAU Gazetteer of Planetary Nomenclature descriptor terms: http://planetarynames.wr.usgs.gov/jsp/append5.jsp
IAU Gazetteer of Planetary Nomenclature feature name request form: http://planetarynames.wr.usgs.gov/jsp/request.jsp
USGS Planetary Interactive GIS on-the-Web Analyzable Database (PIGWAD): http://webgis.wr.usgs.gov/
USGS Map-a-Planet: http://www.mapaplanet.org/

USGS tips and information for preparation of astrogeology maps: http://astrogeology.usgs.gov/Projects/PlanetaryMapping/guidelines/preparationTips.pdf

USGS instructions on building polygons in Illustrator: http://astrogeology.usgs.gov/Projects/PlanetaryMapping/guidelines/layersexample_small.pdf

References

Anand M et al (2012) A brief review of chemical and mineralogical resources on the Moon and likely initial in situ resource utilization (ISRU) applications. Planet Space Sci 74:42–48. https://doi.org/10.1016/j.pss.2012.08.012

Archinal BA, Kirk RL., Duxbury TC, Lee EM, Sucharski R, Cook D (2003) Mars digital image model 2.1 control network. In: 34th lunar and planetary science conference, #1485

Beyer RA (2015) An introduction to the data and tools of planetary geomorphology. Geomorphology 240:137–145. https://doi.org/10.1016/j.geomorph.2014.11.022

Blagg MA, Mueller K (1935) Named lunar formations. International Astronomical Union, Percy Lund, Humphries and Co Ltd, 196 p, London

Brothers TC, Holt JW, Spiga A (2015) Planum Boreum basal unit topography, Mars: Irregularities and insights from SHARAD. J Geophys Res Planets 120:1357–1375. https://doi.org/10.1002/2015JE004830

Bruno DE, Ruban DA (2017) Something more than boulders: a geological comment on the nomenclature of megaclasts on extraterrestrial bodies. Planet Space Sci 135:37–42. https://doi.org/10.1016/j.pss.2016.11.006

Carr MH (2013) Geologic exploration of the planets: the first 50 years. EOS Trans AGU 94(3):29–30. https://doi.org/10.1002/2013EO030001

Carr MH (1975) Geologic map of the tharsis quadrangle of Mars. USGS IMAP 893, scale 1:5,000,000. https://doi.org/10.3133/i893.

Crumpler LS et al (2011) Field reconnaissance geologic mapping of the Columbia Hills, Mars, based on Mars Exploration Rover Spirit and MRO HiRISE observations. J Geophys Res 116: E00F24. https://doi.org/10.1029/2010je003749

Crumpler L (2016) The first field geologic maps on another planet. In: Annual planetary geologic mappers meeting 2016, abstract 7035

De Hon, RA, Mouginis-Mark PJ, Brick EE (1999) Geologic map of the Galaxias quadrangle (MTM 35217) of Mars. USGS IMAP 2579, scale 1:500,000, https://doi.org/10.3133/i2579.

Dohm JM, Tanaka KL, Hare TM (2001) Geologic map of the thaumasia region, Mars. U.S. Geological Survey Geologic Investigations Series I–2650, 3 sheets, scale 1:5,000,000. https://pubs.usgs.gov/imap/i2650/.

Edwards CS, Nowicki KJ, Christensen PR, Hill J, Gorelick N, Murray K (2011) Mosaicking of global planetary image datasets: 1. Techniques and data processing for Thermal Emission Imaging System (THEMIS) multi-spectral data. J Geophys Res: Planets 116:E10008. https://doi.org/10.1029/2010je003755

El-Maarry MR et al (2016) Regional surface morphology of comet 67P/Churyumov-Gerasimenko from Rosetta/OSIRIS images: the southern hemisphere (corrigendum). Astron Astrophys 593, A110. https://doi.org/10.1051/0004-6361/201628634, https://doi.org/10.1051/0004-6361/201628634e

Eppler DB, Bleacher J, Evans C, Feng W, Gruener J, Hurwitz D, Janoiko B, Skinner J, Whitson P (2014) Comparing geologic data sets collected by planetary analog traverses and by standard geologic field mapping: implications for planetary exploration planning. In: 45th lunar and planetary science conference, LPI contribution no. 1777, #2078

FGDC (2006) Digital Cartographic Standard for Geologic Map Symbolization. Federal Geographic Data Committee prepared by the U.S. Geological Survey (FGDC-STD-013-2006)

Greeley R, Guest JE (1987) Geologic map of the eastern equatorial region of Mars. U.S. Geological Survey IMAP 1802-B, scale 1:15,000,000. https://doi.org/10.3133/i1802B.

Greeley R, Batson RM (eds) (1990) Planetary mapping. Cambridge University Press, Cambridge

Greeley R (2013) Introduction to planetary geomorphology. Cambridge University Press, Cambridge

Grotzinger JP, Arvidson RE, Bell JF, Calvin W, Clark BC, Fike DA, Golombek M et al (2005) Stratigraphy and sedimentology of a dry to wet eolian depositional system, Burns formation, Meridiani Planum, Mars. Earth Planet Sci Lett 240(1):11–72. https://doi.org/10.1016/j.epsl.2005.09.039

Gwinner K, Jaumann R, Hauber E et al (2016) The High Resolution Stereo Camera (HRSC) of Mars Express and its approach to science analysis and mapping for Mars and its satellites. Planet Space Sci 126:93–138. https://doi.org/10.1016/j.pss.2016.02.014

Hackman RJ (1961) Photointerpretation of the Lunar Surface. Photogram Eng 27(3):377–386

Hansen VL (2000) Geologic mapping of tectonic planets. Earth Planet Sci Lett 176:527–542. https://doi.org/10.1016/S0012-821X(00)00017-0

Hare TM, Skinner JA, Kirk RL (2018) Cartography tools. In: Rossi A-P, van Gasselt S (eds) Planetary geology. Springer Praxis Books, Chichester, pp 55–70

Hargitai H, Wang J, Stooke PJ, Karachevtseva I, Kereszturi A, Gede M (2017) Map projections in planetary cartography. In: Lapaine M, Usery E (eds) Choosing a map projection. Lecture notes in geoinformation and cartography. Springer, Cham, pp 177–202. https://doi.org/10.1007/978-3-319-51835-0_7

Hargitai H, Willner K, Buchroithner M (this volume) Methods in planetary topographic mapping: a Review. In: Hargitai H (ed) Planetary cartography and GIS. Lecture notes in geoinformation and cartography. Springer

Heiken G, Vaniman D, French BM (eds) (1991) Lunar sourcebook. Cambridge University Press, Cambridge

IAU (2017) International Astronomical Union. https://www.iau.org/

INSPIRE (2013) D2.8.II.4 Data Specification on Geology - Technical Guidelines. INSPIRE – Infrastructure for Spatial Information in Europe, Thematic Working Group Geology, European Commission Joint Research Centre. http://inspire.ec.europa.eu/id/document/tg/ge

Ivanov MA, Head JW (2011) Global geological map of Venus. Planet Space Sci 59(13):1559–1600. https://doi.org/10.1016/j.pss.2011.07.008

Kneissl T, van Gasselt S, Nekum G (2010) Measurement of strike and dip of geologic layers from remote sensing data—new software tool for ArcGIS. In: 41th lunar and planetary science conference, LPI contribution no. 1533, 1640

Kneissl T, van Gasselt S, Neukum G (2011) Map-projection-independent crater size-frequency determination in GIS environments—New software tool for ArcGIS. Planet Space Sci 59:1243–1254. https://doi.org/10.1016/j.pss.2010.03.015

Kneissl T, Michael GG, Platz T, Walter SHG (2015) Age determination of linear surface features using the Buffered Crater Counting approach—Case studies of the Sirenum and Fortuna Fossae graben systems on Mars. Icarus 250:384–394. https://doi.org/10.1016/j.icarus.2014.12.008

Kotelnikov VA (ed) (1989) Atlas Poverhnosti Venery (eng.: Atlas of the Surface of Venus) (in Russian). Moscow

Kumar PS, Head JW (2013) Geologic map of the Lada Terra quadrangle (V–56), Venus. U.S. Geological Survey Scientific Investigations Map 3249, scale 1:5,000,000, 11 p. https://dx.doi.org/10.3133/sim3249

Le Deit L, Hauber E, Fueten F, Pondrelli M, Rossi AP, Jaumann R (2013) Sequence of infilling events in Gale Crater, Mars: Results from morphology, stratigraphy, and mineralogy. J Geophys Res Planets 118:2439–2473. https://doi.org/10.1002/2012JE004322

Lillesand T, Kiefer RW, Chipman J (2015) Remote sensing and image interpretation, 7th edn. Wiley, Hoboken

McSween HY (2015) Mapping the planets—geology stakes its claim. GSA Today 25(1):4–9. https://doi.org/10.1130/GSAT-14PresAdrs.1

Michael G (2013) Planetary surface dating from crater size–frequency distribution measurements: multiple resurfacing episodes and differential isochron fitting. Icarus 226:885–890. https://doi.org/10.1016/j.icarus.2013.07.004

MIIGAiK (ed) (1992) Atlas Planet Zemnoj Gruppy i ih Sputnikov (eng.: Atlas of the terrestrial planets and their satellites) (in Russian). Moscow

Milazzo MP, Kestay L, Dundas C (2017) The challenge for 2050: cohesive analysis of more than one hundred years of planetary data. In: Planetary science vision 2050 workshop 2017, LPI contribution no. 1989, abstract 8070

Molina A, Ángel de Pablo MA, Hauber E, Le Deit L, Fernández-Remolar D (2014) Geology of the Ariadnes Basin, NE Eridania quadrangle, Mars—1:1Million. J Maps 10(3):487–499. https://doi.org/10.1080/17445647.2014.888018

Mouginis-Mark PJ (2015) Geologic map of tooting crater, amazonis planitia region of mars. U.S. Geological Survey Scientific Investigations Map 3297, scale 1:200,000, pamphlet 4 p. https://doi.org/10.3133/sim3297.

NADM Steering Committee (2015) North American geologic map data model (NADM) – version 1.0.4. Technical Report U. S. Geological Survey, revised: 05-Feb-2015

Nass, A., and the Dawn Mapping Team (2017) One GIS-based data structure for Geological mapping using 15 map sheets – Dawn at Ceres. In: 48th lunar and planetary science conference, LPI contribution no. 1533, 1892

Nass A, van Gasselt S, Jaumann R (2010) Map description and management by spatial metadata: requirements for digital map legend for planetary geological and geomorphological mapping. ISPRS Technical Commission IV & AutoCarto in conjunction with ASPRS/CaGIS 2010, Orlando Florida. http://www.isprs.org/proceedings/XXXVIII/part4/

Nass A, van Gasselt S, Jaumann R, Asche H (2011) Implementation of cartographic symbols for planetary mapping in geographic information systems. Planet Space Sci 59:1255–1264. https://doi.org/10.1016/j.pss.2010.08.022 Special Issue: Planetary Mapping

Nass A, Hargitai H (this volume a) Cartography—Its role and interdisciplinary character as intersection in planetary science. In: Hargitai H (ed) Planetary cartography and GIS. Lecture notes in geoinformation and cartography. Springer

Nass A, Hargitai H (this volume b) Participants and initiatives in planetary cartography. In: Hargitai H (ed) Planetary cartography and GIS. Lecture notes in geoinformation and cartography. Springer

Niedermaier G et al (2002) A topographic image map of the MC-18 Quadrangle «Coprates» at 1: 2,000,000 using data obtained from the Mars Orbiter Camera and the Mars Orbiter Laser Altimeter of Mars Global Surveyor. In: 33rd annual lunar and planetary science conference, #1529

Nichols G (2009) Sedimentology and Stratigraphy, 2nd edn. Wiley, Chichester

Ody A et al (2014) OMEGA global maps of non-hydrous key minerals on Mars now available. In: Eighth international conference on Mars, LPI contribution no. 1791, 1191

Okubo CH (2014) Bedrock geologic and structural map through the western Candor Colles Region of Mars. U.S. Geological Survey Scientific Investigations Map 3309, scale 1:18,000, pamphlet 8 p. https://dx.doi.org/10.3133/sim3309.

Okubo CH, Gaither TA (2017) Bedrock and structural geologic maps of eastern Candor Sulci, western Ceti Mensa, and southeastern Ceti Mensa, Candor Chasma, Valles Marineris region of Mars. U.S. Geological Survey Scientific Investigations Map 3359, scale 1:18,000, pamphlet 12 p. https://doi.org/10.3133/sim3359

Pacifici A (2008) Geomorphological map of areas Vallis Mars (ASI Planetary Map Series – Map n °1). Boll. Soc. Geol. It. (Ital. J. Geosci.) 127(1):75–92

Pondrelli M et al (2015) Equatorial layered deposits in Arabia Terra, Mars: facies and process variability. GSA Bull. 127(7/8):1064–1089. https://doi.org/10.1130/b31225.1

Pondrelli M, Baker VR, Hauber E, Rossi A P (2018) Geologic tools. In: Rossi A-P, van Gasselt S (eds) Planetary geology. Springer Praxis Books, Chichester, pp 15–31

Portree DSF (2013) 1961: USGS Astrogeology's First Published Map. https://astrogeology.usgs.gov/rpif/exhibits/astros-first-lunar-map

Rossi AP, van Gasselt S (eds) (2018) Planetary geology. Springer Praxis Books, Chichester

Rothery DA (2010) Planets – a very short introduction. Oxford University Press, Oxford

Sabins FF (2007) Remote sensing principles and interpretation, 3rd edn. Waveland Press, Long Grove 494 p

Salvador A (ed) (1994) International stratigraphic guide. A guide to stratigraphic classification, terminology and procedure, 2nd edn. In: IUGS international subcommission on stratigraphic classification; published by International Union of Geological Sciences and Geological Society of America, Boulder, Colorado

Savinykh VP (ed) (2015) The phobos atlas. MIIGAiK, Moscow 220 p

Scott DH, Carr MH (1978) Geologic map of Mars. Scale 1:25,000,000. USGS Map IMAP-1083

Scott DH, Tanaka KL (1986) Geologic map of the western equatorial region of Mars. Scale 1:15,000,000. USGS Map I-1802-A

Sheehan WP, Dobbins TA (2001) Epic moon – a history of lunar exploration in the age of the telescope. Willmann-Bell, Richmond

Shoemaker EG, Hackman RJ (1962) Stratigraphic basis for a lunar time scale. In: Kopal Z, Mikhailov ZK (eds) The Moon, IAU Symposium 14, pp 289–300. Academic Press, New York

Skinner JA (2015) The challenges of standardized planetary geologic mapping. In: Second planetary data workshop (2015), abstract 7071

Skinner JA (2017a) The role of planetary geologic mapping in establishing a common framework for scientific investigation and exploration. In: 48th lunar and planetary science conference, LPI contribution no. 1964, abstract 2759

Skinner JA (2017b) Evolving planetary geologic mapping efforts in future decades by implementing a common framework for scientific investigation, exploration, and policy. In: Planetary science vision 2050 workshop, LPI contrib. no. 1989, abstract 8243

Smith DE, Zuber MT, Frey HV, Garvin JB, Head JW, Muhleman DO, Phillips GH, Solomon SC, Zwally HJ, Banerdt WB, Duxbury TC, Golombek MP, Lemoine FG, Neumann GA, Rowlands DD, Aharonson O, Ford PG, Ivanov AB, Johnson CL, McGovern PJ, Abshire JB, Afzal RS, Sun X (2001) Mars Orbiter Laser Altimeter—Experiment summary after the first year of global mapping of Mars. J Geophys Res 106(E10):23689–23722. https://doi.org/10.1029/2000JE001364

Snyder JP (1987) Map Projections – A working Manual. U.S. Geological Survey Professional Paper 1395, Washington

Stack KM, Edwards CS, Grotzinger JP, Gupta S, Sumner DY, Calef FJ, Edgar LA, Edgett KS, Fraeman AA, Jacob SR, Le Deit L, Lewis KW, Rice MS, Rubin R, Williams RME, Williford KH (2016) Comparing orbiter and rover image-based mapping of an ancient sedimentary environment, Aeolis Palus, Gale crater, Mars. Icarus 280:3–21. https://doi.org/10.1016/j.icarus.2016.02.024

Strobell ME, Masursky H (1990) Planetary nomenclature. In: Batson RM, Greeley R (eds) Planetary mapping. Cambridge University Print, Cambridge, pp 96–140

Tanaka K (1986) The stratigraphy of Mars. J Geophys Res 91(B13):E139–E158

Tanaka KL (1994) The venus geologic mappers' handbook. US Geological Survey Open-File Report 94-438

Tanaka KL, Skinner JA (2003) Mars: updating geologic mapping approaches and the formal stratigraphic scheme. In: Sixth International Conference on Mars, 3129

Tanaka KL, Anderson R, Dohm JM, Hansen V, Pappalardo R, Watters TR, Schultz RA (2010) Planetary structural mapping. In: Watters TR, Schultz RA (eds) Planetary tectonics. Cambridge University Press, Cambridge, pp 351–396

Tanaka KL, Skinner JA, Hare TM (2011) Planetary Geologic Mapping Handbook – 2011. US Geological Survey. https://planetarymapping.wr.usgs.gov/

Tanaka KL (2014a) The new geologic map of Mars: guiding research and education. EOS Trans AGU 95(38):341. https://doi.org/10.1002/2014EO380001

Tanaka KL, Skinner JA, Dohm JM, Irwin RP, Kolb EJ, Fortezzo CM, Platz T, Michael GG, Hare TM (2014b) Geologic Map of Mars. U.S. Geological Survey Scientific Investigations Map 3292, scale 1:20,000,000, pamphlet 43 p. https://dx.doi.org/10.3133/sim3292.

USGS (1995) Map Projections. U.S. Geological Survey, Paper Poster, Product No 16573, ISBN: 978-0-607-73058-6, Version 01/01/1993. https://store.usgs.gov/product/16573

van Gasselt S, Nass A (2011) Planetary mapping: the datamodel's perspective and GIS framework. Planet Space Sci 59:1231–1242. https://doi.org/10.1016/j.pss.2010.09.012 Special Issue: Planetary Mapping

van Gasselt S, Nass A (2014) On the concept and integration of geologic time in planetary mapping. In: Buchroithner M et al (eds) Cartography from pole to pole. Lecture notes in geoinformation and cartography. Springer, Heidelberg, pp 271–280. https://doi.org/10.1007/978-3-642-32618-9_20

Wilhelms DE (1972) Geologic mapping of the second planet. U.S. Geol. Surv. Interagency Report, Astrogeology 55:36

Wilhelms DE (1976) Geologic Map of the Oxia Palus quadrangle of Mars. USGS IMAP-895

Wilhelms DE (1987) The geologic history of the Moon. Washington, DC, USGS Professional Paper 1342.

Wilhelms DE (1990) Geologic mapping. In: Greeley R, Batson RM (eds) Planetary mapping. Cambridge University Press, Cambridge, pp 208–260

Williams DA (2016) NASA's Planetary Geologic Mapping Program: Overview. ISPRS - International Archives of the Photogrammetry, Remote Sensing and Spatial Information Sciences, Volume XLI-B4, pp. 519–520. https://doi.org/10.5194/isprs-archives-xli-b4-519-2016

Winchester S (2009) The map that changed the world. HarperCollins, New York

Methods in Planetary Topographic Mapping: A Review

Henrik Hargitai, Konrad Willner and Manfred Buchroithner

Abstract Elevation data can characterize geology, from global to local scales. For centuries, however, the only planetary topographic data were those of lunar peaks and craters. In the last few decades, several independent techniques have been developed to extract topographic information from diverse types of planetary datasets, which provide key information for the distinction and geologic interpretation of surface features. In this chapter, we discuss techniques to obtain, reconstruct, and visualize elevation data.

Keywords Topography · Hypsometry · Relief · Terrain · Elevation

1 Introduction

Topographic maps on Earth are complex, large-scale reference maps, which historically evolved from military maps. Additional to contour lines, these maps usually show/contain information of roads, landmarks, vegetation types, and detailed nomenclature. Planetary topographic maps, however, are thematic maps that primarily show information on surface relief.

The first planetary topographic maps used contour lines (Franz 1899; Wu 1976, 1978) suitable to display height differences on a plane surface. Modern "topographic maps" are often distributed as gridded digital elevation models (DEMs) that contain a height value for each grid point.

H. Hargitai (✉)
Eötvös Loránd University, Budapest, Hungary
e-mail: hargitaih@caesar.elte.hu

K. Willner
German Aerospace Center (DLR), Institute of Planetary Research, Berlin, Germany

M. Buchroithner
TU DRESDEN, Dresden, Germany

© Springer Nature Switzerland AG 2019
H. Hargitai (ed.), *Planetary Cartography and GIS*, Lecture Notes in Geoinformation and Cartography, https://doi.org/10.1007/978-3-319-62849-3_6

We note that in addition to “topographic map,” the terms hypsometric *(“height measurement”)* and altimetric map are also used in planetary cartography, although hypsometric is more commonly used by Russian planetary mappers.

1.1 Applications of Topographic Data

Datasets of planetary surface topography are tools to derive geodetic, geophisical, geologic and other mission critical information. The long-wavelength (more than hundreds of km) planetary topography contains information on the interior and thermal history of the planet (Smith et al. 1997) as long-wavelength features are influenced by convection or lithospheric thickness variation, in contrast to short-wavelength features (“landforms”) that are supported elastically (Nimmo et al. 2011; Barnett et al. 2000).

Global surface models are applied to establish reference surfaces such as best-fitting ellipsoids and spheres that are used to map project image data and as a height reference to derive controlled planetary maps.

On a regional scale, topographic information supports the geologic interpretation of the surface features. The number of geologic applications is only limited by the creativity of the researchers. Previously unnoticed surface features may be revealed based on topographic information where these are visually not apparent due to their large size and small relief (e.g., large basins, Frey et al. 2002, or large volcanoes, Spudis et al. 2013). Likewise, landscape or landform types may be directly extracted from topographic information (Stepinski et al. 2009, Bue and Stepinski 2005). In hydrologic modeling, topographic data analysis can be utilized to delineate watersheds, determine (paleo) flow directions (Smith et al. 2001), discharge rates, volumes of water, eroded and deposited material (Kleinhans 2005). Analysis of the compass direction of a slope (aspect) reveals connections between landforms and microclimate (Glines and Gulick 2014).

In landing site selection topographic analysis is applied for a small region. Important requirements here are the visibility between the instrument and Earth or a relay spacecraft, and the absence of high slopes and strong surface roughness. Engineering constraints prohibit landing in rock or boulder fields or in an area that is heavily cratered (Pajola et al. this volume, Golombek et al. 2012).

2 Techniques of Topographic Mapping

With the geodetic control established, large numbers of points, also called *point clouds,* to reconstruct the surface can be correctly located within the geodetic reference frame.

Point clouds can be determined by various observational methods. In practice, planetary missions carry a certain set of instrumentation and experiments that are designed to *primarily* meet *other* scientific objectives besides the determination of topographic information. Instruments *specifically* designed to measure topography include radar and laser *altimeters*. The High-Resolution Stereo Camera (HRSC) on the European Mars Express mission is also specifically designed to derive topographic information of Mars through stereophotogrammetry (Jaumann et al. 2007; Neukum and Jaumann 2004).

When a radar or laser altimeter is not part of a mission, surface relief can still be extracted from the analysis of data from active or passive imaging sensors. Where stereo (overlapping) image pairs are available (either from planned stereo image acquisition or by examining the existing image catalog), *stereogrammetry* can be used to extract topographic information from *geometric* data. Another technique, *photoclinometry*, can derive topographic information from the *radiometric* data of single images. Stereogrammetry and photoclinometry are complementary techniques. Stereogrammetry prefers sharp boundaries and large variations in inherent surface reflectance (albedo) and ignores topographic shading, whereas in photoclinometry (shape-from-shading) it is the albedo variations that are avoided or subtracted from the image. While photoclinometry can be used to determine the shape of irregular small objects (Gaskell et al. 2008; Gaskell 2011; Ernst et al. 2015; and Ermakov et al. 2014), this method is limited to smaller regions on planetary surfaces. Topographic maps of global coverage are usually generated from altimetric data (where they are available), or they can also be constructed with the well-established procedure of stereogrammetry (Preusker et al. 2015, 2017b; Willner et al. 2014).

We can distinguish different dimensionalities of height data. Shadow measurement, the first technique to identify elevations on the Moon, yields short, isolated topographic profiles; laser or radar altimetry instruments produce one-dimensional (point) data along lines (tracks). Point data can be connected to form topographic profiles (as first-order products, in 2D); stereogrammetry and radar interferometry lead to *three-dimensional* models of the surface, e.g., one height value per lateral location. In Earth-based applications, such boundary representation of relief is also frequently referred to as *2.5-dimensional* (2.5D) data, as only one height per lateral location can be assigned. As the data derived by stereogrammetry and radar interferometry have three-dimensional coordinates, we further refer to it as three-dimensional data.

The target body may also pose limitations of the potential measurement methods. An opaque atmosphere, such as that of Venus or Titan, cannot be investigated by optical imagery, but radar waves can reach their surfaces.

In the following sections, we give an overview of the techniques to derive topographic information.

2.1 Interpretation of Shadows and Lights

2.1.1 Cast Shadow Measurements

Measuring the length of a cast shadow is historically the first method to determine elevations on the Moon (Schröter 1791). The length of the shadow that is cast by an elevated feature is measured in relation to the known elevation of the Sun above the local horizon.

The height of a landform can be calculated using

$$h = l * \tan\alpha \quad (1)$$

where l is the shadow length on flat ground, α is the Sun's angle of elevation, and h is the unknown height.

For measuring the depths of small craters, shadow length measurement still may yield the most accurate results. Chappelow and Sharpton (2002) refined the shadow length measurement technique and extended it to crater shapes between cone-like and parabolic (hyperbolic) (Chappelow 2013). Previous methods yielded correct values only for parabolic craters and when the shadow boundary reaches near to the center of the crater.

On Io, where topographic information is difficult to obtain, Schenk et al. (2001) measured mountain heights applying several techniques, including shadow lengths. This study showed that cast shadow measurements are still viable in certain situations.

2.1.2 Terminator (Twilight)

Using this method, the position of the tip of a mountain behind the terminator line is measured when the Sun still or already illuminates it while the base of the mountain remains in shadow. Twilight measurements provided limited data on the minimum heights of some of the mountains of Io (Schenk et al. 2001).

2.1.3 Shape-from-Shading

The method of shape-from-shading that includes *photoclinometry* and *radarclinometry* (see below) obtains slopes from brightness variations in a single image (e.g., Beyer et al. 2003). It assumes that all brightness variations on the image are caused by topography (van Diggelen 1951). Put another way, brightness variations all represent different levels of shading and not inherent radiometric variations of the surface material. Image brightness is converted into slopes and from those into relative elevations by the integration of slopes. This method requires accurate photometric modeling. Potential sources of errors are any inherent brightness variations (albedo for photoclinometry, the electric properties of the surface for radar) and uncalibrated haze (for photoclinometry). Errors accumulate over long distances. Shape-from-shading yields only relative elevations, and the resulting DEM has the same pixel resolution as the input image. This single-pixel resolution is higher than that obtained by stereogrammetry (see below), but additional information is needed to determine a correct scale of the model.

2.1.4 Photoclinometry

Photoclinometry produces relative height data from a single image or multiple images, assigning elevation and slope values for each pixel (Kirk et al. 2003b). Photometric functions link brightness values to possible slope angles under different incidence and emission angles (McEwen 1991).

Shape-from-shading measurements may come in three different dimensionalities: *Point photoclinometry* can determine slope angles from point measurements; *profiling photoclinometry* produces an elevation profile from image brightness values along a line; and *area photoclinometry* builds a DEM from a series of profiles where cross-Sun elevation information is derived from iterative interpolation (Kirk et al. 2003b).

In *single-image* approaches, it is assumed that surface albedo (inherent brightness) is uniform. Variations of surface reflectivity will be misinterpreted as shading; therefore, a constant albedo should ideally characterize the data on which the method is applied. Topographic shading should exceed the contrast from albedo variations; therefore, areas with a relatively uniform albedo and images with large incidence angle should be selected (Kirk et al. 2003a). Effects of albedo could be identified when they correlate with color if color images are available.

However, results from single-image analysis are ambiguous even if albedo is constant over the area because in each pixel a single grayscale value represents two slope variables, the inclination of a surface in two directions (strike and dip/azimuth) (Lohse et al. 2006, Beyer et al. 2003).

For the photometric modeling, both surface and the atmosphere have to be modeled and corrected for surface albedo and atmospheric haze (Day et al. 1992). Haze can shift the height values. The brightness of atmospheric haze can be determined in shadowed regions where the radiance of the shadowed areas is subtracted from all pixels. However, sky illumination may vary and some images may not have sufficient shadowed pixels (Kirk et al. 2003b). On bodies where solar illumination is diffuse due to atmospheric haze, such as Titan, optical images would have too weak contrast for photoclinometry as shadows are not appreciable and shading is ambiguous (Radebaugh et al. 2007).

The usage of *multiple images* refines the results. Using multiple images, albedo variations may be separated from shading.

Several methods use a combination of image matching and shape-from-shading (Wöhler 2004; Dorrer et al. 2004; Kirk et al. 2003a; and Lohse et al. 2006). It is an option when multiple images are available but they are not suitable for stereo processing because of unfavorable base (separation of camera viewpoints)-to-height ratio, significantly different viewing directions (Lohse et al. 2006), or when the surface texture is too subdued for stereo matching. One such example is that of modeling small-scale features in, e.g., planetary rover images or when stereogrammetry fails due to featureless scenery (O'Hara and Barnes 2012).

In *photometric stereo analysis*, radiometric information of multiple images is used and no stereo parallax is measured. Although many methods use overlapping images, the term *stereophotoclinometry* is specifically applied to those techniques that reconstruct the surface relief or the shape of small, irregular bodies. In the former, multiple (>3) images with different illumination and viewing conditions are used to determine slope and albedo (Gaskell et al. 2008). *Multiresolution stereophotoclinometry* (Capanna et al. 2013) has been developed to reconstruct the shape of irregular bodies in an iterative process.

Conversely, when DEM, camera, and Sun angle information is available, shading and shadows can be removed and the inherent albedo of the surface can be reconstructed (Nefian et al. 2013).

2.1.5 Radarclinometry

Shape-from-shading can also be applied to radar image data (Thomas et al. 1991, Radebaugh et al. 2007), assuming that the strength of the radar echo is proportional to the inclination of the terrain toward the antenna. Shading is created by the radar illumination process. Since the electric properties of the surface are not known, this method can be applied only to small areas with can only be applied to small areas under the assumption of homogenneous properties. Radarclinometry can be combined with stereo radargrammetry to refine a low-resolution DEM (Thomas et al. 1991; Leberl 1993).

2.2 Planetary Limb Measurements

2.2.1 Limb Profiles

Limb images show the boundary between the planetary body and space and thus its topographic profile along this line. It is similar to a surface track obtained through e.g. light detection and ranging (LIDAR) (Elgner et al. 2012). Such observations are important to determine the global shape (Oberst et al. 2011) and long-wavelength topography (swells and basins) (Nimmo and Parsons 2011) in cases where no other relief information is available. Limb profiles can also provide more detailed information such as location, minimum height, and shape (silhouette) of mountains (e.g., Io: Schenk et al. 2001) or other features such as fault scarps (e.g., Miranda: Pappalardo et al. 1997) depending on the image quality and data reduction method.

The orientation of the limb tracks depends on the observation geometry of a certain image that depicts the limb and is arbitrarily oriented with respect to the the longitude and latitude grid of a body. In the case of Io observation, the predominant observation geometry led to north–south-oriented limb tracks, so latitudinal variations were more apparent from the profiles (Thomas et al. 1998). Localized depressions such as paterae are not detectable in limb profiles. "Off-limb" mountains and other high relief features (White et al. 2014) that appear from beyond the limb (below the horizon) should be ignored in constructing topographic profiles. However, they are useful to identify previously unidentified surface features such as peaks (see also Sect. 2.1.2).

2.2.2 Occultations

The shape and marginal elevation profile of a planetary disk also can be determined from occultation studies. Early studies used stellar occultations (Chugunov 1979; Rizvanov et al. 2007). Spacecraft observations use radio occultations where the radius of a body can be determined by recording the time immediately before and after the spacecraft orbital occultation entry (ingress) and exit (egress) points viewed from the Earth. After several orbits, these observations eventually result in data points scattered on the surface (Cain et al. 1972; Wu 1976, 1978; and Perry et al. 2011).

2.3 Atmospheric Pressure

Inferring topographic heights from atmospheric pressure is a *non-imaging* spectrum-based technique. In the seventeenth-century terrestrial observations, air pressure and later temperature of boiling water (Cajori 1929) were used as proxies for elevation. The large-scale topography of Mars was first determined from atmospheric pressure data calculated from Mariner 9 infrared (Herr et al. 1970) and ultraviolet (Barth and Hord 1971; Cintala et al. 1976; and Hord et al. 1972) spectral observations. Conversely, in TES datasets, local air pressure can be calculated using topographic data (Christensen et al. 2006). Air pressure on Mars is not a reliable parameter to determine absolute elevations because it changes seasonally and the air pressure reference surface shifts 1.5–2.5 km vertically over a Martian year. The former reference surface, at 6.1 mb air pressure, corresponds to −1600 m MOLA elevation at $L_s = 0°$ (Smith and Zuber 1998).

2.4 Altimetry

Altimetry is the *most direct method* of elevation measurement (Leberl et al. 1991). While visual limb observations result in continuous profiles at the margin of the body, altimetry techniques collect profiles from all over the surface in the form of a series of elevation point measurements along a line (track). The altimetry dataset can be transformed into a continuous representation of the surface relief by the interpolation of the profiles. Laser or radar pulses similarly can be used for altimetry. Laser and radar altimetry differ in frequency, but both use electromagnetic waves ($\sim$1000 nm and $\sim$10 cm wavelength, respectively).

Elevation data are calculated from the radar or laser signal time delay measurement of points, in other words, from the round-trip time of flight of individual electromagnetic wave pulses between the instrument and the surface (Zuber et al. 1992, Wu 1978). The range from the instrument to the surface (R) is

$$\mathrm{R} = \mathrm{c}\ (\Delta T/2) \tag{2}$$

where ΔT is the time of flight of the pulse and c is the speed of light (Zuber et al. 1997).

The precision of the range measurement is a function of the precision of the time measurements of the observation (time tags) (Wu 1978, Pettengill et al. 1980, Shan et al. 2005), which contains noise in the signal. For remotely obtained data, the time measurement also creates a connection between the altimetry measurement and the corresponding orbital location of the spacecraft that is determined from Earth-based tracking of the telemetry carrier signal (Pettengill et al 1980).

The precision of the topographic data obtained also depends on the accuracy of the location in latitude and longitude. This value is limited by several factors, including the knowledge of the instrument pointing (orientation), and the instrument position (e.g., spacecraft orbit, orbital altitude) and the knowledge of the shape of the target body (Smith et al. 1999, Zuber et al. 1997).

Altimetry data can be adjusted using height differences at orbit crossover points (e.g., Aharonson et al. 2004) or differenced altimeter technique where different altimeter measurements are calibrated against each other (Shum et al. 2012).

2.4.1 Laser Altimetry

A laser altimeter sends a pulse toward the target surface, and when the reflected pulse is received, the range counter stops. The resolution of laser altimetry is limited by the laser shot spacing along track, the spot size on the surface, and their orbit-to-orbit variations.

The backscattered *laser pulse* will have lower amplitude, and it will be spread in time compared to the transmitted pulse. The energy of the reflected laser pulse depends on *albedo* and *surface roughness*. Pulse spreading is caused by instrument effects, pointing jitter, RMS surface roughness, slope effects, and surface albedo variations (Zuber et al. 1997 and Harding et al. 1994). Ranging is also affected by the dispersion of the laser pulses by clouds (Neumann et al. 2001).

The Apollo program was the first to use laser altimetry extensively to investigate the selenodetic figure (Wollenhaupt and Sjogren 1972; Wollenhaupt et al. 1973;

and Wu et al. 1972a). Clementine LIDAR laser altimeter data generated a model of the lunar shape, including center of mass vs. center of figure difference and polar flattening (Zuber and Smith 1996 and Smith et al. 1997).

The Mars Orbiter Laser Altimeter (MOLA) project was proposed by Zuber et al. (1992) to provide a global, uniform, geodetically referenced topographic dataset of Mars. At that time, the highest precision global Mars DEM was determined from combined IR, UV, radar, occultation, and photogrammetry studies (Wu et al. 1986 and USGS 1989) and had contour lines of 1 km spacing with vertical errors of 500–2500 m. The MOLA project (Smith et al. 1999) provided two end products. (1) The original data are point (track) data (Precision Experiment Data Record, PEDR) with a 100 m horizontal and 3 m vertical precision, which add together as two-dimensional topographic profiles (Okubo et al. 2004, Neumann et al. 2001). (2) The three-dimensional product is a 128 px/deg (463 m/px) resolution gridded DEM (Experiment Gridded Data Record, MEGDR) that contains binned median PEDR altimetry values with gaps up to 12 km (commonly 1–2 km between PEDR tracks) that are filled by spline interpolation. However, different interpolation methods will yield different DEMs (Okubo et al. 2004).

The Lunar Orbiter Laser Altimeter (LOLA) is a laser altimeter designed to measure the shape of the Moon and to provide a geodetic grid that can be used as a reference surface (Burns et al. 2012). A specialty of LOLA is that the one laser beam is split into five beams, and five detectors record the time of flight separately. As a result, the ground pattern of the laser is shaped like a five on a dice. This densifies the resolution along the spacecraft ground track but also creates a track of two dimensions—having an extent also across track. Lunar polar areas are well covered by LOLA data tracks, and elevation models based on the altimetry data are much more accurate for the lunar *polar areas* in comparison to stereogrammetric results (Gläser et al. 2014).

Other laser altimeters include the Mercury Laser Altimeter (MLA) (Cavanaugh et al. 2007) and the NEAR–Shoemaker Laser Rangefinder (NLR) (Zuber et al. 1997, 2000). Future missions will also include laser altimeter experiments such as the Ganymede Laser Altimeter (GALA) as part of the Jupiter Icy Moons Explorer (JUICE) mission, or the BepiColombo Laser Altimeter (BELA) launched in 2018.

2.4.2 Radar Altimetry

Radar altimetry uses nadir-looking radar to conduct topographic measurements (Kirk et al. 1992). Earth-based (e.g., Arecibo, Goldstone, Green Bank) time delay radar profiles (Pettengill et al. 1969) have been largely superseded by space-based observations.

For a *radar echo*, a normalized echo power from a smooth surface has different shapes depending on off-nadir angle, progressively delayed and spread out with increasing off-nadir look angle. The strength of the radar echo also depends on surface roughness variations (Zebker et al. 2009).

The size of the footprint (i.e., spatial resolution) is a function of the time and frequency characteristics of the transmitted waveform and the distance between the instrument and the target surface (Pettengill et al. 1980).

For nadir-looking geometries, which are required in conventional radar altimetry, the leading edge of the echo yields the highest surface elevation because this is the initial encounter of the pulse with the terrain and it is returned to the receiver first. The mean value (centroid) of echo delay is more representative of the average elevation of the terrain. The spread of the delay is related to the spread of surface elevations. The return time is different from various off-nadir directions because of the different ranges even for similar elevations. On rough terrain, mean and peak values will diverge (Zebker et al. 2009). These two values, and their differences, can be visualized in radargrams and are useful tools of terrain analysis.

For Earth-based planetary radar altimetry observations, the elevation profile is produced along the locus of the sub-Earth point as the target planet rotates along its axis (Esposito et al. 1992). For orbiting spacecraft observations, the profile is made where the spacecraft points (it may be at nadir or off-nadir) in its orbit around the target body along the flight direction (Smith et al. 1999). Planetary rotations provide different opportunities for successive Earth-based radar measurements, from almost no change in the imaged area (Moon from Earth) to 1.5°/day change of the sub-Earth longitude (Venus) to almost a full rotation per day (Mars).

Photogrammetry or laser altimetry cannot be used on Venus because it is blanketed by an optically opaque atmosphere. Imaging, as well as altimetry, is only possible there using radar. Earth-based radar is able to produce both. However, due to the quasi-resonance between Venus and Earth, at inferior conjunction when Venus is closest to Earth, always the same hemisphere is facing the Earth providing only limited coverage of the surface for Earth-based observations (Pettengill et al. 1980). Global Venus altimetric maps have used data from radar altimeters aboard Venera 15 and 16 (Rzhiga 1987), Pioneer Venus (Pettengill et al. 1980), and aboard Magellan spacecraft (Leberl et al. 1991).

2.4.3 Radar Interferometry

Radar interferometry is an imaging range measurement technique that uses the principles of radar altimetry but results in a 3D model of the target surface.

Radar interferometry was introduced to measure the topography of the Moon, developed simultaneously by Zisk (1972b) and Shapiro et al. (1972).

In point measurements (altimetry), only time delay is analyzed. To obtain two-dimensional radar images, both time delay and Doppler frequency shift are analyzed. The round-trip echo (delay) defines a plane perpendicular to the radar–target body vector, and the Doppler-shift of the echo defines another that is parallel to it (Shapiro et al. 1972). The resulting radar picture is a projection of the target surface onto a plane normal to the Doppler axis and shows the surface reflectivity (backscatter) where brightness values are proportional to the radar echo strength.

Radar interferometry produces a three-dimensional view by the use of a radio interferometer receiver for the radar echo signals. The body is illuminated by the radar, and the echo is received simultaneously by two radar nearby receivers where the difference between the two signals' phase contains the elevation information. The value of phase change is proportional to the elevation.

For spacecraft radar interferometry, two images must be taken from nearly identical positions relative to the target to produce the required phase differences. If images are taken from two orbits several km apart, the two signals will not be coherent enough and phase differences cannot be observed (Leberl 1993).

Earth-based radar interferometry was used to determine the topography of the lunar poles and from simulated solar illumination conditions to identify permanently shadowed cold trap craters that may contain ice (Margot et al. 1999). The terrestrial Shuttle Radar Topography Mission (SRTM) mapped 80% of Earth's land surface. Data were acquired by Interferometric Synthetic Aperture Radar (InSAR). Data gaps occurred in regions with heavy vegetation canopy, flat terrain corresponding to calm water bodies where no meaningful reflection was detected due to scattering, and in radar shadows behind steep (>20°) slopes in mountainous terrain (Mukul et al. 2015 and Luedeling et al. 2007).

2.5 *Stereogrammetry*

Stereogrammetry obtains height values in the overlapping portion of two convergent images (a stereo pair) using their geometric information, with the technique of aerotriangulation. A key criterion is that the scene contain permanent surface features that can be identified on both images. Feature identification is optimal when the image has a strong high-frequency content (i.e., distinct details). Tie points that link the two images are ideally at high contrast boundaries (Becker et al. 2016). When stereogrammetry is applied to passive remote sensing images, it is called *photogrammetry*. This technique can also be used with radar images, where it is called *radargrammetry*, taking into account that the geometry of a radar image differs from that of the optical images.

2.5.1 Photogrammetry

Photogrammetry determines and elevation differences of depicted features based on measurements in images.

A height measurement in photographs is performed using the phenomenon of *relief displacement* (Blasius 1973). For example, the image pixel of a hilltop is situated at a different position relative to the image pixel of a point at the hill's foot on two images taken from different perspectives. One can simulate this effect by looking at close objects with alternating eyes.

The two overlapping image portions are linked using the method of *bundle adjustment*. To perform this least squares adjustment, image coordinates of corresponding points—tie points or conjugate points—need to be provided as observational data. With the increase of planetary image data, sparse automated matching is applied to determine tie point coordinates. Two types of matching routines are commonly used. *Feature-based matching* detects—as the name already suggests—features in images that can be described uniquely. *Area-based matching* on the other hand compares gray value differences between two patches from different images and requires sufficient texture in the image data to deliver reliable results.

The bundle adjustment corrects inconsistencies in the relative orientation of the images to each other and is followed by dense stereo matching to derive point clouds suitable for the generation of DEMs. Additional editing might be needed to remove undetected blunders manually.

A limitation of the photogrammetric method is that it needs to have a stereo pair and it can produce a horizontal resolution 3–5 times lower than the original images (Kirk et al. 2003b).

Stereo pairs can be produced from along-track (same orbit), e.g., HRSC on Mars Express (Jaumann et al. 2007), or across-track tilt. In the latter case, convergence angle between two images is achieved by tilting the spacecraft in one of the two different orbits off-nadir. Prominent examples for such an operation mode are the Mars Reconnaissance Orbiter HiRISE Camera (McEwen et al. 2007) and the Lunar Reconnaissance Orbiter Camera (Robinson et al. 2010).

While changes in the viewing angles are necessary for successful stereogrammetric image analysis, many other kinds of changes, for example severe changes in illumination and strong scale changes, should be avoided. Day et al (1992) and Cook et al. (1996) defined constraints for the selection of a stereo pair and its individual images.

For individual images, these criteria include that the Sun should be sufficiently high above the horizon so that no long shadows are present, and only high contrast images should be used.

High *phase angles* (the angle between the Sun and the spacecraft at a point on the surface) (Fig. 1) should also be avoided in general because Lunar observations show that at phase angles >100° some fresh craters appear dark and some maria appear bright (in contrast to their normal reflectance) and the original albedo cannot be reconstructed (McEwen 1996).

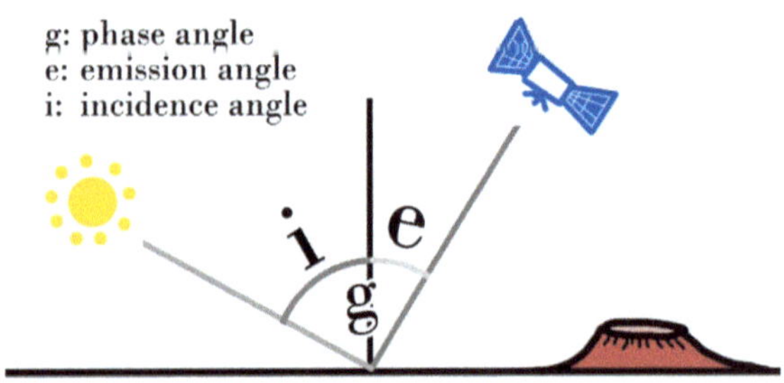

Fig. 1 Illustration of the concepts of different angles used in remote sensing. The spacecraft represents the observer

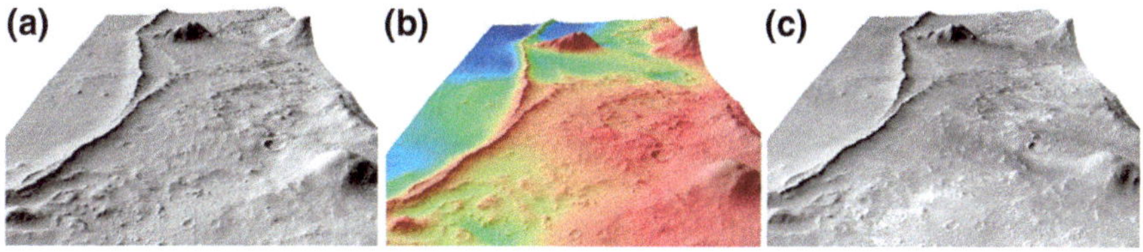

Fig. 2 Perspective views of sinuous ridges in Eastern Peta Crater, Mars, draped on corresponding DEM. **a** Synthetic hillshading produced from DEM (albedo features are not shown); **b** colored DEM with hillshading; **c** orthophotograph showing albedo features. HiRISE ESP_019466_1585. Original map scale is 2 m/px. The ridge is about 30 m high

Io poses specific problems because the photometric behavior of its surface includes contrast reversals and substantial contrast changes with phase angle, which makes stereophotogrammetry very difficult (Simonelli et al. 1997 and Kirk et al. 2003b) but locally possible with gaps between data (White et al. 2014). In addition, Io has low-contrast featureless volcanic plains, and imagery is corrupted by radiation noise, compression artifacts, and data dropouts (White et al. 2014).

For stereo pairs, the two images should have a resolution ratio smaller than 2.5:1; the two images should be taken with an identical spectral filter; the base-to-height (parallax–height) ratio should be between 0.01 and 1.0 (providing a strong stereo effect with not too large and not too small stereo convergence angles); and the difference in solar altitude should be <10° and solar azimuth <45°. Further criteria can be atmospheric clarity (for high image contrast) and minimal seasonal changes, for example, on Mars in polar regions with seasonal frost and defrosting features (Kirk et al. 1999) or due to dust devil tracks (Kirk et al. 2003a).

Preusker et al. (2017a) defined the following criteria for stereo processing of MESSENGER images: 0–10° difference in illumination, 15–45° stereo angle, 0–55° *emission angle*, 5–75° *incidence angle,* 5–180° phase angle. It is to be noted that all of these criteria need to be fulfilled such that phase angles 180°—i.e., directly looking toward the Sun—will never be reached due to the limitations of the other angles.

Another issue with stereo pairs is the so-called *jitter*, or high-frequency across-track spacecraft pointing oscillation that mimic elevation-related parallax and cause washboard-like artifacts (stripes) in DEMs (Kirk et al. 2003a). This effect becomes more apparent with an increasing resolution capability of the imaging sensor.

Potential errors in bundle adjustment may be caused by incorrect camera position and orientation data, too small overlap area, wrong sensor model, too few control points, etc.

The end product may be a stereo pair that can be viewed in 3D, a DEM on which measurements can be made, or a simulated perspective view where an orthophotograph (or the same DEM) is draped over a corresponding DEM with or without vertical exaggeration.

Today, the highest resolution planetary DEMs have one meter per pixel resolution and are generated from HiRISE stereo pairs using stereophotogrammetry

(Fig. 2) (Shan et al. 2005 and Li et al. 2011). This level of resolution can only be achieved locally. The HRSC aims at a global digital elevation model coverage with DEM mosaic resolution of 50 m per pixel (Gwinner et al. 2016).

Photogrammetry has some unconventional applications. A global network of ~94 million tie points were used to directly produce a global DEM of Mercury by interpolation of the irregularly spaced tie point dataset with bundle adjustment (Becker et al. 2016). Feature-based matching may also be paired with shape-from-shading (Lohse et al. 2006).

Because stereo processing is conducted on pixel level, it is inherently computationally demanding, an issue that is aggravated as the high-resolution image dimensions increase. Currently, the production of a high-resolution DEM requires several hours (or even days) of processing on state-of-the-art workstations. As a result, the batch-mode processing on the cloud has recently been suggested as an alternative (Tao et al. 2018).

2.5.2 Radargrammetry

Radargrammetry (Delay-Doppler Stereoscopy, Shapiro et al. 1972) is analogous to stereophotogrammetry (Zisk 1972b). Radar images, unlike optical ones, are not sensitive to solar illumination angles and can be produced on the night side and on surfaces behind visually nontransparent atmospheres.

Radargrammetric analysis uses images obtained from side-looking synthetic aperture radar (SAR, onboard, e.g., Magellan and Cassini spacecraft), where the sensor can look at off-nadir directions (e.g., east, west). Images have to be rectified in a process called dead-reckoning (Leberl et al. 1991). Stereo pairs may be constructed from overlapping same-side or opposite-side radar images (Leberl and Maurice 1922).

Brightness variations in a SAR image may be caused by: (1) surface roughness at scales comparable to radar wavelength: Rough surface have high backscatter; (2) topography: The surface that slopes toward the spacecraft appears brighter than away-facing slopes (radar illumination is received from the spacecraft's direction, and a shading effect is created similar to when a surface is illuminated from an angle); and (3) the variations of surface reflectivity at the radar's wavelength, corresponding to the surface electrical properties. Radar illumination incidence angle also influences brightness but is normalized in mosaicked image products (Kirk et al. 1992).

Sources of stereo matching errors in radar images include shadowing of the radar beam, differences in illumination direction, and radar-specific geometric distortions (layover: Slopes facing the radar are foreshortened, slopes facing away the radar are elongated). Illumination in SAR images changes from flight line to flight line (Thomas et al. 1991).

On Titan, a special technique called SARtopo estimated along-track topographic profile. In SARtopo, simultaneous radar footprints overlap. When processing the images, the overlapping parts match in brightness only when the images are using the correct local mean surface heights; otherwise, they show a banding effect (Fig. 2

in Stiles et al. 2009). On Titan, both radar image and altimetry data are available in narrow swaths. To generate a global map, these sparse data have to be interpolated. Lorenz et al. (2013) used the spline interpolating technique, also known from other geological gridded analyses, to produce a global, interpolated topographic model of Titan.

3 Final Products of Topographic Mapping

3.1 Elevation Models

Elevation models (Table 1) are spatial databases that store elevation data linked to locations, representing the relief of a surface. These models do not contain visualization information.

A *digital terrain model* (DTM) is a dataset of elevation points of the geologic surface (bare ground). The DTM defines body radius or geometric height above the body reference surface as a function of cartographic latitude and longitude (PDS 2008). A DTM may be in the format where sample points are interpolated/resampled into a regularly gridded dataset of similar-sized square pixels (DEM), or (rarely) a *triangulated irregular network* (TIN) that may include any number and arrangement of height points.

A *digital surface model* (DSM) represents the first echo elevations (from airborne altimetric observations) and includes vegetation and man-made objects. The term *digital elevation model* (DEM) is used in general for any data file of elevation values (USGS 2017a). In practice, a DTM or DEM is a georeferenced raster grayscale image, where the gray values represent the height at this geolocation.

On extraterrestrial bodies, in the absence of vegetation or artificial objects, the term DSM is not used as it represents essentially the same surface as a DEM.

On Earth, a DEM is usually limited to data over land while underwater elevations are stored in digital bathymetry models (DBMs), such as the General Bathymetric Chart of the Oceans (GEBCO) (Weatherall et al. 2015). The term for a land plus ocean floor model is *global topography model* or *global relief model* that has two versions: one with ice surface topography and one with bedrock topography (which differs in Greenland and Antarctica). The actual topography of the ocean's water surface can be measured with radar altimetry, and its difference from the geoid, due to ocean currents by about ± 1 m, is called the *mean dynamic ocean topography* (DOT) (Tapey et al. 2003).

Such distinctions of land, time-averaged liquid surface, underwater (bathymetric), and sub-ice relief (Buchroithner and Radig 2017a) may have relevance for planetary polar caps (e.g., Mars), the surface relief under global water oceans of icy moons or the liquid bodies of Titan (e.g., Mastrogiuseppe et al. 2014).

Some useful, how-to guides on DEM production can be found in Öhman 2013, USGS 2015, and USGS 2013.

Table 1 Global DEMs. They are available at USGS (Map-a-Planet, Astrogeology Science Center Websites), except those indicated with an asterisk

Body	Name	Source	Resolution m/px at equator	Method and accuracy	Reference
Mercury	MESSENGER Global DEM 665 m	MESSENGER MDIS NAC and WAC	665	Bundle adjustment	Becker et al. 2016
Venus	GTDR-SINUS.3;2	Magellan radar	4641	Radar altimetry	Magellan Project
Moon ±60 lat	LDEM GDR	LRO LOLA	118	Laser Altimeter ± 20 m horizontal and ± 1 m radius	Neumann et al. 2011
Moon	SLDEM2015	LOLA + Kaguya	59	Laser altimeter, coregistered to stereo 10 m horizontal and 4 m vertical	Barker et al. 2016
Moon ±67.9 lat	LROC WAC GDL100 (Global Lunar DTM 100 m)	LROC WAC	118	Stereogrammetry ±18, horizontal ±2 m vertical	Scholten et al. 2012
Moon*		Chang'E-1 CCD	500	Photogrammetry ±192 m, horizontal ±120 m vertical	Li et al. 2015
Mars	MEGDR	MGS MOLA	463	Laser altimetry ±100, horizontal ±3 m vertical	Smith et al. 2001
Mars	(planned)	HRSC controlled to MGS MOLA	50 (DTM) < 20 (mosaic)		Gwinner et al. 2016
Phobos	Phobos Mars Express HRSC DEM	Mars Express HRSC SRC	100	Stereogrammetry	Willner et al. 2014

(continued)

Table 1 (continued)

Body	Name	Source	Resolution m/px at equator	Method and accuracy	Reference
Ceres	Ceres Digital Terrain Model	Dawn HAMO Framing Camera 2	136.7	Stereogrammetry ±10 m vertical	Roatsch et al. 2016
Vesta	Vesta Dawn HAMO DTM	Dawn HAMO Framing Camera	92	Stereogrammetry	Preusker et al. 2014
Io* (Partial)		Galileo, Voyager, controlled to Galileo limb profiles	140–4080	Stereogrammetry	White et al. 2014
Titan	Titan Cassini Global Topographic Data Record	Cassini RADAR SARTopo and altimetry		Spline interpolation	Lorenz et al. 2013
Mimas, Enceladus, Tethys, Dione, Rhea, Iapetus*	(Data available on request)	Cassini	150–2000	Stereogrammetry and photoclinometry	Schenk 2008, 2010a, 2010b
Pluto (portion)	Pluto New Horizons Global DEM	New Horizons Long-Range Reconnaissance Imager (LORRI) and the Multispectral Visible Imaging Camera (MVIC)	300	Stereogrammetry	Moore et al. 2016
Charon (portion)	Charon New Horizons Global DEM	New Horizons Long-Range Reconnaissance Imager (LORRI) and the Multispectral Visible Imaging Camera (MVIC)	300	Stereogrammetry	Moore et al. 2016
P67 Churyumov Gerasimenko		OSIRIS NAC on Rosetta	5–10	Stereogrammetry	Preusker et al. (2015)

3.2 Elevation Model Visualizations

3.2.1 Shaded Relief

While shaded relief maps, unlike digital terrain models, provide no actual elevation data, at least not directly, they visualize the terrain relief with hillshade of uniform (synthetic) illumination. On a hillshade image, surface coloration and albedo variations are *excluded* and only the effects of topographic variations are shown.

Hillshading had numerous manual cartographic styles (Hargitai and Naß, this volume), all labor intensive, of which twentieth-century USGS/NASA planetary maps utilized the airbrush technique (Inge and Bridges 1976). Today, shaded relief maps are produced directly from DEM data using various automated digital methods (Horn 1981).

3.2.2 Contour Lines

A contour line connects equal elevations on a planar map. To provide the user with more meaningful information than just one elevation value, contour lines in certain intervals are used to indicate elevation differences. Where contour lines are far apart the area changes little in elevation, while a dense pattern of contour lines indicates a steep slope (cf. Figure 3).

Interval contour lines (1 km) were used as early as in the 1963 AMS topographic lunar maps combined with *gradient colors* where contour lines were derived by stereophotogrammetry from telescopic photography (AMS 1963). Post-Apollo lunar maps typically showed 300 m contours over *airbrush hillshading* or 10- to 100-m-interval red contours over Apollo photography on "topographic orthophotograph maps" or "topophotograph maps". A synthesis of various datasets into the first global topographic map of Mars visualized topography with 1-km interval contour lines and *shaded relief* (*Wu* 1976*)*. In early Mars maps, low-accuracy contour lines were modified to be in accordance with shaded relief renderings (Batson 1973). The first digital topographic model of Mars was produced by digitizing and interpolating hardcopy contour data (Kirk et al. 2000).

The first detailed maps of Venus showed 500 m contour lines based on Pioneer Venus, Earth-based (USGS 1984) and additional Venera radar data (GUGK 1988, Rzhiga 1987, with shaded relief background: USGS 1989).

With the introduction of all-digital photogrammetric systems, the primary end products became digital gridded elevation datasets; however, contour line products are still useful in visualizing general topography (e.g., USGS 2003), and special isolines, for example potential paleoshorelines.

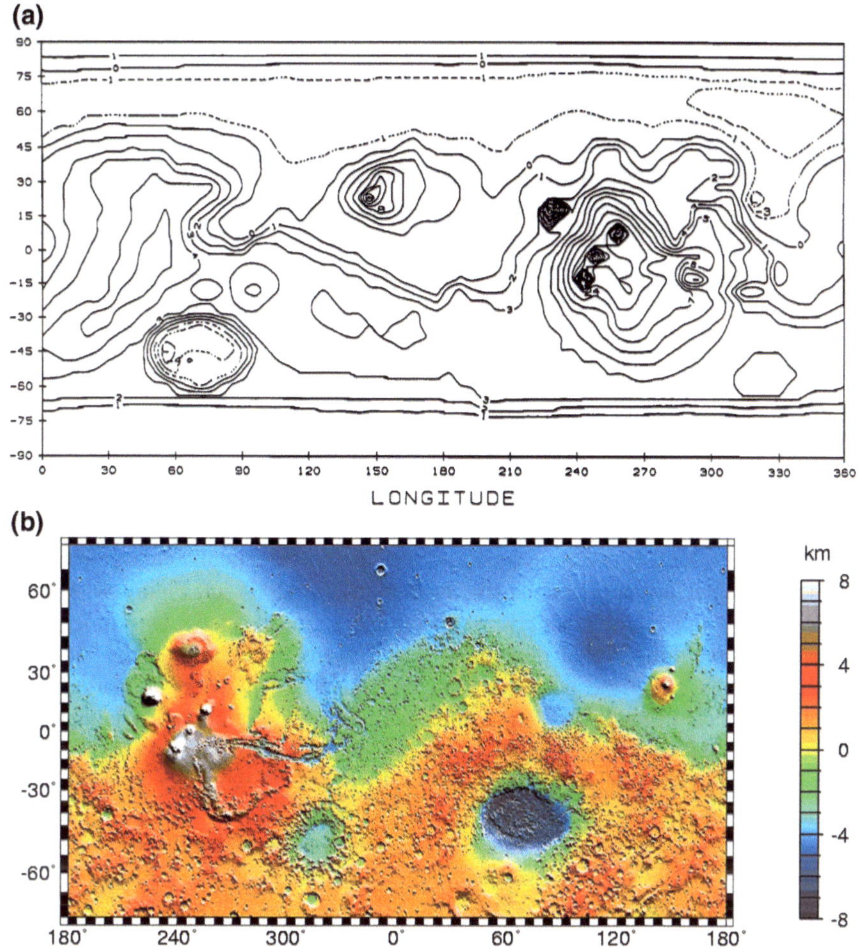

Fig. 3 Digital elevation model of Mars prior to the MOLA mission (a, Wu et al. 1986 cited in Zuber 1992) and the first publication of the MOLA map (b) (Smith 1999). Note change in cartography. Zero longitude is placed in the middle, and contour lines give way to color coding with shaded relief (the 1986 map also had a color-coded version, NASA 1990). Shaded relief is now derived from DEM values, and not manual interpretation of photographs, as before. The MOLA map is based on 333 million measurements (Smith et al. 2001)

3.2.3 Color-Coded Topography

In a color-coded (gradient tint) topographic map, particular colors are assigned to ranges of heights. The colors may reflect particular surface attributes of the given height band as on physical geographic maps of the Earth or suggest in general "a rocky landscape rather than water or vegetation," or they may be chosen to "lead the viewer to intuitively accept elevation information… and automatically see one elevation higher or lower than other elevations" (Aeschliman 1998). Alternatively, colors may be linearly scaled, e.g., rainbow colored. Rainbow coloring, the signature character of the MOLA map, "simply represents elevation changes and is not intended to imply anything about surface characteristics (e.g., past or current presence of water or ice)" (USGS 2017a).

Similar to contrast stretching on individual images, a single color scheme may be used through several different maps, or individual maps in a series may use the same, fully stretched color scheme, using Dynamic Range Adjustment, maximizing the visualization of height differences on each map view, but losing a fixed assignment of certain colors to certain heights.

3.2.4 Lenticular True-3D Cartography

In addition to the above-described, well-established techniques, a novel, analogue, visualization technique was developed within the Research and Development activities of ESA/DLR's *Mars Express* project. True-3D lenticular cartography goes far beyond planetary applications (Buchroithner et al. 2005, cf. Figure 4). Prior to its development, hardcopy displays were confined to mostly static, two-dimensional or pseudo-three-dimensional depictions. The various effects offered by lenticular foil technology allow the generation of flip image effects, short animations, and true-3D displays in hardcopy form. One advantage of this technology is that these effects can be spontaneously viewed with the unaided eye. Figure 4 (Buchroithner et al. 2005) tries to give an oblique view 2D impression of what can be achieved by means of lenticular cartography. The interested reader is kindly referred to view the cover of the October 2005 issue of *Photogrammetric Engineering & Remote Sensing* where the depicted scenes can be seen in true-3D.

The great advantage of the lenticular technology is the spontaneous perception of complex sets of information contained in a single image. Moreover, no additional aids like glasses (e.g., anaglyph approach, polarization, classical stereoscopy) or a special illumination (like for holography) are required for image separation. True-3D lenticular displays, in particular, allow height relations to be directly perceived and interpreted—all this being of importance for the analysis of planetary remote sensing imagery, but, in addition, the capability of the lenticular technology to interlace multiple scenes for the creation of various 2D effects offers a simple possibility for easy comparison of temporally and spatially changing objects and processes. Hence, the main application potential of lenticular displays lies in the fields of science, education, planning, and representation (Buchroithner 2012 (eds.), Buchroithner 2016, Knust and Buchroithner 2014).

The (lenticular) autostereoscopic (= "true-3D") image map of Valle Marineris on Mars, sized approximately 125 cm × 225 cm, was ordered by the Space Museum of the European Space Agency (ESA) in Noordwijk, NL, and put on display there in 2005. The globally first autostereoscopic relief representation of two overlaying

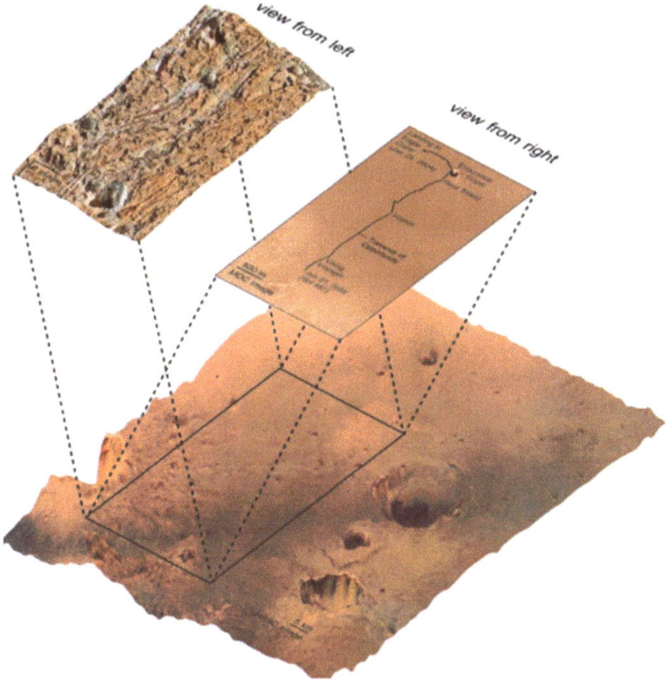

Fig. 4 Principle of perception of true-3D/flip lenticular display. (Buchroithner et al. 2005)

relief surfaces (ice and rockbed) in a lenticular foil map (Antarctica, published by Alfred Wegener-Institute/Germany) in 2015 is also a result of R&D work carried out within *Mars Express* project (Buchroithner and Radig 2017).

Acknowledgements The authors are grateful to R. Kirk and T. Hare for the helpful discussions and suggestions during the planning and reviewing of the manuscript.

References

Aeschliman R (1998) Topographic map of the Guinevere Planitia of Venus. V10 M 30/240 RTK, USGS

Aharonson O, Zuber MT, Smith, DE, Neumann, GA, Feldman WC, Prettyman TH (2004) Depth, distribution, and density of CO2 deposition on Mars. J Geophys Res 109(E05004). https://doi.org/10.1029/2003je002223

AMS (1963) Topographic Lunar Map. Army Map Service, Corps of Engineers, U.S. Army, Washington, D.C.

Barker MK, Mazarico E, Neumann GA, Zuberc MT, Haruyama J, Smith DE (2016) A new lunar digital elevation model from the Lunar Orbiter Laser Altimeter and SELENE Terrain Camera. Icarus 273:346–355. https://doi.org/10.1016/j.icarus.2015.07.039

Barnett DN, Nimmo F, McKenzie D (2000) Elastic thickness estimates for venus using line of sight accelerations from Magellan cycle 5. Icarus 146:404–419
Barth CA, Hord CW (1971) Mariner ultraviolet spectrometer: topography and polar cap. Science 173:197–201
Batson RM (1973) Cartographic Products from the Mariner 9 Mission. J Geophys Res 78 (20):4424–4435
Becker KJ et al (2016) First global digital elevation model of mercury. In: 47th lunar and planetary science conference abstract 2959
Beyer RA, McEwen AS, Kirk RL (2003) Meter-scale slopes of candidate MER landing sites from point photoclinometry. J Geophys Res 108(E12):8085. https://doi.org/10.1029/2003JE002120
Blasius KR (1973) A study of Martian topography by analytic photogrammetry. J Geophys Res 78 (20):4411–4423
Buchroithner M (2016) 3D visualisation of geodata: gimmick, hype or necessity? GIM Int. https://www.gim-international.com/content/article/3d-visualisation-of-geodata-gimmick-hype-or-necessity. Accessed 26 Feb 2018
Buchroithner MF, Gründemann T, Randolph KL, Habermann K (2005) Three in one: multiscale hardcopy depiction of the Mars surface in true-3D. Photogrammetric Eng Remote Sens Okt 71:1105–1108
Buchroithner MF, Radig L (2017) The first depiction of two superimposed geographical surfaces in one autostereoscopic map: Antarctica's topography and the southern ocean seafloor. Cartographic J World Mapp 1–10. http://dx.doi.org/10.1080/00087041.2017.1323153
Buchroithner M (eds) (2012) True-3D in Cartography: Autostereoscopic and Solid Visualization of Geodata. Lecture Notes in Geoinformation and Cartography, Springer, Heidelberg. https://doi.org/10.1007/978-3-642-12272-9
Bue BD, Stepinski TF (2005) Automated classification of landforms on Mars. Comput Geosci https://doi.org/10.1016/j.cageo.2005.09.004
Burns KN, Speyerer EJ, Robinson MS, Tran T, Rosiek MR, Archinal BA, Howington-Kraus E, et al (2012) Digital Elevation Models And Derived Products from LROC NAC Stereo Observations.International Archives of the Photogrammetry, Remote Sensing and Spatial Information Sciences, Volume XXXIX-B4
Cain DL, Kliore AJ, Seidel BL, Sykes MJ (1972) The shape of Mars from the Mariner 9 occultations Icarus 7:517–524
Cajori F (1929) History of determinations of the heights of mountains. Isis 12(3):482–514
Capanna C, Gesquière G, Jorda L, Lamy P, Vibert D (2013) Three-dimensional reconstruction using multiresolution photoclinometry by deformation. Vis Comput 29(6–8):825–835
Cavanaugh JF et al (2007) 2007 the mercury laser altimeter instrument for the MESSENGER mission. Space Sci Rev 131:451–480. https://doi.org/10.1007/s11214-007-9273-4
Chappelow JE, Sharpton VL (2002) An improved shadow measurement technique for constraining the morphology of simple impact craters. Meteorit Planetary Science 37:479–486
Chappelow JE (2013) Simple impact crater shape determination from shadows. Meteorit Planetary Sci 48(10):1863–1872. https://doi.org/10.1111/maps.12201
Christensen PR, Thorpe T, Fletcher Y (2006) Mars Global Surveyor Thermal Emission Spectrometer Data Processing User's Guide. http://tes.asu.edu/mgst/document/process.pdf
Chugunov IG (1979) System of absolute elevations in the marginal zone of the Moon derived from stellar occultations. Moon Planets 21:277–281
Cintala MJ, Head JW, Mutch TA (1976) Martian crater depth/diameter relationships: comparison with the Moon and Mercury. In: Proceedings of 7th lunar science conference, pp 3575–3787
Cook AC et al (1996) Clementine imagery: selenographic coverage for cartographic and scientific use. Planet Space Sci 44(10):1135–1148
Day T, Cook AC, Muller JP (1992) Automated digital topographic mapping techniques for Mars. In: International Archives of Photogrammetry and Remote Sensing, vol 29, no B4. American Society of Photogrammetry and Remote Sensing, Washington, DC, pp 801–808

Dorrer E, Mayer H, Ostrovskiy A, Reznik S, Neukum G et al (2004) De- and re-shading of Mars express HRSC image data for homogenization of map relief shading. Int Arch Photogrammetry Remote Sens Spatial Inform Sci 35(B4):1299–1303

Elgner S, Oberst J, Perry ME, Zuber MT, Robinson MS, Solomon SC (2012) Updates to Mercury limb topography from MESSENGER images. In: European planetary science congress, 23–28 September 2012, Madrid, Spanien

Ermakov A, Zuber MT, Smith DE, Raymond CA, Gaskell RW, Preusker F (2014) Assessment of Systematic Differences Between the SPC and SPG Vesta Shape Models Derived from the Dawn Mission. AGU Fall Meeting Abstracts 41

Ernst CM, Gaskell RW, Kahn EG, Barnouin OS, Roberts JH, Wilcomb KK (2015) Updated shape models of Phobos and Deimos from stereophotoclinometry. In: Lunar and planetary science conference, p 2753

Esposito PB, Banerdt WB, Lindal GF, Sjogren WL, Slade MA, Bills BG, Smith DE, Balmino G (1992) Gravity and Topography. In: Keiffer HH, Jakosky BM, Snyder CW, Matthews MS (eds) Mars. The University of Arizona Press, Tucson, pp 209–249

Franz J (1899) Die Figur des Mondes. Astron, Beobachtungen, Königsberg, Bd, p 38

Frey HV, Roark JH, Shockey KM, Frey EL, Sakimoto SEH (2002) Ancient lowlands on Mars. Geophys Res Lett 29(10):1384. https://doi.org/10.1029/2001GL013832

Gaskell RW, Barnouin-Jha OS, Scheeres DJ, Konopliv AS, Mukai T, Abe S, Saito J et al (2008) Characterizing and navigating small bodies with imaging data. Meteorit. Planetary Sci 43 (6):1049–1061

Gaskell RW (2011) Phobos Shape Model V1.0. NASA Planetary Data System 154

Gläser P, Scholten F, De Rosa D, Marco Figuera R, Oberst J, Mazarico E, Neumann GA, Robinson MS (2014) Illumination conditions at the lunar south pole using high resolution digital terrain models from LOLA. Icarus 243:78–90. https://doi.org/10.1016/j.icarus.2014.08.013

Glines NH, Gulick VC (2014) Comparative study of Gullies in Kaiser Crater on Mars. In: 45th lunar and planetary science conference, #2926

Golombek M, Grant J, Kipp D, Vasavada A, Kirk R, Fergason R, Bellutta P, Calef F, Larsen K, Katayama Y, Huertas A, Beyer R, Chen A, Parker T, Pollard B, Lee S, Sun Y, Hoover R, Sladek H, Grotzinger J, Welch R, Noe Dobrea E, Michalski J, Watkins M (2012) Selection of the Mars science laboratory landing site. Space Sci Rev 170:641–737. https://doi.org/10.1007/s11214-012-9916-y

GUGK (1988) 27 photomap sheets of Venus. Map series of Venus based on Venera 15 and Venera 16 observations. 1:5 M. GUGK SSSR, Moscow

Gwinner K, Jaumann R, Hauber E, Hoffmann H, Heipke C, Oberst J, Neukum G, Ansan V, Bostelmann J, Dumke A, Elgner S, Erkeling G, Fueten F, Hiesinger H, Hoekzema NM, Kersten E, Loizeau D, Matz KD, McGuire PC, Mertens V, Michael G, Pasewaldt A, Pinet P, Preusker F, Reiss D, Roatsch T, Schmidt R, Scholten F, Spiegel M, Stesky R, Tirsch D, van Gasselt S, Walter S, Wählisch M, Willner K (2016) The high resolution stereo camera (HRSC) of Mars express and its approach to science analysis and mapping for Mars and its satellites. Planetary Space Sci 126:93–138. https://doi.org/10.1016/j.pss.2016.02.014

Harding DJ, Bufton JL, Frawley JJ (1994) Satellite laser altimetry of terrestrial topography: vertical accuracy as a function of surface slope, roughness, and cloud cover. IEEE Trans Geosci Remote Sens 32:329–339

Herr KC, Horn D, McAfee JM, Pimentek GC (1970) Martian topography from the Mariner 6 and 7 infrared spectra. Astron J 75(8):883–894

Hord CW (1972) Mariner 6 and 7 ultraviolet spectrometer experiment: photometry and topography of Mars. Icarus 16:253–280

Horn BKP (1981) Hill-shading and the reflectance map. Proc. IEEE 69(1):14–47. http://people.csail.mit.edu/bkph/papers/Hill-Shading.pdf

Inge JL, Bridges PM (1976) Applied photointerpretation for airbrush cartography. Photogrammetric Eng Remote Sens 42(6):749–760

Jaumann R, Neukum G et al (2007) The high-resolution stereo camera (HRSC) experiment on Mars express: instrument aspects and experiment conduct from interplanetary cruise through the nominal mission. Planetary Space Sci (PSS) 55:928–952

Kirk R, Barrett J, Howington-Kraus E (1999) A Database of Viking Orbiter Image Coverage of Mars for Cartographic and Scientific Use LPS XXX #1857

Kirk RL, Horwington-Kraus E, Rosiek M (2000) Recent planetary topographic mapping at the USGS, Flagstaff: Moon, Mars, Venus, and beyond. Int Arch Photogrammetry Remote Sens XXXIII(Part B4)

Kirk RL, Soderblom LA, Lee EM (1992) Enhanced Visualization for Interpretation of Magellan Radar Data: Supplement to the Magellan Special Issue. J Geophys Res 97(El0):16371–16380

Kirk RL, Howington-Kraus E, Redding B, Galuszka D, Hare TM, Archinal BA, Soderblom LA, Barrett JM (2003a) High-resolution topomapping of candidate MER landing sites with Mars Orbiter Camera narrow-angle images. J Geophys Res 108(E12):8088. https://doi.org/10.1029/2003JE002131

Kirk RL, Barrett JM, Soderblom LA (2003a) Photoclinometry made simple. . .?, 2003b paper presented at ISPRS-ET Working Group IV/9 Workshop "Advances in Planetary Mapping 2003,"Int. Soc. for Photogramm. and Remote Sens., Houston, Tex., 2003a. http://astrogeology.usgs.gov/Projects/ISPRS/MEETINGS/Houston2003/ abstracts/Kirk_isprs_mar03.pdf

Kleinhans MG (2005) Flow discharge and sediment transport models for estimating a minimum timescale of hydrological activity and channel and delta formation on Mars. JGR 110(E12003). https://doi.org/10.1029/2005JE002521

Knust C, Buchroithner M (2014) Principles and terminology of true-3D geovisualisation. Cartographic J World Mapp. 51(3):191–202. https://doi.org/10.1179/1743277413Y.0000000038

Leberl F, Maurice K (1922) Stereo-mapping of planet venus from Magellan SAR images: a status repoRT. In: ISPRS XXIX, pp 795–800

Leberl F, Maurice K, Thomas J, Kober W (1991) radargrammetric measurements from the initial Magellan coverage of the planet venus. Photogrammetric Eng Remote Sens 57(l2):1561–1570

Leberl FW (1993) How to extract topographic information from Magellan radar images. SPIE Vol. 1943 State-of-the-Art Mapping. https://doi.org/10.1117/12.157164

Li C, Ren X, et al (2015) A new global and high resolution topographic map product of the Moon from Chang'E-2 image data. In: LPSC Conf., LPI No. 1903, #1638

Li R, Hwangbo J, Chen Y, Di K (2011) Rigorous photogrammetric processing of HiRISE stereo imagery for mars topographic mapping. IEEE Trans Geosci Remote Sens 49(7):2558–2572

Lohse V, Heipke C, Kirk RL (2006) Derivation of planetary topography using multi-image shape-from-shading. Planetary Space Sci 54:661–674

Lorenz RD et al (2013) A global topographic map of Titan. Icarus 225:367–377

Luedeling E, Siebert S, Buekert A (2007) Filling the voids in the SRTM elevation model — A TIN-based delta surface approach. ISPRS J Photogrammetry Remote Sens. https://doi.org/10.1016/j.isprs.jprs.2007.05.004

Margot JL, Campbell DB, Jurgens RF, Slade MA (1999) Topography of the lunar poles from radar interferometry: a survey of cold trap locations. Science 284:1658–1660

Mastrogiuseppe M et al (2014) The bathymetry of a Titan sea. Geophys Res Lett 41:1432–1437. https://doi.org/10.1002/2013GL058618

McEwen AS (1991) Photometric functions for photoclinometry and other applications. Icarus 92:298–311

McEwen AS (1996) A precise lunar photometric function (abs.). In: Lunar and Planetary Science, XXVII, Lunar and Planetary Institute, Houston, pp. 841–842

McEwen AS, Eliason EM, Bergstrom JW, Bridges NT, Hansen CJ, Delamere WA, Grant JA, Gulick VC, Herkenhoff KE, Keszthelyi L, Kirk RL, Mellon MT, Squyres SW, Thomas N, Weitz CM (2007) Mars Reconnaissance Orbiter's High Resolution Imaging Science Experiment (HiRISE). J Geophys Res 112, E05S02. https://doi.org/10.1029/2005JE002605

Moore JM et al (2016) The geology of pluto and charon through the eyes of new horizons. Science 351(6279):1284–1293. https://doi.org/10.1126/science.aad7055

Mukul M, Srivastana V, Mukul M (2015) Analysis of the accuracy of Shuttle Radar Topography Mission (SRTM) height models using international global navigation satellite system service (IGS) network. J Earth Syst Sci 124(6):1343–1357
Nefian AV, Alexandrov O, Kim T, Moratto Z, Beyer R (2013) Albedo reconstruction of the apollo metric camera zone. In: 44th Lunar and Planetary Science Conference, abs. 1649
Neukum G, Jaumann R (2004) HRSC: the high resolution stereo camera of mars express. In: Wilson A, Chicarro A (eds) ESA Special Publication, pp 17–35
Neumann GA (2011) Lunar Reconnaissance Orbiter Lunar Orbiter Laser Altimeter Reduced Data Record and Derived Products Software Interface Specification, version 2.42. http://imbrium.mit.edu/DOCUMENT/RDRSIS.PDF
Neumann GA, Rowlands DD, Lemoine FG, Smith DE, Zuber MT (2001) Crossover analysis of Mars Orbiter laser altimeter data. J Geophys Res 106(E10):23753–23768
Nimmo F, Parsons RA (2011) Long wavelength satellite topography from limb profiles: geophysical implications. In: 42nd LPSC 1523
O'Hara R, Barnes D (2012) A new shape from shading technique with application to Mars Express HRSC images. ISPRS J Photogrammetry Remote Sens 67:27–34. https://doi.org/10.1016/j.isprsjprs.2011.07.004
Oberst J et al (2011) Radius and limb topography of Mercury obtained from images acquired during the MESSENGER flybys. Planet Space Sci. https://doi.org/10.1016/j.pss.2011.07.003
Öhman T (2013) A beginner's guide to stereo-derived DEM production and analysis using ISIS, ASP, and ArcMap. http://www.lpi.usra.edu/lunar/tools/dems/Ohman_2013_ISIS-ASP-ArcMap_workflow.pdf
Okubo CH, Schultz RA, Stefanelli GS (2004) Gridding Mars orbiter laser altimeter data with GMT: effects of pixel size and interpolation methods on DEM integrity. Comput Geosci 30:59–72
Pappalardo RT, Reynolds SJ, Greeley R (1997) Extensional tilt blocks on Miranda: evidence for an upwelling origin of Arden Corona. J Geophys Res 102(E6):13369–13379
PDS (2008) PDS Standards Reference, Chapter 2. Cartographic Standards. Draft: v. 4.3, 12.10.08. https://pds.jpl.nasa.gov/documents/sr/stdref3.7/Chapter2_20081210_v4_3_final_rev.pdf
Perry ME et al (2011) Measurement of the radius of Mercury by radio occultation during the MESSENGER flybys. Planet Space Sci. https://doi.org/10.1016/j.pss.2011.07.022
Pettengill GH, Counselman CC, Rainville LP et al (1969) Radar measurements of martian topography. Astron J 74(3):461–482
Pettengill GH, Eliason E, Ford PG, Loriot GB, Masursky H, McGill GE (1980) Pioneer venus radar results: altimetry and surface properties. J Geophys Res 85(A13):8261–8270
Preusker F, Scholten F, Matz K-D, Roatsch T, Kersten E, Raymond CA, Russell CT (2014) Global shape of vesta from dawn FC stereo images. In: Lunar planetary science conference, Houston, TX, USA 2027
Preusker F, Scholten F, Matz KD, Roatsch T, Willner K, Hviid SF, Knollenberg J, Jorda L, Gutiérrez PJ, Kührt E, Mottola S, A'Hearn MF, Thomas N, Sierks H, Barbieri C, Lamy P, Rodrigo R, Koschny D, Rickman H, Keller HU, Agarwal J, Barucci MA, Bertaux JL, Bertini I, Cremonese G, Da Deppo V, Davidsson B, Debei S, De Cecco M, Fornasier S, Fulle M, Groussin O, Güttler C, Ip WH, Kramm JR, Küppers M, Lara LM, Lazzarin M, Lopez Moreno JJ, Marzari F, Michalik H, Naletto G, Oklay N, Tubiana C, Vincent JB (2015) Shape model, reference system definition, and cartographic mapping standards for comet 67P/Churyumov-Gerasimenko—Stereo-photogrammetric analysis of Rosetta/OSIRIS image data*. A&A 583. https://doi.org/10.1051/0004-6361/201526349
Preusker F et al (2017a) The global meter level shape model of comet 67P/Churyumov-Gerasimenko. Astronomy Astrophysica 607. https://doi.org/10.1051/0004-6361/201731798
Preusker F, Stark A, Oberst J, Matz K.-D, Gwinner K, Roatsch T, Watters TR (2017b) Toward high-resolution global topography of Mercury from MESSENGER orbital stereo imaging: A prototype model for the H6 (Kuiper) quadrangle. Planet Space Sci 142:26–37. https://doi.org/10.1016/j.pss.2017.04.012

Radebaugh J, Lorenz RD, Kirk RL, Lunine JI, Stofan ER, Lopes RMC, Wall SD, the Cassini Radar Team (2007) Mountains on Titan observed by Cassini Radar. Icarus 192(1):77–91
Rizvanov NG, Nefed'ev YA, Kibardina MI (2007) Research on selenodesy and dynamics of the moon in Kazan. Solar Syst Res 41(2):140–149
Roatsch T, Kersten E, Matz K-D, Preusker F, Scholten F, Jaumann R, Raymond CA, Russell CT (2016) High-resolution ceres high altitude mapping orbit atlas derived from dawn framing camera images. Planetary Space Sci (in press)
Robinson MS, Brylow SM, Tschimmel M, Humm D, Lawrence SJ, Thomas PC, Denevi BW, Bowman-Cisneros E, Zerr J, Ravine MA, Caplinger MA, Ghaemi FT, Schaffner JA, Malin MC, Mahanti P, Bartels A, Anderson J, Tran TN, Eliason EM, McEwen AS, Turtle E, Jolliff BL, Hiesinger H (2010) Lunar Reconnaissance Orbiter Camera (LROC) Instrument overview. Space Sci Rev 150:81–124. https://doi.org/10.1007/s11214-010-9634-2
Rzhiga ON (1987) Venera-15 and -16 spacecraft—Images and maps of Venus. COSPAR, IAU, Plenary Meeting, 26th vol 7, no 12, pp 269–278
Schenk P, Hargitai H, Wilson R, McEwen A, Thomas P (2001) The mountains of Io' Global and geological perspectives from Voyager and Galileo. J Geophys Res 106(E12):33201–33222
Schenk P (2008) Cartographic and topographic mapping of the icy satellites of the outer solar system. In: ISPRS XXXVII, Commission IV, WG IV/7
Schenk P (2010a) Global topographic mapping of saturn's midsize icy satellites: system-wide thermal and impact effects. Amer. Astron. Soc., D.P.S. meeting 42, abstr. 9.16
Schenk P (2010b) New Moons – First Global Topographic Maps of (Saturn's) Icy Moons. http://stereomoons.blogspot.com/2010/11/new-moons.html
Scholten F, Oberst J, Matz K-D, Roatsch T, Wählisch M, Speyerer EJ, Robinson MS (2012) GLD100: the near-global lunar 100 m raster DTM from LROC WAC stereo image data. J Geophys Res 117:E00H17. https://doi.org/10.1029/2011je003926
Schröter JH (1791) Selenotopographische Fragmente, vol 1. Lilienthal, J.G. Rosenbusch, Univ. Buchdr
Shan J, Yoon J, Lee DS, Kirk RL, Neumann GA, Acton CH (2005) Photogrammetric analysis of the mars global surveyor mapping data. Photogrammetric Eng Remote Sens 71:97–108
Shapiro II, Zisk SH, Rogers AEE, Slade MA, Thompson TW (1972) Lunar topography: global determination by Radar. Science 178:939–948
Shum CK et al (2012) Lunar topography model determined by integrating laser altimetry from multiple orbiters. LPS XLIII:2407
Simonelli DP, Veverka J, McEwen AS (1997) Io: Galileo evidence for major variations in regolith properties. Geophys Res Lett 24:2475–2478
Smith DE, Zuber MT (1998) The relationship between MOLA northern hemisphere topography and the 6.1-Mbar atmospheric pressure surface of Mars. Geophys Res Lett 25(24):4397–4400
Smith DE, Zuber MT, Neumann GA, Lemonie FG (1997) Topography of the Moon from the Clementine lidar. J Gephys Res 102(E1):1591–1611
Smith DE, Zuber MT, Solomon SCRJ, Phillips JW, Head JB, Garvin WB, Banerdt DO, Muhleman GH, Pettengill GA, Neumann FG, Lemoine JB, Abshire O, Aharonson CD, Brown SA, Hauck AB, Ivanov PJ, McGovern HJ, Zwally TC Duxbury (1999) The global topography of Mars and implications for surface evolution. Science 284:1495–1503
Smith DE, Zuber MT, Frey HV, Garvin JB, Head JW, Muhleman DO, Pettengill GH, Phillips RJ, Solomon SC, Zwally HJ, Banerdt WB, Duxbury TC, Golombek MP, Lemoine FG, Neumann GA, Rowlands DD, Aharonson O, Ford PG, Ivanov AB, Johnson CL, McGovern PJ, Abshire JB, Afzal RS, Sun X (2001) Mars Orbiter Laser Altimeter—Experiment summary after the first year of global mapping of Mars. J Geophys Res 106 (E10):23689–23722
Spudis PD, McGovern PJ, Kiefer WS (2013) Large shield volcanoes on the Moon. J Geophys Res Planets 118:1063–1081. https://doi.org/10.1002/jgre.20059
Stepinski TF, Mendenhall MP, Bue BD (2009) Machine cataloging of impact craters on Mars. Icarus 203. https://doi.org/10.1016/j.icarus.2009.04.026

Stiles BW, Hensley S, Gim Y, Bates DM, Kirk RL, Hayes A, Radebaugh J, Lorenz RD, Mitchell KL, Callahan PS, Zebker H, Johnson WTK, Wall SD, Lunine JI, Wood CA, Janssen M, Pelletier F, West RD, Veeramacheneni C, The Cassini RADAR Team (2009) Determining Titan surface topography from Cassini SAR data. Icarus 202:584–598
Tao Y, Muller J-P, Sidiropoulos P, Xiong S-T, Putri ARD, Walter SHG, Veitch-Michaelis J, Yershov V (2018) Massive stereo-based DTM production for Mars on cloud computers. Planetary Space Sci 154:30–58
Tapley BD, Chambers DP, Bettadpur S, Ries JC (2003) Large scale ocean circulation from the GRACE GGM01 Geoid Geophys Res Lett 30(22). https://doi.org/10.1029/2003gl018622
Thomas J, Kober W, Leberl F (1991) Multiple image SAR Shape-from-Shading. Photogrammetric Eng Remote Sens 57(1):51–59
Thomas P, Davies M, Colvin T et al (1998) The shape of Io from Galileo limb measurements. Icarus 135:175–180
USGS (1984) Topographic map of Venus. VRM Planning Chart. Atlas of Venus. I-1562. V50 M 6/60RT Scale 1:50 M
USGS (1989) Maps of Part of the Northern Hemisphere of Venus. Map I-2041. Miscellaneous Investigations Series. 1:15 M
USGS (2003) Color-Coded Contour Map of Mars. Geologic investigations series I–2782
USGS (2013) Stereo Processing of Planetary Stereo Imagery using ISIS3 and SOCET SET® aPrimer. https://astropedia.astrogeology.usgs.gov/download/Docs/Photogrammetry/PGF_Primer.pdf
USGS (2015) Stereo Processing of HiRISE Imagery Using SOCET SET® Astrogeology Science Center, USGS. ftp://pdsimage2.wr.usgs.gov/pub/pigpen/tutorials/SS4HIRISE_JULY2015/HiRISE_StereoProcessing_Tutorial_July_2015.pdf
USGS (2017a) Mimas Voyager Image Control Network (RAND)
USGS (2017b) Control Networks. https://astrogeology.usgs.gov/maps/control-networks
van Diggelen J (1951) A photometric investigation of the slopes and heights of the ranges of hills in the maria of the Moon. Netherlands Astron Inst Bull 11:283–289
Weatherall P, Marks KM, Jakobsson M, Schmitt T, Tani S, Arndt JE, Rovere M, Chayes D, Ferrini V, Wigley R (2015) A new digital bathymetric model of the world's oceans. Earth Space Sci 2:331–345. https://doi.org/10.1002/2015EA000107
White OL, Schenk PM, Nimmo F et al (2014) A new stereo topographic map of Io: implications for geology from global to local scales. J Geophys Res Planets 119:1276–1301. https://doi.org/10.1002/2013JE004591
Willner K, Shi X, Oberst J (2014) Phobos' shape and topography models. Planet Space Sci 102:51–59. https://doi.org/10.1016/j.pss.2013.12.006
Wöhler C (2004) Shape from shading under coplanar light sources. In: Rasmussen CE, Bülthoff HH, Giese MA, Schölkopf B (eds) Proceedings of the 26th DAGM symposium, pattern recognition, Tübingen. Springer, Heidelberg pp 278–285
Wollenhaupt WR, Sjogren WL (1972) Apollo 16 Preliminary Science Report. NASA 30-1–30-5
Wollenhaupt WR, Sjogren WL, Lingenfelter RE et al (1973) Apollo 17 Preliminary Science Report. NASA SP-330 33-41
Wu SSC (1976) Mars synthetic topographic mapping. Dissertation, University of Arizona
Wu SSC (1978) Mars synthetic topographic mapping. Icarus 22:417–440
Wu SSC, Schafer FJ, Jordan R et al (1972a) Photogrammetry of Apollo 16 Photography. Apollo 15 Preliminary Science Report. 25-36–25-48
Wu SSC, Jordan R, Schafer FJ (1986) Mars global topographic map 1:15,000,000 scale, NASA Tech. Memo., TM-88383, 614-617
Zebker HA, Gim Y, Callahan P, Hensley S, Lorenz R, The Cassini Radar Team (2009) Analysis determining titan surface topography from titan radar altimeter echoes. Icarus 200, 240–255
Zisk SH (1972) A new, earth-based radar technique for the measurement of lunar topography. Moon 4(3–4):296–306

Zuber MT, Smith DE, Cheng AF, Cole TD (1997) The NEAR laser ranging investigation. J Geophys Res 102:23761–23773

Zuber MT et al (1992) The mars observer laser altimeter investigation. J Geophys Res 97(E5): 7781–7797

Zuber MT et al (2000) The Shape of 433 Eros from the NEAR-Shoemaker Laser Rangefinder Science 289, 2097–2101. https://doi.org/10.1126/science.289.5487.2097

Zuber MT, Smith DE (1996) Topographic mapping of the Moon. Int Arch Photogrammetry Remote Sens 31(B4):1011–1015

Planetary Mapping for Landing Sites Selection: The Mars Case Study

Maurizio Pajola, Sandro Rossato, Emanuele Baratti and Alexandre Kling

Abstract The selection of a landing site on a planetary body is a multistep process that involves both the fulfillment of several engineering constraints and the accomplishment of scientific requirements. In this chapter, we will show how the simultaneous production and exploitation of different GIS maps depicting these criteria are pivotal in the landing site selection. Indeed, all of such constraints are presently evaluated through the use of GIS-based software. To show this, we will focus on the Martian site identification outline, providing multiple real examples taken from two ongoing study cases, i.e., the Simud Vallis landing site proposed by Pajola et al. (Icarus 268:355–381, 2016a) for the ESA ExoMars rover and the Eridania landing site proposed by Pajola et al. (Icarus 275:163–182, 2016b) for the NASA Mars 2020 landing site selection.

Keywords Planetary mapping · Landing site · Mars · Engineering constraints · Robotic exploration

M. Pajola (✉)
National Institute for Astrophysics, Astronomical Observatory of Padova, Vicolo Osservatorio 5, 35122 Padova, Italy
e-mail: maurizio.pajola@gmail.com

M. Pajola · A. Kling
NASA Ames Research Center, Moffett Field, CA 94035, USA

S. Rossato
University of Padova Geosciences Department, 35131 Padova, Italy

E. Baratti
Department of Civil, Chemical, Environmental and Material Engineering, University of Bologna, 40136 Bologna, Italy

A. Kling
Universities Space Research Association, Mountain View, CA 94043, USA

© Springer Nature Switzerland AG 2019

H. Hargitai (ed.), *Planetary Cartography and GIS*, Lecture Notes in Geoinformation and Cartography, https://doi.org/10.1007/978-3-319-62849-3_7

1 Introduction

Selecting a landing site on the surface of Mars, and in general on planetary surfaces, is a long (>4 years (Golombek et al. 2012)) and difficult task that is characterized by several engineering and scientific issues. The engineering requirements include all the constraints related to the landing and roving safety, such as the latitude and the elevation of the site, the ellipse landing dimensions, the surface slopes, the rock abundances, the thermal inertia, the albedo, the radar reflectivity, and the atmospheric properties. The scientific constraints' goal, on the contrary, is to answer to those questions raised by the scientific community such as, for the case of Mars, to search for past habitats where organisms could have spread, leaving signatures on the present-day geology and mineralogy. While scientists aim to observe specific locations, maximizing the scientific return of the mission, on the other hand, engineers have to figure out how to reach them. Therefore, it is not unusual in the site selection process that scientifically compelling regions must be ruled out after the quantification of unacceptable risks that could arise during the landing or the roving phase. Eventually, the final selected landing site is the best attainable compromise between scientific and engineering requirements.

2 The Engineering Constraints

Engineering criteria that have to be fulfilled during the landing and the roving phase are many and may vary from mission to mission depending on the entry descent and landing (EDL) parameters, as well as on the rover design. In Table 1, we present those criteria that ESA and NASA defined for the ExoMars 2020 and the Mars 2020 rovers.

Here below, we describe each of these criteria providing GIS maps depicting them:

(a) **Latitude and longitude** (i.e., the geographical coordinates): If rovers are powered by solar panels, the landing site areas are always defined in a narrow latitude band located around the equator, i.e., where the maximum day insolation occurs. However, when the rover is powered by a radioisotope thermoelectric generator (RTG), a wider latitude band is conceivable, given that the insolation conditions are no more restricting (Fig. 1). Moreover, the surface platform and rover design drive the latitude band thickness, given that the illumination conditions resulting from the latitude band choice can result in challenging thermal conditions.

(b) **Elevation:** It is required for sufficient atmosphere braking during EDL; see Table 1. The elevation values always refer to the Mars Global Surveyor-Mars Orbiter Laser Altimeter (MOLA) geoid (Smith et al. 2001).

Table 1 Summary of the ExoMars 2020 and the Mars 2020 surface/terrain engineering constraints as from the ExoMars landing site call (http://exploration.esa.int/mars/53462-call-for-exomars-2018-landing-site-selection/) and from Golombek et al. (2012)

	ExoMars 2020	Mars 2020
Landing latitude	5°S–25°N	30°S–30°N
Landing elevation	⩽−2 km MOLA	⩽+ 0.5 km MOLA
Landing ellipse dimensions	104 × 19 km	25 × 20 km (nominal)
(major × minor axis)		18 × 14 km (range trigger)
		13 × 7 km (range trigger)
Landing ellipse orientation	83–127°	Roughly east–west
Slopes at 2 to 10 km length scale	⩽3.0°	⩽5.71°
Slopes at 333 m length scale	⩽8.6°	⩽5.71°
Slopes at 7 m length scale	⩽12.5°	⩽25.0–30.0°
Slopes at 2 m length scale	⩽15.0°	⩽25.0–30.0°
Rock abundance	K ⩽7%	K ⩽ 12%, locally K ⩽ 20%
Thermal inertia	⩾150 J m^{-2} $s^{-0.5}$ K^{-1}	⩾ 100 J m^{-2} $s^{-0.5}$ K^{-1}
Albedo	0.1 ⩽ Albedo ⩽ 0.26	Albedo ⩽ 0.25
Radar reflectivity	−15 dB ⩽ RR ⩽ 27.5 dB	−20 dB ⩽ RR ⩽ 15.0 dB

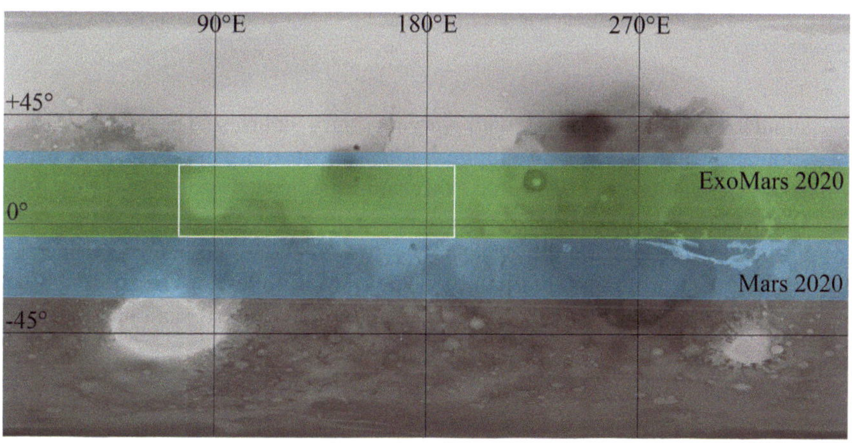

Fig. 1 Latitude bands requested for the ExoMars (green) and the Mars 2020 (blue) rovers overlaid on a Mars MOLA map. *The evaluation here shows only insolation requirements*. The Mars 2020 rover is powered by an RTG, while ExoMars is not: hence the wider permitted latitude range containing the ExoMars' one. The white box shows the areal extent of Fig. 3

(c) **Landing ellipses dimensions and orientation:** They are dictated by the spacecraft entry angle into the planet's atmosphere, the atmospheric density, its drag, and the entry mass (Fig. 2). Given the unavoidable uncertainties in their estimation, it is impossible to previously know exactly the spacecraft's final landing point. For this reason, in order to predict the landing area, numerical

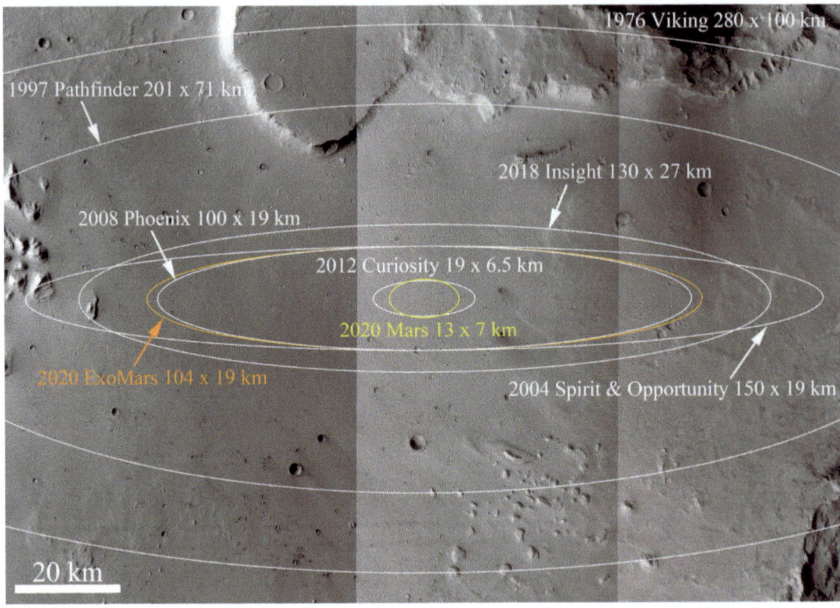

Fig. 2 Different sizes of the landing ellipses from Viking Landers to the next NASA Mars 2020 rover overlaid on the Simud landing area proposed by Pajola et al. (2016a) for the ExoMars mission

simulations are performed with varying entry courses. The result of such simulation processes is an elliptical footprint on the planet's surface with a specific azimuth orientation, i.e., the landing ellipse. Inside this area, all the engineering constraints must be verified.

(d) **Slopes at different length scales:** They are generally computed on the GIS environment from digital terrain models (DTMs), following the Burrough and McDonnell (1998) method. They are required to ensure (i) slant and incidence

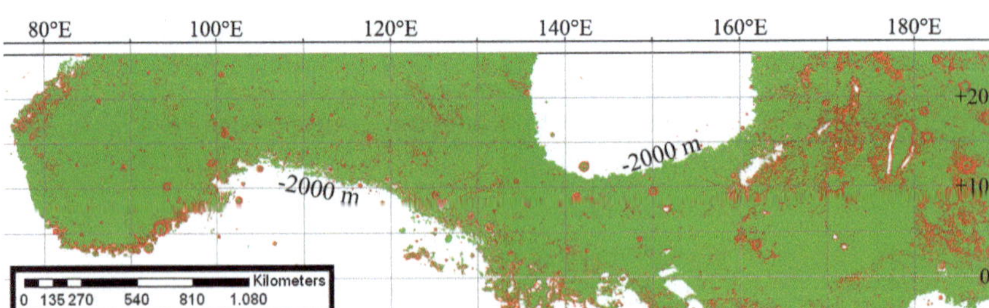

Fig. 3 A combined map prepared for the ExoMars landing site selection showing the Mars surface below the −2000 m MOLA constraint (white areas do not fulfill such requirement) and that does not exceed 3° slope at 2 km length scale (in red the no-go areas). The extent of this image is presented in the white box in Fig. 1

compatible with the onboard radar (2–10-km length scale (Fig. 3)), (ii) proper fuel consumption during the powered descent (330-m length scale), (iii) proper altitude error in the touchdown phase (7-m length scale), and (iv) stability during landing (2-m length scale).

(e) **Rock abundances:** They are needed to evaluate the probability of crashing on a rock during landing and drive the rover traversability performance once on the ground. In order to extract the rock abundances, commonly called *K*, the Viking Infrared Thermal Mapper (IRTM) data products (Christensen 1986a; 1986b) are first used. Despite being the IRTM estimates at a spatial resolution with 1 degree bins (60 km), comparisons with rock counts at the surface at much smaller scales have been successful (Moore and Jakosky 1989; Christensen et al. 1992; Golombek et al. 2003, 2005; Golombek 2008) both in terms of the total area covered by rocks and their diameter distribution, matching the exponential model for rocks greater than 10 cm. Clearly, when available, more detailed rock abundance maps based on the Mars Global Surveyor-Thermal Emission Spectrometer (TES, 7.4 km resolution) data products are used (Nowicki and Christensen 2007), Fig. 4a. The resulting cumulative fraction of surface covered by rocks with a height $\geq$ H is then derived, Fig. 4b (Golombek and Rapp 1997; Golombek et al. 2003; Golombek 2008). Eventually, when the High Resolution Imaging Science Experiment (HiRISE) images (scale of 0.25 m/pixel, McEwen et al 2007) of a landing site are taken, it is possible to perform manual or automatic boulder counting on test areas (Pajola et al. 2017; Golombek et al. 2008) to have an extremely precise evaluation of the *K* values (Fig. 5).

(f) **Dust coverage:** It is required to have free loose material/dust deposits for the rover not to be stuck while moving. For this purpose, it is currently used the 3.5 km resolution map of the Dust Cover Index (DCI) produced by Ruff and Christensen (2002) from the TES data. The lowest quantities of surface dust appear with a DCI close to 0.99, while the maximum quantity of surface dust is present when the DCI approaches 0.89. Generally, dust-free terrains are identifiable when the DCI value is $\geq$0.97.

(g) **Thermal inertia:** It is driven by the rover thermal constraints and by the need to have a load-bearing surface. Following Nowicki and Christensen (2007), Mars can be separated into three primary thermophysical components: dust (<150 TIU, where TIU is thermal inertia units, i.e., J m^{-2} $s^{-0.5}$ K^{-1}), a variable moderate inertia component (up to ~300 TIU), and rocks (>1250 TIU). In order to avoid landing on a dusty surface, the proposed landing area must have a thermal inertia larger than 100 (Mars 2020)-150 TIU (ExoMars). The MGS-TES global night-side and day-side seasonal thermal inertia maps (3 km resolution) presented by Putzig et al. (2005) and Putzig and Mellon (2007) are used for this purpose (Fig. 6).

(h) **Albedo:** It is driven by the rover thermal design constraints and the consequent surface temperatures that the rover has to face on the Mars surface. It is evaluated by using the MGS-TES (Christensen et al. 2001) map that has a resolution of 7.4 km (Fig. 7).

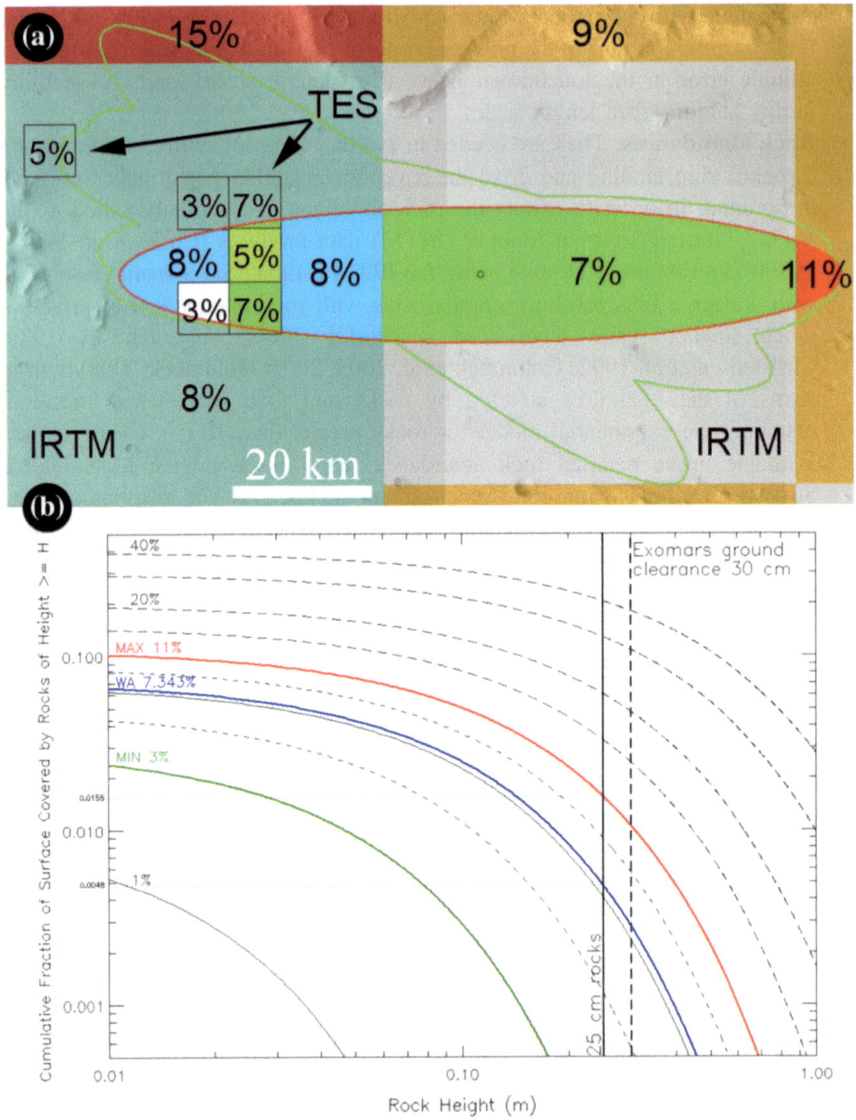

Fig. 4 **a** Rock abundance map based on IRTM (larger squares) and TES (smaller squares) data of the Simud ExoMars proposed landing ellipses (Pajola et al., 2016a). **b** Curves representing the cumulative fraction of surface covered by rocks of height $\geq$ H. The red and green curves represent the maximum and minimum rock abundances observed on the Simud landing site, while the blue curve represents the computed weighted average

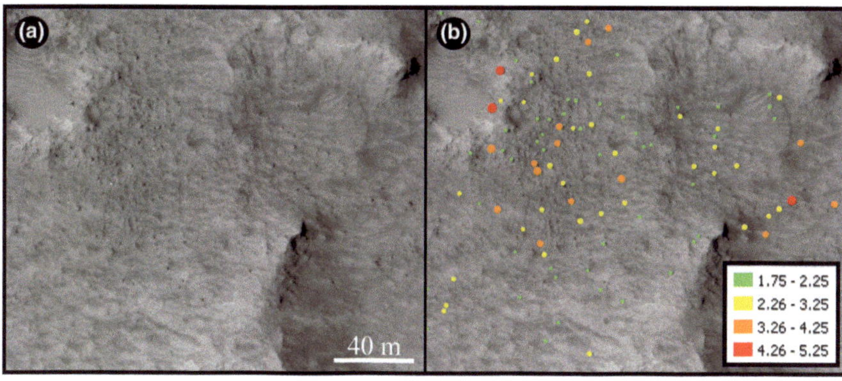

Fig. 5 Identification of the boulders located over the final Oxia Planum ExoMars landing ellipse (Quantin et al. 2016; Pajola et al. 2017). The boulders are manually identified on HiRISE 0.25 m/px scale images and are classified on their different sizes

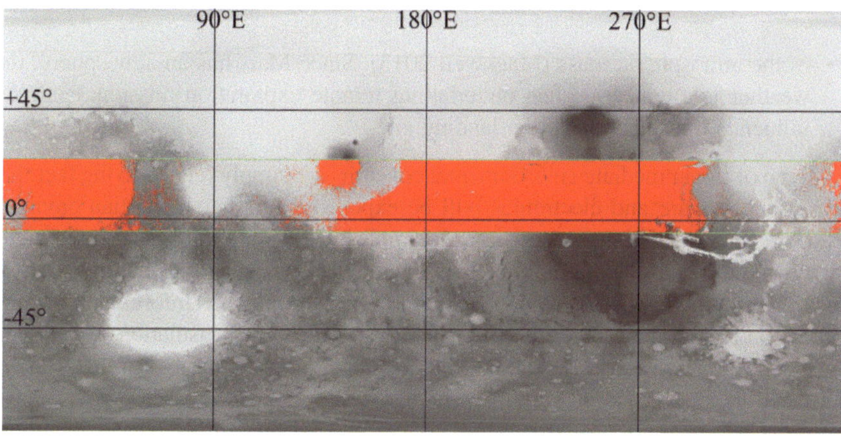

Fig. 6 Thermal inertia map derived using the Putzig et al. (2005) and Putzig and Mellon (2007) TES dataset. The red areas within the ExoMars permitted latitude band do not fulfill the ≥ 150 TIU criterion. The background is the MOLA Mars global map

(i) **Radar reflectivity:** It is fundamental to the proper functioning of the radar on the rover and relevant to nadir backscatter during data relay to the orbiters. The required values are derivable from the MEX-MARSIS (Mars Advanced Radar for Subsurface and Ionosphere Sounding) global reflectivity map presented in Mouginot et al. (2010).

(j) **Atmospheric circulations and slope winds at the landing site:** Mars has a relatively thin atmosphere, with an annually averaged global mean surface pressure of approximately 6 mbar and a global mean surface temperature approaching 215 K (Showman 2002). Carbon dioxide (CO_2) represents ~95%

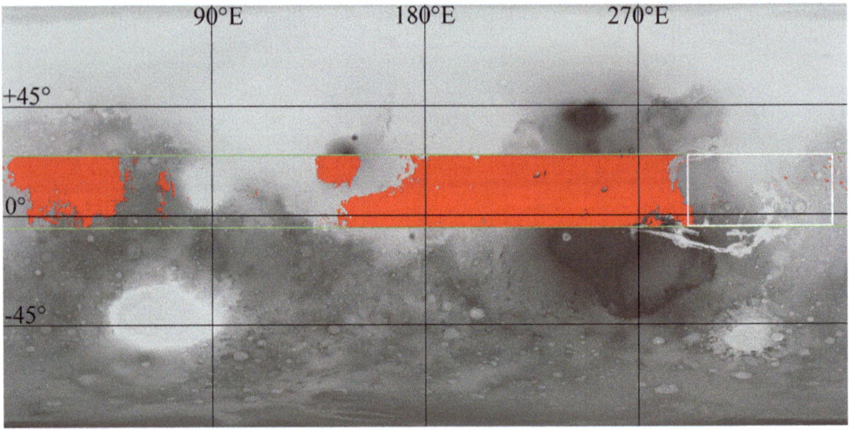

Fig. 7 Albedo map derived using the Christensen et al. (2001) TES dataset. The red areas within the ExoMars permitted latitude band do not fulfill the $0.1 \leq$ albedo ≤ 0.26 criterion. The white box shows the extent of Fig. 10. The background is the MOLA Mars global map

of the atmospheric mass (Mackwell 2013). Since Mars has an atmosphere, the weather is a concern when undertaking remote exploration missions and can influence the selection of the landing site.

Some of the immediate environmental concerns are simply related to the fact that sensitive hardware and electronics will be exposed to the harsh Martian environment for an extended period of time. Examples of design factors that have to be considered are extreme thermal cycling of both electronics and mechanical parts or dust intrusion and adhesion to the solar panels or instruments. Unfortunately, some mitigation techniques to these issues, such as better thermal insulation, are limited by the reduced resources available and the total mass of hardware that can be delivered to the surface of Mars.

The EDL phase is the one when the local environment can have the most dramatic consequences on the success of a robotic mission. The main atmospheric factors to consider are:

1. **Sensitivity to the air density variations** (Vasavada et al. 2012): This is due to the fact that the aerodynamic forces on the descent stage vary linearly with the atmospheric density. Large-scale dynamics and local atmospheric circulations, e.g., due to a mountain range or a crater, can result in significant regional variations of the air density (Rafkin et al. 2003). Also, about 20% of the atmospheric mass condenses at the poles of Mars during winter and sublimates in spring, resulting in a CO_2 cycle and large seasonal variations of the atmospheric density (Showman 2002).
2. **Sensitivity to the mean horizontal winds:** As the descent stage is decelerating under a parachute, it will move at a speed close to the mean horizontal wind speed. Therefore, uncertainties on the winds will tend to broaden the size of the

landing ellipse, possibly making a geologically interesting target no longer reachable to the EDL system. Moreover, high slope wind speeds can be hazardous during touchdown, especially if passive decelerating systems are used, such as airbags (Rafkin et al. 2003).

3. **Sensitivity to the wind shear:** As the EDL system falls under a parachute, certain frequencies in the horizontal wind shear and variations in the vertical structure of the wind fields can trigger hazardous oscillations of the EDL system (Kass et al. 2003).
4. **Sensitivity to the vertical winds:** Landers and rovers that use rockets to brake before touchdown are particularly sensitive to the vertical winds when initiating the power descent (Vasavada et al. 2012).

For these reasons, a good understanding of the environmental constraints at the landing site is critical when designing a spacecraft. Overall, little meteorological data has been available at Mars: Clouds and dust storms have been imaged from orbit, and examples of remote sensing techniques are temperature profile retrievals from spectrometers onboard Mars orbiters (Conrath et al. 2000) or Earth-based radio occultation experiments (Tellmann et al. 2013). Nonetheless, meteorological measurements near the surface, which are the most relevant to the selection of a landing site, are not accessible from orbit and have to be performed in situ. To date, only seven missions have made it safely to the surface of Mars. Between them, meteorological data have been available at only five locations scattered both geographically across the planet and temporally over several decades of Mars exploration history (Viking Landers 1 and 2, Pathfinder, Phoenix and Mars Science Laboratory).

The lack of extensive datasets to assist the design of Mars missions has fostered the development of Mars global circulation models (GCMs, Haberle et al. 1999; Kahre et al. 2006; Hollingsworth and Kahre 2010; Kahre et al. 2015) in order to better understand and predict the Martian climate. The spatial resolution of the GCMs is typically a few degrees of latitude/longitude; thus, they are particularly well equipped to capture the large-scale components of the atmospheric circulations and for long-term (>1 Martian year) climate sensitivity studies. Their high-resolution counterparts, called mesoscale models, can achieve spatial resolution down to the km scale over a specific area of interest and resolve regional circulations over complex topography (Fig. 8). Both families of models are originally derived from the numerical models used on Earth for weather forecasting and have been adapted to simulate the Martian atmosphere. The global and mesoscale models are extremely useful (i) for atmospheric safety assessment (for instance, it was revealed that in certain locations, the horizontal wind speed near the surface can be as high as 50 m/s and the vertical velocities up to 40 m/s due to intense thermals and mesoscale circulations (Rafkin et al. 2003)); (ii) to support the scientific observations from the instruments on the Mars rovers and landers, (Haberle et al. 2014); and (iii) to predict the prevailing wind directions to understand the aeolian processes such as erosion or transport and deposition of material that have shaped a specific Martian region of interest (Greeley 2008). An example of a statistical analysis of wind speed at the Eridania Basin landing site is shown in Fig. 9.

Fig. 8 Regional circulation and slope winds over the Eridania Basin during the northern hemisphere winter solstice as predicted by the Mars Regional Atmospheric Modeling System (MRAMS, Rafkin et al. 2001). This image shows the magnitude (m/s) and direction of the slope winds over the Eridania landing site at local time 5 pm

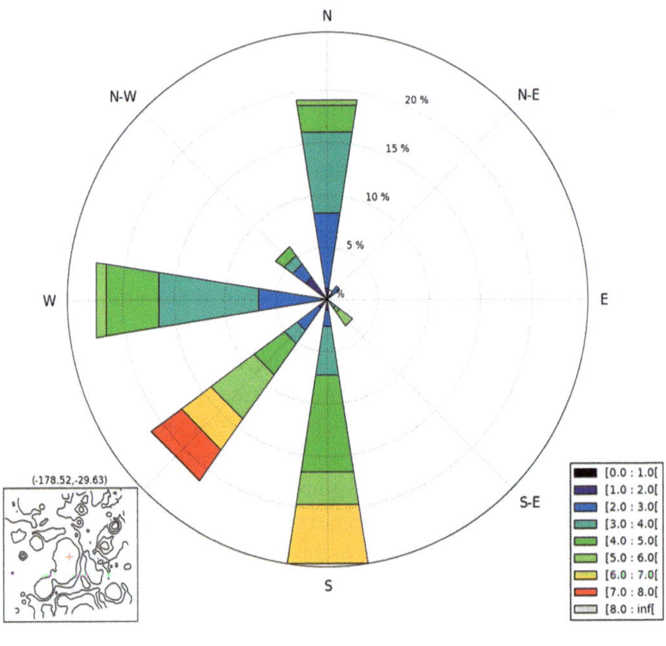

Fig. 9 "Wind rose" that shows the dominant wind directions and magnitudes (m/s) at the Eridania Basin. Single-day predictions from MRAMS (Rafkin et al. 2001) at the solstices and equinoxes are used to extrapolate the winds to the full Martian year. This is important in order to understand where the winds are blowing from, and their magnitudes, during the landing day

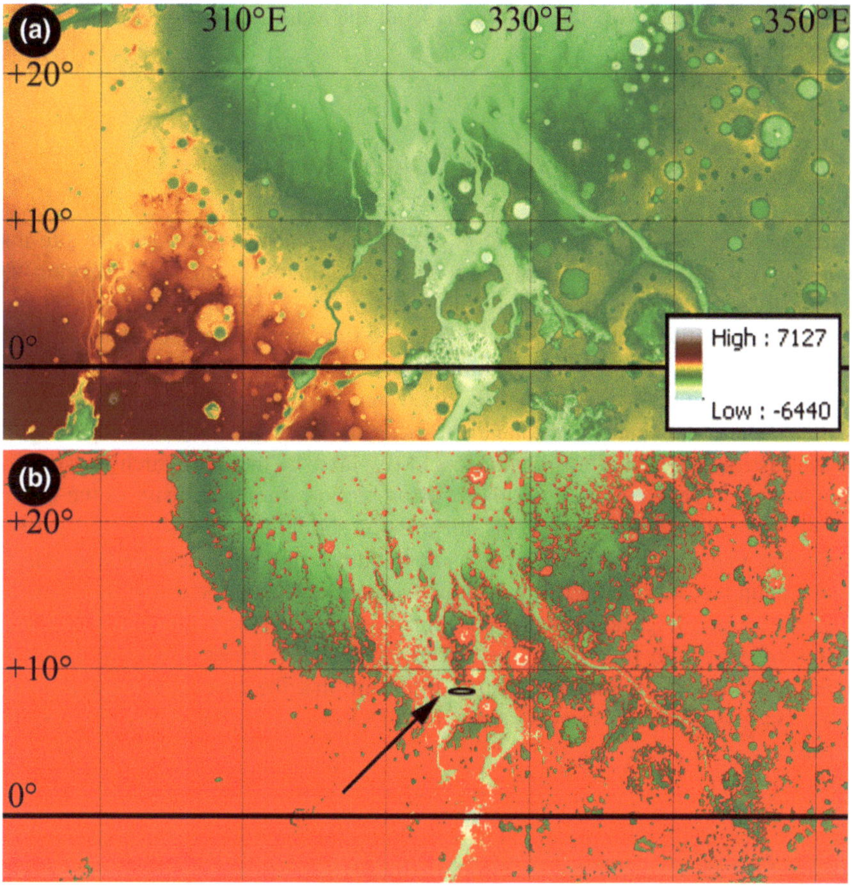

Fig. 10 **a**: MOLA-based elevation map of a sector of Mars (see Fig. 7 for location). **b**: Resulting constraint map showing in red those areas that do not satisfy the elevation, slope, thermal inertia, and albedo requirements for the ExoMars 2020 landing site. The arrow indicates a proposed landing site on Simud Vallis (104 × 19 km ellipse)

When all the above-mentioned constraints are evaluated, only specific areas of the surface of Mars can be proposed as a final landing site (see Fig. 10 depicting an example of such multi-criteria evaluation steps).

3 The Scientific Requirements

The ongoing ESA and NASA missions clearly differ on the engineering constraints and technicalities, but scientifically they are both targeted to explore sites that are likely to preserve signs of the past life (see the ExoMars 2018 landing site selection call (ESA 2014) and Mustard et al. (2013) for the Mars 2020 landing site scientific constraints).

3.1 The ESA ExoMars 2020 Landing Site

The ExoMars 2020 mission has to target a geologically diverse, ancient site that could have had strong potential for both past habitability and for preserving the physical and chemical signs of life (Kereszturi et al. 2016) and organic matter (including abiotic/prebiotic organics). The ExoMars rover is foreseen to (i) analyze the local geology at km to sub-mm scales and down to $\sim$2 m depth, being equipped with a new-generation drill, (ii) search for and evaluate the nature of past habitable environments at the landing site, (iii) investigate favorable geological materials for preserving biosignatures, and (iv) analyze them to search for signs of life as well as seek evidence of abiotic or prebiotic carbon chemistry in the 0–2 m depth range. To achieve these goals, the mission should land in a site with a surface age older than 3.6 Ga, when Mars was potentially habitable. The landing site must show abundant morphological and mineralogical evidence for a long duration, or frequently recurring, aqueous activity, including numerous sedimentary rock outcrops covered with little or no dust (Kereszturi 2012). These outcrops should be distributed all over the landing ellipse to ensure that the rover can reach some of them. Indeed, the rover traverse range is 4–15 km—during the mission's nominal duration.

3.2 The NASA Mars 2020 Landing Site

The landing site for the Mars 2020 rover must show the presence of subaqueous or hydrothermal sediments, as well as hydrothermally altered rocks coupled with the presence of outcrops characterized by minerals indicative of aqueous phases, such as phyllosilicates, carbonates, or sulfates. A Noachian/Early Hesperian (i.e., older than 3.5 Ga) surface age based on stratigraphic relations and crater counts and a free path access to unaltered igneous rocks are also required. Moreover, it is pivotal for such site to present morphological criteria for standing bodies of water and/or fluvial activity, such as deltaic deposits or shorelines; the presence of former water ice, glacial activity, or its deposits; igneous rocks of Noachian age of known stratigraphic relation (the igneous rocks with a well-constrained age are of prime consideration for the Mars 2020 aim of sampling and caching: Indeed derived radiometric ages of igneous rocks provide the formation age, while sedimentary rocks provide a combination of component ages), volcanic unit of Hesperian or Amazonian age well-defined by crater counts and well-identified by morphology and/or mineralogy; potential for resources for future human missions.

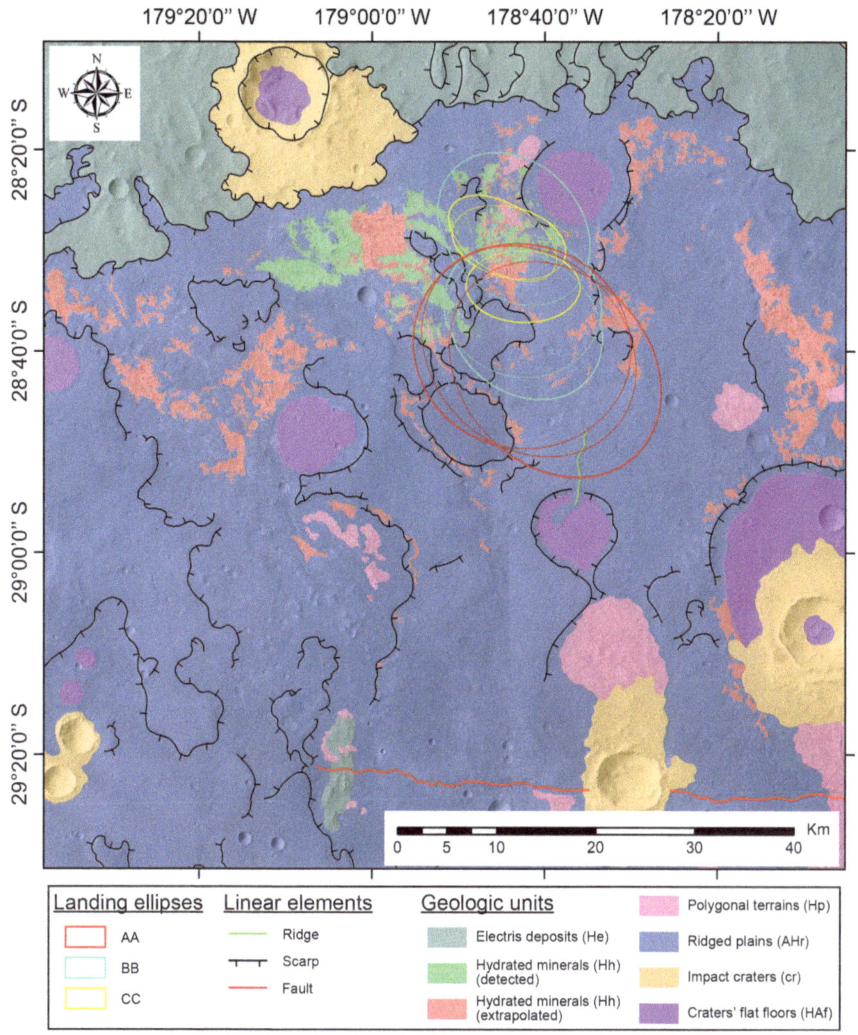

Fig. 11 Geological context map presenting the different units identified in this sector of the Eridania paleolake floor (adapted from Pajola et al. 2016b)

In order to fulfill the scientific requirements, specific GIS maps have to be prepared showing both the geological units and the mineralogical occurrence. An example is the analysis presented by Pajola et al. (2016b) on the Eridania ancient paleolake floor (Irwin et al. 2002) as a possible landing site for Mars 2020 rover. Indeed, on this area, uneven high-albedo patches of material characterized by the absence of dust are present in the local geology (Fig. 11), and when analyzed with orbital imaging spectroscopy, they return signatures of a set of aqueous minerals in the stratigraphy. This is visible thanks to erosional windows in the first several tens of meters of the sedimentary sequence (Fig. 12).

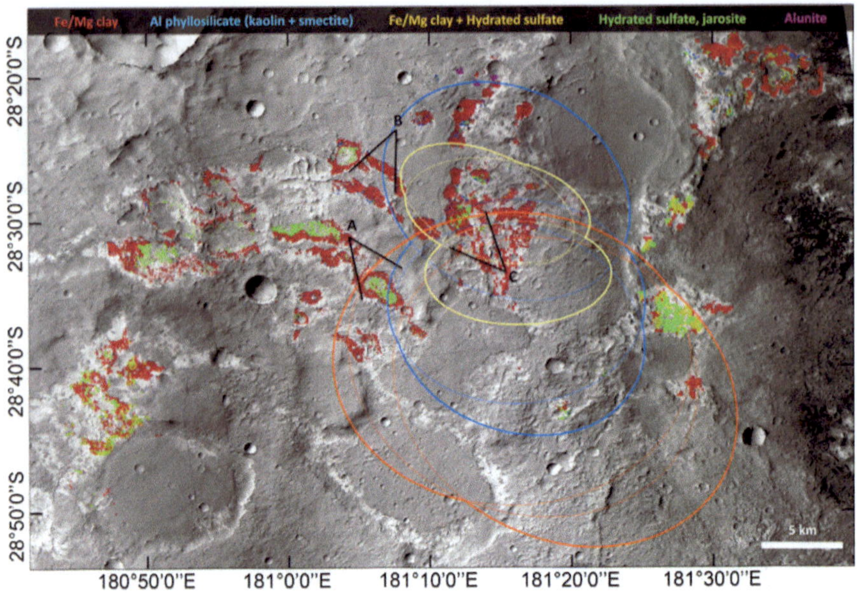

Fig. 12 High-resolution mapping of the NE light-toned basin deposit (taken from Pajola et al. 2016b) based on CRISM data (Murchie et al. 2007). Fe/Mg clays are mapped in red, Al-rich phyllosilicates are mapped in blue, polyhydrated sulfates with occasional jarosite are mapped in green, and alunite is mapped in purple. The proposed Mars 2020 landing ellipses of Fig. 11 are superposed

These maps (Figs. 11 and 12) underline all the scientific benefits that landing on such a spot would bring. When the scientific criteria evaluation and accomplishment are overlain with the engineering cartographic maps, e.g., Fig. 10, the synthesis of all such data results in a complete and consistent landing site proposal that can be both objectively evaluated and thoroughly compared with other landing competitors.

4 Conclusion

We have presented and discussed the engineering and scientific criteria that have to be fulfilled when a landing site is proposed on the surface of Mars. The widespread production and use of GIS maps both for representation and evaluation of such constraints have become a conditio sine qua non in this research field, as indicated by the ESA ExoMars and the NASA Mars 2020 ongoing study cases and detailed by the multiple maps showed in this chapter. The superimposition of the different engineering cartographic products and data, as indicated in Fig. 10, results in the safest landing spots attainable on the surface of Mars, but have eventually to be overlain with the geological and mineralogical maps (Figs. 11 and 12) in order to derive the complete and consistent study site that can be quantitatively evaluated and compared with other proposed landing areas.

Acknowledgements We thank Dr. Henrik Hargitai and an anonymous reviewer for important comments and suggestions that highly improved the book's chapter.

References

Burrough PA, McDonnell RA (1998) Principles of Geographical Information Systems. Oxford University Press, New York

Christensen PR (1986a) The spatial distribution of rocks on mars. Icarus 68:217–238

Christensen PR (1986b) Regional dust deposits on mars. J Geophys Res 91(B3):3533–3545

Christensen PR et al (1992) Thermal emission spectrometer experiment: the mars observer mission. J Geophys Res 97:7719–7734

Christensen PR et al (2001) Mars global surveyor thermal emission spectrometer experiment: investigation description and surface science results. J Geophys Res 106(E10):23823–23871

Conrath BJ et al (2000) Mars global surveyor thermal emission spectrometer (TES) observations: atmospheric temperatures during aerobraking and science phasing. J Geophys Res 105(E4): 9509–9519

ESA (2014). http://exploration.esa.int/mars/53462-call-for-exomars-2018-landing-site-selection/

Golombek MP, Rapp D (1997) Size-frequency distributions of rocks on mars and earth analog sites: implications for future landed missions. J Geophys Res 102(E2):4117–4129

Golombek MP et al (2003) Rock size-frequency distributions on Mars and implications for Mars exploration rover landing safety and operations. J Geophys Res 108(E12):8086

Golombek MP et al (2005) assessment of Mars exploration rover landing site predictions. Nature 436:44–48

Golombek MP et al (2008) Size-frequency distributions of rocks on the northern plains of mars with special reference to Phoenix landing surfaces. J Geophys Res 113 (E00A09)

Golombek M et al (2012) Selection of the mars science laboratory landing site. Space Sci Rev 170(1–4):641–737

Greeley R et al (2008) Columbia hills, mars: aeolian features seen from the ground and orbit. J Geophys Res 113

Haberle RM et al (1999) General circulation model simulations of the mars pathfinder atmospheric structure investigation/meteorology data. J Geophys Res 104:8957–8974

Haberle RM et al (2014) Preliminary interpretation of the REMS pressure data from the first 100 sols of the MSL mission. J Geophys Res Planets 119:440–453

Hollingsworth JL Kahre MA (2010) Extratropical cyclones, frontal waves, and mars dust: modeling and considerations. Geophys Res Lett 37

Irwin RP et al (2002) A large paleolake basin at the head of Ma'adim Vallis, Mars. Science 296 (5576):2209–2212

Kass DM et al (2003) Analysis of atmospheric mesoscale models for entry, descent, and landing. J Geophys Res 108:8090

Kahre MA (2006) Modeling the martian dust cycle and surface dust reservoirs with the NASA ames general circulation model. J Geophys Res 111

Kahre MA (2015) Coupling the mars dust and water cycles: the importance of radiative-dynamic feedbacks during northern hemisphere summer. Icarus 260:477–480

Kereszturi A (2012) Landing site rationality scaling for subsurface sampling on mars—case study for exomars rover-like missions. Planet Space Sci 72:78–90

Kereszturi A, Bradak B, Chatzitheodoridis E, Ujvari G (2016) Indicators and methods to understand past environments from exomars rover drills. Orig Life Evol Biosph 46:435–454
Mackwell et al (2013) Comparative climatology of terrestrial planets. University of Arizona Press, pp 55–89
McEwen AS et al (2007) Mars reconnaissance orbiter's high resolution imaging science experiment (HiRISE). J Geophys Res 112:E05S02
Moore HJ, Jakosky BM (1989) Viking landing sites, remote-sensing observations, and physical properties of martian surface materials. Icarus 81(1):164–184
Mouginot J et al (2010) The 3–5 MHz global reflectivity map of mars by MARSIS/Mars express: implications for the current inventory of subsurface H_2O. Icarus 210(2):612–625
Murchie S et al (2007) Compact reconnaissance imaging spectrometer for Mars (CRISM) on Mars reconnaissance orbiter (MRO). J Geophys Res 112:E05S03
Mustard JF, Beaty D, Bass D (2013) Mars 2020 science rover: science goals and mission concept. In: American astronomical society, DPS meeting No. 45, No. 211.17
Nowicki SA, Christensen PR (2007) Rock abundance on Mars from the thermal emission spectrometer. J Geophys Res 112:E05007
Pajola M et al (2016a) The Simud-Tiu valles hydrologic system: a multidisciplinary study of a possible site for future Mars on-site exploration. Icarus 268:355–381
Pajola M et al (2016b) Eridania basin: an ancient paleolake floor as the next landing site for the Mars 2020 rover. Icarus 275:163–182
Pajola M et al (2017) Boulder abundances and size-frequency distributions on oxia Planum-Mars: scientific implications for the 2020 ESA ExoMars rover. Icarus 296:73–90
Putzig NE, Mellon MT (2007) Apparent thermal inertia and the surface heterogeneity of Mars. Icarus 191:68–94
Putzig NE et al (2005) Global thermal inertia and surface properties of mars from the MGS mapping mission. Icarus 173:325–341
Quantin C, Carter J, Thollot P et al (2016) Oxia planum, the landing site for exomars 2018. In: 47th lunar and planetary science conference 2016
Rafkin SCR, Michaels TI (2003) Meteorological predictions for 2003 Mars exploration rover high-priority landing sites. J Geophys Res 108(E12):809
Rafkin SCR, Haberle RM, Michaels TI (2001) The mars regional atmospheric modeling system: model description and selected simulations. Icarus 151:228–256
Ruff SW, Christensen PR (2002) Bright and dark regions on Mars: particle size and mineralogical characteristics based on thermal emission spectrometer data. J Geophys Res 107(E12):5119
Showman AP (2002) Planetary atmospheres: Mars Encyclopedia of Atmospheric Sciences. Academic Press, pp 1745–1755
Smith DE et al (2001) Mars orbiter laser altimeter: experiment summary after the first year of global mapping of Mars. J Geophys Res 106(E10):23689–23722
Tellmann S et al (2013) The structure of mars lower atmosphere from mars express radio science (MaRS) occultation measurements. J Geophys Res Planets 118:306–320
Vasavada AR et al (2012) Assessment of environments for Mars science laboratory entry descent, and surface operation. Space Sci Rev 170:793

Mapping Irregular Bodies

Philip Stooke and Maurizio Pajola

Abstract Map projecting small, irregular bodies present in the Solar System is not a trivial task. The first 3D models attempting to reconstruct the Martian satellite Phobos occurred at the end of the 1980s. After that, an increasing number of high-resolution observations of asteroids, comets and planets' satellites lead to the identification of specific standards used to both shape model the targets as well as to map project them. In this chapter, we will present an excursus of the early 3D model reconstruction of irregular bodies, as well as the shape modelling and the retrieval of illumination conditions methodology, and the specific 3D shape reconstruction of comet 67P/Churyumov–Gerasimenko.

Keywords Minor bodies mapping · Irregular objects · 3D shape modelling · Map projections · Comet 67P

1 Introduction

Solar System objects (asteroids, comets and natural satellites) smaller than c. 500 km across (varying with composition and thermal history) have insufficient mass to collapse into roughly spheroidal shapes and may be very irregular in shape. Many have been imaged by spacecraft since the first images of Phobos and Deimos, the small satellites of Mars, were taken in 1971–1972 by Mariner 9, (e.g. Duxbury et al. (2004), Massironi et al. (2012), Robinson et al. (2002), Stooke (1996), Thomas (1979), Thomas et al. (1994), Thomas et al. (1995), Thomas et al. (1996), Keller et al. (2010), Magrin et al. (2012), Pajola et al. (2012), Pajola et al. (2013)). Many of them have very irregular shapes, and mapping them involves challenges

P. Stooke (✉)
Department of Geography, University of Western Ontario,
London, ON N6A5C2, Canada
e-mail: pjstooke@uwo.ca

M. Pajola
NASA Ames Research Center, Moffett Field, CA 94035, USA

© Springer Nature Switzerland AG 2019

H. Hargitai (ed.), *Planetary Cartography and GIS*, Lecture Notes in Geoinformation and Cartography, https://doi.org/10.1007/978-3-319-62849-3_8

not encountered in conventional cartography. These include 3D shape modelling, new map projections and special cases such as "overhanging cliffs", where a radius vector passes through a body more than once, creating non-unique latitude–longitude values. Here we discuss these points and illustrate them for the recent case of comet 67P/Churyumov–Gerasimenko, a bilobate object which poses special difficulties for mapping.

2 The First Maps of Irregular Objects

The first attempt to map any irregular body was made by Duxbury (1974). He measured the positions of small craters in multiple Mariner 9 images of Phobos to estimate their locations in three dimensions and fitted a triaxial ellipsoid surface to those points. A coordinate system drawn on that ellipsoid was overlain on each image to locate major surface features, which could then be transferred grid cell by grid cell to a map. Duxbury used a conventional map grid based on a sphere, not a new map projection, for his map, but Turner (1978) created the first map projection for an irregular body, again for Phobos. He made a globe, a physical model of Phobos, by modifying an initial ellipsoid until its outline matched all available images, and plotted features on pole-centred azimuthal projections which were elongated to represent the elongated outline of Phobos. Following these simple beginnings, many other approaches have been devised. Maps of over 30 irregular bodies (Table 1) have been published by 2017 (the number depends on many assumptions about what to include, including how much detail should be included before a sketch is called a map. Specifically, in this table sketches of possible locations of cometary active regions or visualizations of shape models with no additional information are not included as maps).

2.1 *Early Soviet Models of Phobos*

In the late 1980s, a digital model of Phobos was constructed for the solution of navigational tasks in the project "Phobos" by IKI RAS and MSU on the basis of Turner's map. It was necessary to ensure the accuracy of the definitions of the centre of mass of Phobos, the distance to it and the orientation of Phobos in space. In the laboratory of aerospace methods of the Faculty of Geography of Moscow State University, an irregular globe model was created for Phobos at a scale of 1:100,000 using Ralph Turner's maps. This work was carried out by L. Vinnikov, I. Indichenko, A. Koshitz and B. Serapinas. The Phobos model was used to referencing images by photographing and modelling various imaging conditions (Avanesov et al. 1994).

Table 1 Published maps of irregularly shaped bodies

Name	Dimensions (km)[a]	Citation
Planetary satellites		
Phobos	9.9 by 13.5	Wählisch M, et al. (2010). EPSL. doi: 10.1016/j.epsl.2009.11.003
Deimos	5.5 by 7.7	Thomas, P. (1979). Icarus. doi: 10.1016/0019-1035(79)90069-1
Amalthea	64 by 125	Thomas, P. et al. (1998). Icarus. doi: 10.1006/icar.1998.5976
Thebe	42 by 58	Simonelli et al. (2000). Icarus. doi: 10.1006/icar.2000.6474
Prometheus	29.7 by 67.8	Stooke, P. (1993). Earth Moon Plan. 62(3), 199–221
Pandora	32 by 52	Stooke, P. (1993). Earth Moon Plan. 62(3), 199–221
Janus	76.3 by 102	Stooke, P., Lumsdon, M. (1993). Earth Moon Plan. 62 (3), 223–237
Epimetheus	53 by 65	Stooke, P. (1993). Earth Moon Plan. 63(1), 67–83
Methone	1.2 by 1.94	Thomas P. et al. (2013). 44th LPSC, abstract no. 1598
Hyperion	103 by 180	Stooke, P. (1996). Earth Moon Plan. 74(1), 61–83
Phoebe	102 by 110	Roatsch T. et al. (2006). Plan. Space Sci. 54(12), 1137-45
Proteus	195 by 212	Stooke, P. (1994). Earth Moon Plan. 65(1), 31–54
Larissa	84 by 108	Stooke, P. (1994). Earth Moon Plan. 65(1), 31–54
Asteroids		
2 Pallas	238 by 275	Carry et al. (2010). Icarus. doi: 10.1016/j.icarus.2009.08.007
4 Vesta	230 by 285	Jaumann et al. (2012). Science. doi: 10.1126/science.1219122
21 Lutetia	37.5 by 60.5	Sierks et al. (2011). Science. doi: 10.1126/science.1207325
243 Ida	9.3 by 29.9	Thomas et al. (1996). Icarus. doi: 10.1006/icar.1996.0033
243 Ida 1 Dactyl	0.6 by 0.8	Veverka et al. (1996). Icarus. doi: 10.1006/icar.1996.0045
253 Mathilde	c. 22 by 33	Veverka et al. (1999). Icarus. doi: 10.1006/icar.1999.6120
433 Eros	5.6 by 17.2	Bussey et al. (2002). Icarus. doi: 10.1006/icar.2001.6771
951 Gaspra	4.4 by 9.1	Stooke (1996). Earth, Moon, Planets. doi.org/10.1007/BF00056410
2867 Steins	2.2 by 3.3	Leyrat et al. (2010). Plan. Space Sci. doi: 10.1016/j.pss.2010.04.003
4179 Toutatis	0.9 by 2.3	Stooke (1996). 27th LPSC abstracts, pp. 1283-1284.
4769 Castalia	0.4 by 0.9	Stooke, P. J. (1998). Can. Geog. doi: 10.1111/j.1541-0064.1998.tb01553.x
5535 Annefrank	1.7 by 3.3	Stryk and Stooke (2016). https://www.hou.usra.edu/meetings/lpsc2016/eposter/1148.pdf
Comet nuclei		
1/P Halley	8.0 by 15.0	Stooke and Abergel (1991). Astron. Astrophys. 248 (2), 656-668.
9/P Tempel	4.4 by 8.0	Veverka et al., (2013). Icarus. doi: 10.1016/j.icarus.2012.03.034
19/P Borrelly	4.0 by 8.0	Britt et al. (2004). Icarus. doi: 10.1016/j.icarus.2003.09.004

(continued)

Table 1 (continued)

Name	Dimensions (km)[a]	Citation
67/P Churyumov - Gerasimenko	2.3 by 4.3	Preusker et al. (2017). Astron. Astrophys. DOI: 10.1051/0004-6361/201731798
81/P Wild	3.3 by 5.5	Sekanina et al. (2004). Science. doi: 10.1126/science.1098388
103/P Hartley	1.0 by 2.3	Thomas et al. (2012). Icarus. doi: 10.1016/j.icarus.2012.05.034.

[a]Dimensions are the minimum and maximum radii of the body or its triaxial ellipsoid approximation, provided only as an indication of deviation from a sphere. See the cited sources for more detailed information

3 Shape Modelling Methods

Duxbury (1974) calculated the positions in 3D space of small features on Phobos, using multiple images, adapting control network analysis methods used previously for the Moon and Mars. With a large number of control points, a detailed shape model can be constructed. Ideally we could use stereoscopic pairs of images, but this was precluded in early work by low resolution, differing illumination, small numbers of images and very different viewing angles.

Stooke (1988) took a different approach, creating a digital model, initially a triaxial ellipsoid, which consisted of a matrix of radii at regular intervals of planetocentric latitude and longitude. This could be viewed from the same direction as an image of a body and "sculpted" by iteratively adjusting its radii until the shape matched the limb outline and terminator position of the image. This process would be repeated iteratively until the shape matched all available images. This approach was suited to the low-resolution images then available for small satellites of the outer planets, often less than 100 pixels across and inadequate for stereoscopic viewing or control point definition. Simonelli et al. (1993) effectively merged the Duxbury and Stooke methods in a software called "Spud" developed at Cornell University. Other approaches have included defining a shape by means of intersecting prisms, each prism formed by the outline of an image projected along its view direction, for the nucleus of Halley's Comet (Bertaux and Abergel 1986), and fitting control points with spherical harmonic analysis (Duxbury 1991) for Phobos.

Analysis of rotational lightcurves taken under different viewing and illumination conditions allows astronomers to create rough shape models of numerous asteroids not visited by spacecraft (e.g. Kaasalainen and Torppa 2001). Another essentially astronomical approach is the use of radar returns from asteroids, pioneered by Ostro (1993) to create images of small bodies and extended by Hudson and Ostro (1995) to derive shape models. The lightcurve technique provides a shape resembling the convex hull of the true shape (omitting depressions), and the radar method produces high-quality shape models if the data are sufficient in terms of resolution, signal/noise and number of observations. Only rarely can these models be tested with a spacecraft fly-by, one good example being the Chinese fly-by of 4179 Toutatis by Chang'E 2 on 13 December 2012 (Ji et al. 2015). The radar shape was generally correct but gave poor results in the most concave areas.

In more recent years, higher quality datasets, stereoscopic imaging and in some cases laser altimeters have resulted in high-resolution and high-quality shape models. Examples include asteroids 433 Eros and 25143 Itokawa and comet 67P/Churyumov–Gerasimenko (Miller et al. 2002; Demura et al. 2006; Preusker et al. 2015; Jorda et al. 2016). An analysis of the shapes of small tiles derived from multiple overlapping images and combined to cover a body (Gaskell et al. 2008) has been applied to large and small bodies alike, including asteroids Eros, Itokawa and 4 Vesta. Future approaches are expected to include the use of LIDAR on the OSIRIS-REx mission which will reach asteroid 101955 Bennu in 2018. The LIDAR data should result in very high-quality 3D shape models.

4 Retrieval of the Observation Geometries on Irregular Bodies

Objects such as asteroids are generally observed with multiple images having different observing angles, illumination conditions, and resolutions (Fig. 1). In order to determine the shape and derive the spectral properties of these irregular surfaces, such as reflectance and albedo, it is fundamental to retrieve the viewing and illumination geometry (incidence, emission and phase angles) for each observation. These are calculated from knowledge of the spacecraft trajectory and positions of the target and the Sun, and are available from NASA's Navigation and Ancillary Information Facility (NAIF) as SPICE datasets ("kernels") (Acton 1996).

Once all these parameters have been considered, the illumination conditions of a specific image can be reproduced and calculated. An example is the Phobos observation performed by the High Resolution Stereo Camera onboard ESA Mars

Fig. 1 Asteroid 21 Lutetia, observed by the Rosetta spacecraft on 10 July 2010 with different observing geometries, illumination conditions and resolutions. *Image credit* ESA/Rosetta/MPS for OSIRIS Team MPS/UPD/LAM/IAA/SSO/INTA/UPM/DASP/IDA

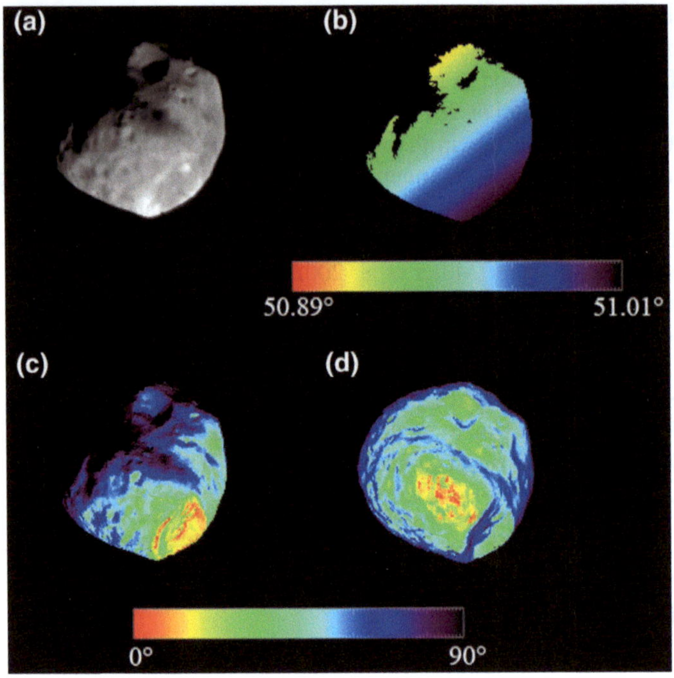

Fig. 2 **a** Original image of Phobos taken on 25 February 2007 from the HRSC imager onboard Mars Express spacecraft. **b**, **c** and **d** Phase angle (i.e. the observer-object-Sun angle), incidence angle (i.e. the angle between the local normal vector and the direction of the Sun) and emission angle (i.e. the angle between the local normal vector and the direction of the observer) calculations generated by using the NAIF SPICE system, as done in Pajola et al. (2018). *Image (a) credit* ESA/DLR/FU Berlin (G. Neukum)

Express on 25 February 2007 (Fig. 2). In this specific case, the 3D shape model of Phobos (Gaskell 2011) has been oriented following the SPICE kernels; consequently, each facet has been illuminated and the phase angle (i.e. the observer-object-Sun angle, Fig. 2b), incidence angle (i.e. the angle between the local normal vector and the direction of the Sun, Fig. 2c) and emission angle (i.e. the angle between the local normal vector and the direction of the observer, Fig. 2d) maps have been computed.

5 Map Projections for Irregular Bodies

A map projection is a method for representing a three-dimensional body such as the Earth on a flat surface. Methods for representing the spherical Earth on a flat map are well established and could be applied to other spherical worlds such as the Moon and Mars without difficulty. This process first encountered serious problems when images of the irregularly shaped satellites of Mars, Phobos and Deimos were acquired by Mariner 9 in 1971 and 1972. Further exploration revealed objects such as asteroids Eros, Ida, Toutatis and Itokawa whose minimum and maximum radii vary by 500% or more. It was soon apparent that new approaches to mapping were required. If a body is not spherical, the geometry of a map projection applied to it would have to be modified to accommodate the shape.

The simplest approach to mapping a non-spherical body is to use a conventional projection for a sphere. This necessarily produces distortions, becoming more severe as the shape departs further from a sphere. Despite this problem, such projections can be useful because they can be produced in standard mapping software without special modifications. A common example is the equirectangular (simple cylindrical, or plate carrée) projection which produces a rectangular map (Figs. 3b, 4c, 7a).

Turner (1978) created the first map projection modified for an irregular object (Phobos) by elongating an initially circular azimuthal projection to match the elongated shape of Phobos. Stooke (1988) extended this concept by substituting the local radius at any point for the conventional radius constant in azimuthal map projection equations. Each bulge or hollow in the shape is depicted by warping of the grid (Stooke and Keller 1990), and analysis showed that distortions were minimized if the greatest deviations from a sphere were placed at the boundary of the modified azimuthal ("morphographic", meaning "shape-drawing") projection.

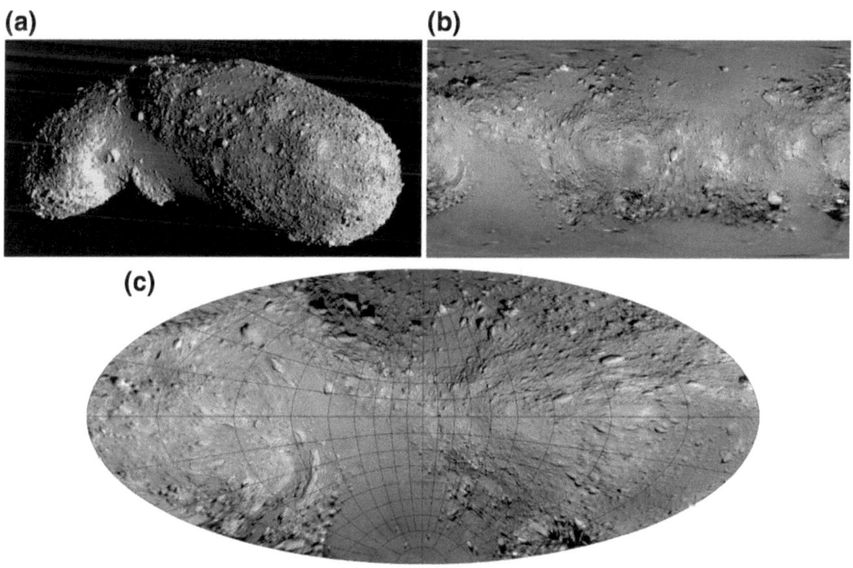

Fig. 3 Mapping asteroid Itokawa. **a** Hayabusa image of Itokawa (JAXA). **b** Cylindrical projection photomosaic of Itokawa (Stooke 2015). **c** One side of Itokawa corresponding to **a**, projected onto a triaxial ellipsoid and portrayed in a morphographic projection (Stooke 2015)

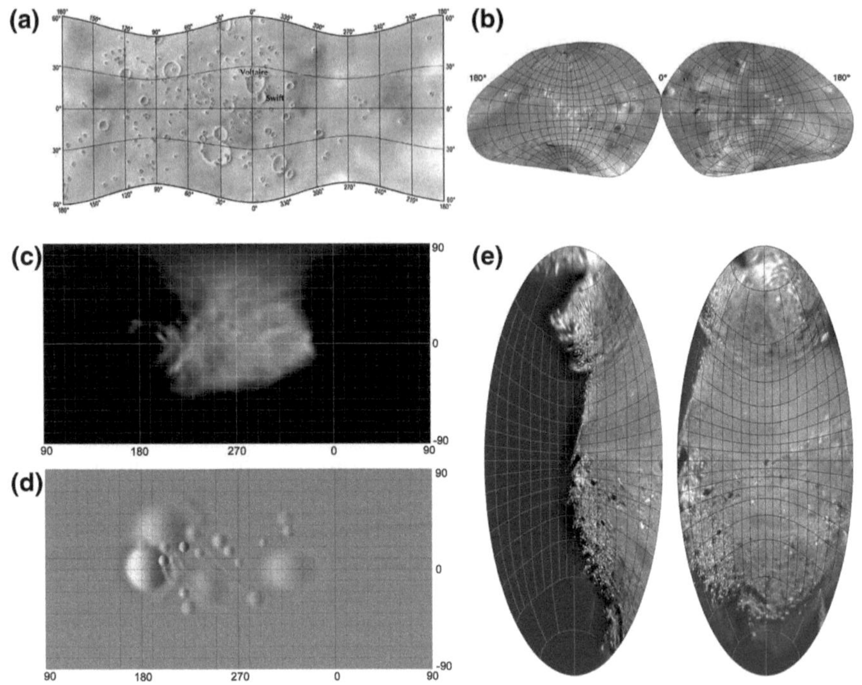

Fig. 4 Maps of several bodies illustrating different map projections. **a** Deimos in Bugaevsky's projection (Hargitai et al. 2006). **b** Deimos in a morphographic projection (Stooke and Keller 1990). **c, d** Asteroid 5535 Annefrank in cylindrical projections (Stryk and Stooke 2016). **e** Comet Hartley 2, two morphographic projections of an ellipsoid showing its elongation along the rotation axis (Stooke 2015)

The shape model used to distort the grid can be actual topography, a convex hull (Fig. 4b), a triaxial ellipsoid (Fig. 3c) or any other version of the shape.

Russian cartographers have modified other projections, work pioneered by Bugaevsky at MIIGAiK in Moscow and continued by Nyrtsov. Bugaevsky's approach was to modify a cylindrical projection, causing it to expand around regions with larger radii at each end of an elongated body (Fig. 4a). Nyrtsov has extended this to many other projections, e.g. Nyrtsov et al. (2014). Some other approaches have resulted in interesting projections such as an equal area projection by Berthoud (2005), but they remain experimental and have not been widely used.

Another approach to mapping is to use multiple images or rendered views of an object to display its surface in sections, and this can be extended to dynamic mapping in which map data are draped over a shape model which can be rotated and viewed from any direction. A good example of this, readily available to all, is the Small Body Mapping Tool which can be downloaded from the Applied Physics Laboratory at Johns Hopkins University.

6 Compiling Maps of Irregular Objects

When a shape model has been derived, it alone can form the basis for a variety of maps. Variations in radius, or elevation above a defined surface such as an equipotential, can be colour-coded to form topographic maps (e.g. Fig. 7). Slopes can be calculated and portrayed similarly, and illumination maps can reveal permanently shaded or illuminated regions based on analysis of the shape. All these may be plotted on orthographic or perspective views of the object or transferred to a map projection for a global map.

Other maps may be rendered over the shape model, including photomosaics and geological interpretations of the surface. The shape is used to manipulate the geometry of individual images so they can be represented in a standard map projection, and multiple images can be combined to create global or broad regional coverage. Figure 4b is a global photomosaic of Itokawa compiled by Stooke (Stooke 2015) in a simple cylindrical projection. Figure 4c is a morphographic (azimuthal) projection of one side of Itokawa using a triaxial ellipsoid shape for this elongated object, which reduces distortions considerably.

7 The Comet 67P/Churyumov–Gerasimenko Case

On 6 August 2014, after a ten-year journey that included three fly-bys of the Earth, one of Mars and two close encounters with the asteroids (2867) Steins and (21) Lutetia, the European Space Agency (ESA) Rosetta spacecraft reached its main target: Comet 67P/Churyumov–Gerasimenko (hereafter 67P). Contrarily to a predicted rugby ball (Lamy et al. 2006) shape for 67P (Fig. 5a, b), this comet appeared (Sierks et al. 2015) as one of the most complex and irregular shape ever observed in the Solar System (Fig. 5c). Indeed, 67P is characterized by a wide diversity of surface morphology (Thomas et al. 2015; Vincent et al. 2015; Massironi et al. 2015; Pajola et al. 2015; Pommerol et al. 2015; El-Maarry et al. 2015) occurring on the

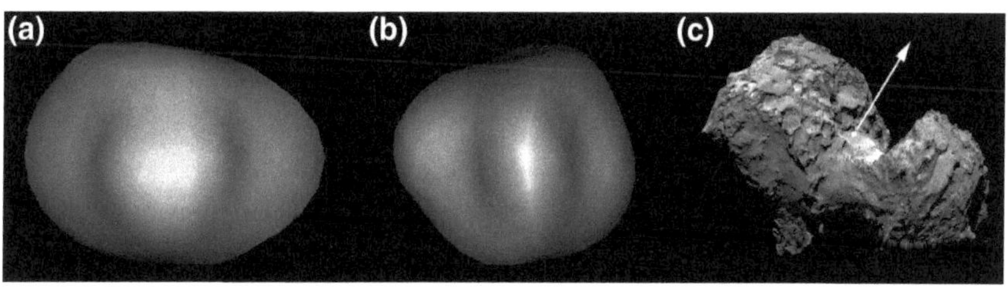

Fig. 5 **a** and **b** Side and lateral view of the pre-Rosetta 3D shape model of 67P. *Image credit* NASA, ESA and Philippe Lamy. **c** The true shape of 67P. The white dot indicates the north pole of 67P, while the arrow its rotation axis. *Image credit* ESA/Rosetta/MPS for OSIRIS Team MPS/UPD/LAM/IAA/SSO/INTA/UPM/DASP/IDA

two main lobes: the larger one, hereafter called the "body", with dimensions of 4.10 × 3.52 × 1.63 km, and a smaller lobe, hereafter called the "head", with dimensions of 2.50 × 2.14 × 1.64 km (Jorda et al. 2016). The region located between the head and the body, hereafter called the "neck" (Sierks et al. 2015; Thomas et al. 2015), is 2.2. km long and roughly 0.8 km wide.

While producing the shape model through stereophotogrammetry (SPG, Preusker et al. 2015) and stereophotoclinometry techniques (SPC, Jorda et al. 2016) by exploiting the high-resolution Optical Spectroscopic and Infrared Remote Imaging System (OSIRIS, Keller et al. 2007) images, it became clear that the extreme irregularity of 67P's shape resulted in a loss of uniqueness in conventional spherical coordinates (latitude and longitude) in regions where individual radius vectors intersect with the surface more than once. This was particularly true in the concave neck area close to the north pole (Fig. 5c). For this reason, Preusker et al. (2015) decided to subdivide the bilobate shape of 67P into separate entities (i.e. the body, the head and the neck) in order to prepare separate maps of each entity, each with its own unique latitude and longitude coordinates. This ingenious solution is so far unique to 67P.

Another important aspect that was considered while defining control points on 67P was the possible occurrence of major surface changes on the cometary nucleus due to its intrinsic activity that could make individual control points or even large areas of the surface disappear (see, e.g. Pajola et al. 2017). For this reason, three different boulders on opposite sides of the body and head of 67P were considered as reference landmarks with fixed coordinates (assuming that at least one of them would not have moved or disappeared over the entire duration of the Rosetta mission). The consequent reference system for 67P was therefore called "Cheops reference frame", due to the biggest of these three boulders dubbed "Cheops" (a 50-m-size block (Preusker et al. 2015)), Fig. 6.

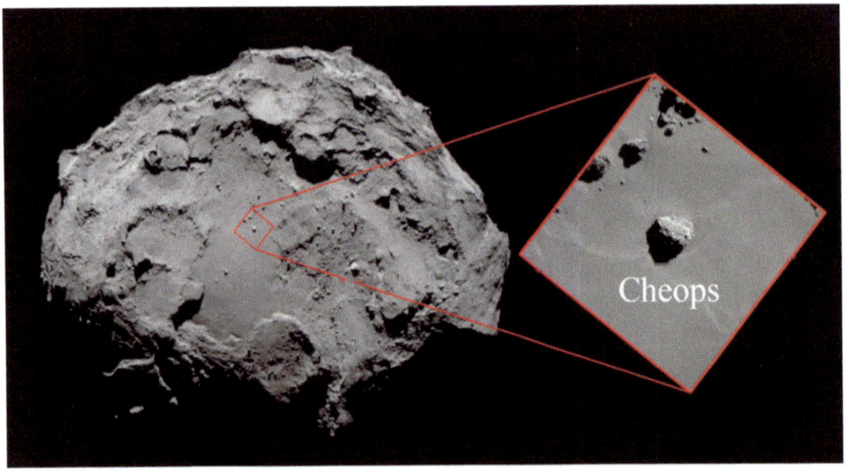

Fig. 6 Cheops boulder which gives its name to 67P's coordinate reference frame. *Image credit* ESA/Rosetta/MPS for OSIRIS Team MPS/UPD/LAM/IAA/SSO/INTA/UPM/DASP/IDA

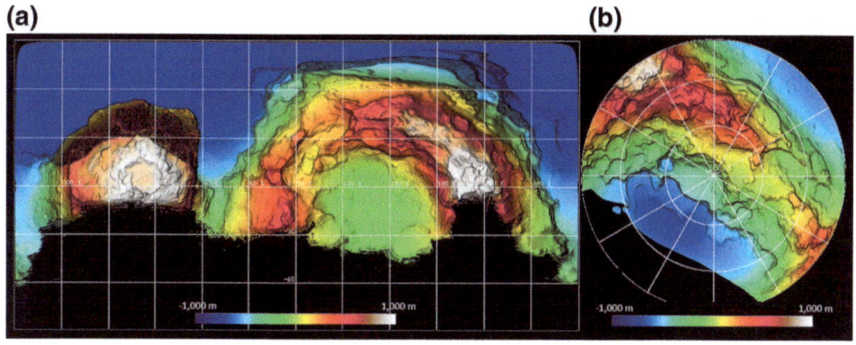

Fig. 7 **a** Simple (equidistant) cylindrical map for the entire shape of 67P. The colour coding shows the height above the 1500 m reference sphere: blue = −1000 m to red/white = 1000 m black areas are regions on the southern hemisphere of 67P not illuminated when this map was prepared (adapted from Preusker et al. 2015). **b** Stereographic projection only for the body (main lobe) of 67P (centred on its north pole). *Image credit* ESA/Rosetta/MPS for OSIRIS Team MPS/UPD/LAM/IAA/SSO/INTA/UPM/DASP/IDA

Figure 7 depicts two examples, out of the several products produced for 67P (Preusker et al. 2015) that have been used for different geomorphological applications, such as the detection of a pervasive onion-like layering structure (Massironi et al. 2015), and a boulder analysis located on one terrace of the comet (Pajola et al. 2016). Therefore, for the specific case of the irregularly shaped 67P, the cartography produced is a fundamental tool not only to locate the visible geomorphic structures, by inserting them into a wider context, but also to retrieve and distinguish different geomorphological units otherwise indistinguishable when merely looking at the unprojected images.

8 Conclusion

Cartography of irregularly shaped objects in the Solar System is necessary because maps remain essential tools for integrating and displaying observations, and because numerous such objects have been explored by spacecraft. These objects, often with varied shapes of great complexity, challenge conventional mapping methods and have necessitated the development of new shape modelling methods, new map projections and sometimes wholly unique approaches to mapping.

Acknowledgements We are grateful to Nyrtsov, M.V., who provided additional text regarding Russian mapping.

References

Acton CH Jr (1996) Ancillary data services of NASA's navigation and ancillary information facility. Planet Space Sci 44:65

Avanesov GA, Zhukov BS, Ziman YaL et al (1994) Televisionniye issledovaniya Fobosa. Nauka publishing, Moskva (in Russian)

Bertaux JL, Abergel A (1986) Some physical characteristics of Halley's nucleus as inferred from VEGA and Giotto pictures. In: ESLAB symposium on the exploration of Halley's Comet, vol 2. ESA SP-250, pp 341–345

Berthoud MG (2005) An equal-area map projection for irregular objects. Icarus 175(2):382–389

Demura H et al (2006) Pole and global shape of 25143 Itokawa. Science 312(5778):1347–1349

Duxbury TC (1974) Phobos: control network analysis. Icarus 23(2):290–299

Duxbury TC (1991) An analytic model for the Phobos surface. Planet Space Sci 39(1):355–376

Duxbury TC, Newburn RL, Brownlee DE (2004) Comet 81P/Wild 2 size, shape, and orientation. J Geophys Res: Planet 109:E12

El-Maarry MR et al (2015) Fractures on comet 67P/Churyumov-Gerasimenko observed by Rosetta/OSIRIS. Geophys Res Lett 42:5170–5178

Gaskell R, et al (2008) Gaskell Itokawa Shape Model V1.0. HAY-A-AMICA-5-ITOKAWASHAPE-V1.0. NASA Planetary Data System

Gaskell RW (2011) Gaskell Phobos Shape Model V1.0. VO1-SA-VISA/VISB-5-PHOBOSSHAPE-V1.0. NASA Planetary Data System

Hargitai H, Krasnopevtseva BV, Shingareva KB (eds) (2006) Moons of Mars. 1:100k. Eötvös Loránd University Cosmic Materials Space Research Group, Budapest-Moscow

Hudson RS, Ostro SJ (1995) Shape and non-principal axis spin state of asteroid 4179 Toutatis. Science 270(5233):84

Ji J et al (2015) Chang'e-2 spacecraft observations of asteroid 4179 Toutatis. Proc Int Astron Union 10(S318):144–152

Jorda L et al (2016) The global shape, density and rotation of Comet 67P/Churyumov-Gerasimenko from preperihelion Rosetta/OSIRIS observations. Icarus 277:257–278

Kaasalainen M, Torppa J (2001) Optimization methods for asteroid lightcurve inversion: I. Shape determination. Icarus 153(1):24–36

Keller HU et al (2007) OSIRIS - the scientific camera system onboard Rosetta. Space Sci Rev 128:433

Keller HU et al (2010) E-type Asteroid (2867) Stein as imaged by OSIRIS on board Rosetta. Science 327:190–193

Lamy PL et al (2006) Hubble space telescope observations of the nucleus and inner coma of comet 67P/Churyumov-Gerasimenko. Astron Astrophys 458:669–678

Magrin S et al (2012) Spectrophotometry from Rosetta-OSIRIS images and comparison to ground based observations. Planet Space Sci 66:43–53

Massironi M et al (2012) Geological map and stratigraphy of asteroid 21 Lutetia. Planet Space Sci 66(1):125–136

Massironi M et al (2015) Two independent and primitive envelopes of the bilobate nucleus of comet 67P. Nature 526:402

Miller JK et al (2002) Determination of shape, gravity, and rotational state of asteroid 433 Eros. Icarus 155(1):3–17

Nyrtsov MV, Fleis ME, Borisov MM, Stooke PJ (2014) Jacobi conformal projection of the triaxial ellipsoid: new projection for mapping of small celestial bodies. In: Buchroithner M, Prechtel N, Burghardt D (eds) Cartography from Pole to Pole. Springer, Heidelberg, pp 235–246. https://doi.org/10.1007/978-3-642-32618-9_17

Ostro SJ (1993) Planetary radar astronomy. Rev Mod Phys 65(4):1235

Pajola M et al (2012) Spectrophotometric investigation of Phobos with the Rosetta OSIRIS-NAC camera and implications for its collisional capture. Mon Not R Astron Soc 427:3230–3243

Pajola M et al (2013) Phobos as a D-type asteroid, spectral modeling from 0.25 to 4.0 μm. Astrophys J 777(127):6
Pajola M et al (2015) Size-frequency distribution of boulders ≥ 7 m on comet 67P/Churyumov-Gerasimenko. Astron Astrophys 583(A37):17
Pajola M et al (2016) Aswan site on comet 67P/Churyumov-Gerasimenko: morphology, boulder evolution, and spectrophotometry. Astron Astrophys 592(A69):17
Pajola M et al (2017) The pristine interior of comet 67P revealed by the combined Aswan outburst and cliff collapse. Nat Astron 1:0092
Pajola M et al (2018) Phobos MRO/CRISM visible and near-infrared (0.5–2.5 μm) spectral modeling. Planet Space Sci 154:63–71
Pommerol A et al (2015) OSIRIS observations of meter-sized exposures of H2O ice at the surface of 67P/Churyumov-Gerasimenko and interpretation using laboratory experiments. Astron Astrophys 583:A25
Preusker F et al (2015) Shape model, reference system definition, and cartographic mapping standards for comet 67P/Churyumov-Gerasimenko – Stereo-photogrammetric analysis of Rosetta/OSIRIS image data. Astron Astrophys 583(A33):19
Preusker F et al (2017) The global meter-level shape model of comet 67P/Churyumov-Gerasimenko. Astron Astrophys 607(L1):5
Robinson MS et al (2002) The geology of 433 Eros. Meteorit Planet Sci 37(12):1651–1684
Sierks H et al (2015) On the nucleus structure and activity of comet 67P/Churyumov-Gerasimenko. Science 347:1044
Simonelli DP et al (1993) The generation and use of numerical shape models for irregular solar system objects. Icarus 103(1):49–61
Stooke PJ (1988) Cartography of non-spherical worlds. Dissertation, University of Victoria
Stooke PJ (1996) The surface of asteroid 951 Gaspra. Earth Moon Planet 75(1):53–75
Stooke P (2015) Stooke Small Bodies Maps V3.0. MULTI-SA-MULTI-6-STOOKEMAPS-V3.0. NASA Planetary Data System
Stryk T, Stooke, PJ (2016) The surface of asteroid 5535 annefrank. In: 47th lunar and planetary science conference. Abstract no. 1148
Stooke PJ, Keller CP (1990) Map projections for non-spherical worlds/the variable-radius map projections. Cartographica 27(2):82–100
Thomas P (1979) Surface features of Phobos and Deimos. Icarus 40(2):223–243
Thomas N, et al (2015) The morphological diversity of comet 67P/Churyumov-Gerasimenko. Science 347(6220), article id. aaa0440
Thomas PC et al (1994) The shape of Gaspra. Icarus 107(1):23–36
Thomas PC, Black GJ, Nicholson PD (1995) Hyperion: Rotation, shape, and geology from Voyager images. Icarus 117(1):128–148
Thomas PC et al (1996) The shape of Ida. Icarus 120(1):20–32
Turner RJ (1978) A model of Phobos. Icarus 33(1):116–140
Vincent J-B et al (2015) Large heterogeneities in comet 67P as revealed by active pits from sinkhole collapse. Nature 523:63

Part III
Cartographic Approaches

Multi-mapper Projects: Collaborative Mercury Mapping

Valentina Galluzzi

Abstract Taking up the challenge of mapping the entire surface of a planetary body may present different levels of difficulty. The effort and time required for such a project depends mainly on the available data quality and workforce. The resolution and coverage of the basemaps provided as data sets by the space missions determine the highest acceptable mapping scale and the possible extent of a project, respectively. The larger the mapping scale, the longer the work. If many mappers are involved, this can considerably decrease the time needed for completing a global map by producing a series of regional maps. However, this also increases the risk of mismatches between the mapped regions. In order to better analyse the complexity of such a plan, here we examine the case of the Mercury 1:3M-scale global mapping project.

Keywords Mercury (planet) · Planetary geologic mapping · Group mapping · Mapping symbology · Mapping methods

1 Introduction

Hitherto, Mercury has been the target of several multi-mapper projects and represents a good example of how planetary geologic mapping can be undertaken and how planetary geologic maps can evolve over time. By the end of the NASA Mariner 10 mission (1973–1975), 45% of Mercury's surface had been imaged by the M10 Television Experiment (Murray et al. 1974) and over 2000 useful pictures were available at a resolution better than 2 km, up to 100 m (Davies et al. 1978). These results led to the production of 1:5M geologic maps of nine of the fifteen quadrangles of Mercury (Spudis and Guest 1988 and references therein). The NASA MESSENGER (MErcury Surface, Space ENvironment, GEochemistry

---

V. Galluzzi (✉)
INAF, Istituto di Astrofisica e Planetologia Spaziali, Via del Fosso del Cavaliere, 100, 00133 Rome, Italy
e-mail: valentina.galluzzi@inaf.it

© Springer Nature Switzerland AG 2019

H. Hargitai (ed.), *Planetary Cartography and GIS*, Lecture Notes in Geoinformation and Cartography, https://doi.org/10.1007/978-3-319-62849-3_9

and Ranging; 2004–2015) mission filled the gap by imaging 100% of the planet with a frame resolution up to 8 m/pixel at the north pole, and a global average resolution of 200 m/pixel, enabling preparation of a new global 1:15M geological map (Prockter et al. 2016).

Despite the quality gap between Mariner 10 and MESSENGER images, no global geologic mapping project with a scale larger than 1:5M—i.e. the same scale used by the Mariner 10 geological maps—has been released yet. However, a complete global series of 1:3M global maps of Mercury is being prepared in support of the BepiColombo joint mission of the European Space Agency (ESA) and the Japan Aerospace Agency (JAXA) (Benkhoff et al. 2010). This project was born from individual geologic quadrangle maps (Galluzzi et al. 2016; Mancinelli et al. 2016; Guzzetta et al. 2017), then it has evolved into a coordinated global mapping plan (Galluzzi et al. 2018), and carried on with the aim of exploiting MESSENGER images at the best resolution available (i.e. global average resolution). This will set up the context for BepiColombo operations and help redefine mission goals as appropriate. What follows is not a step-by-step guide on how to produce a Mercury geological (i.e. morpho-stratigraphical) map, but rather a description of the main issues encountered during the main phases of the project, which is still ongoing today.

2 Coordination of the Work

The planetary geologic map is a tool designed for addressing specific scientific needs, thus different scientific goals may result in different map outputs. Deciding the main goal of a map is therefore a crucial step before starting a project. Hence, a good question to start from is *what will this map be used for?* Answering this question determines the kind of methods that the mappers will use to describe what they see on a basemap. Our main goal is to reconstruct the global stratigraphy of Mercury by means of morpho-stratigraphical classification of the units, corrobortated by the morphological classification of craters that helps understanding the relative timing of the volcanic events that shaped the planet. However, even when the main goal of a map is established, one has to deal with the fact that many mappers have to reach a result that is compatible with the results of their colleagues.

Creating a common and compatible output is probably the biggest difficulty in a multi-mapper project. The ideal case would be that an entire team of mappers works in the same institution at the same time, on the same target, so that everybody can ask for advice, compare and map in a real-time condition. This circumstance seldom occurs. Most of the time, like in our 1:3M-scale Mercury mapping project case, the mappers work during different time spans and in different institutions. Moreover, in the early part of the project, the situation was complicated by the still updating data from the ongoing MESSENGER mission. For comparison, the global mapping projects advanced for Vesta (Williams et al. 2014; Yingst et al. 2014) and Ceres (Mest et al. 2016; Williams et al. 2016), share with this project the same kind

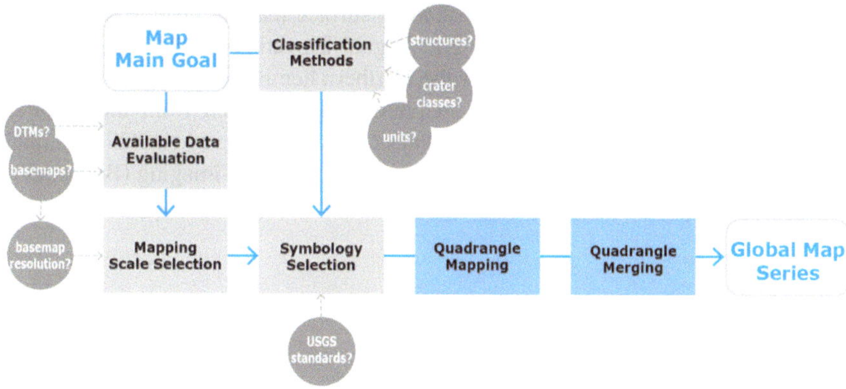

Fig. 1 Workflow scheme followed for defining the mapping steps of the 1:3M-scale Mercury geologic mapping project. The 'Map Main Goal' and the 'Global Map Series' rectangles indicate the start and the end of the workflow, respectively. Grey rectangles indicate the main global rules to follow. Dark grey circles indicate the inputs needed for defining the rules connected by dashed grey arrows. Light blue rectangles indicate actions taken on each quadrangle. Further information on each step can be found in the text

of challenges (i.e. mapping during the acquisition of data); it follows that this kind of difficulties should not prevent any mapper from starting a global mapping plan.

In fact, most difficulties can be overcome by establishing sound mapping rules and a main workflow to follow. Figure 1 summarizes the workflow underlying the 1:3M-scale Mercury mapping project.

2.1 *Evaluating the Available Data*

The evaluation of the available data sets is fundamental before starting a mapping project. Above all other issues, coverage and illumination conditions are the first things to be considered. The MESSENGER mission and its Mercury Dual Imaging System (MDIS) instrument provided several global basemaps at 166 m/pixel: a moderate incidence angle basemap (BDR), two high-incidence angle basemaps with illumination from East and West (HIE and HIW) and a low-incidence angle basemap (LOI). The different illumination conditions lower the bias in detecting the surface features and permit a thorough study of their morphology.

However, when this project started, the MESSENGER mission was still releasing data sets, and this caused basemap coverage and resolution to change frequently. Therefore, the main problem was to select a basemap with a sufficient resolution and a homogeneous coverage, which could be used as a reference work-base by the mappers. Since most basemaps were initially released as 250 m/pixel mosaics, the best data set that could be used as a reference basemap was the 166 m/pixel BDR.

Topography is also a key to understanding surface features. The Mercury Laser Altimeter (MLA) provided a 500-m grid Digital Elevation Model (DEM) for the North Pole and a 665-m grid DEM for the Northern hemisphere (Zuber et al. 2012). This was the main reason why the mapping project started from the Northern hemisphere. More recently, the United States Geological Survey (USGS) has made available a 665-m global coverage DEM derived from stereo-imaging (Becker et al. 2016), providing a better understanding also of the Southern hemisphere.

2.2 *Defining the Mapping Scale*

The USGS usually recommends choosing the map output scale (i.e. the print scale) before selecting the mapping scale (i.e. the scale used to draw), which should be two to five times larger than the output scale (Tanaka et al. 2011). Our main goal required us to make a geological survey at the best resolution possible defined by the available basemaps. Thus, we also considered the cartographic rule by Tobler (1987), which says that the mapping scale should be 2000 times the basemap resolution.

Taking into account both recommendations (the USGS guidelines and the cartographic rule), we finally opted for a 1:3000,000 output scale, which was the best compromise for obtaining a readable and detailed output. The chosen output scale of 1:3M is compatible with the available basemap resolutions that vary between 166 m/pixel and 250 m/pixel. In fact, these resolutions would permit mapping at larger scales, up to ~1:300,000 to ~1:600,000.

Eventually, this choice led to the publication of the first mapped quadrangles as a series of 1:3M-scale quadrangle maps (Galluzzi et al. 2016; Mancinelli et al. 2016; Guzzetta et al. 2017), instead of a unique merged global output. The quadrangles are those defined in Davies et al. (1978), with their limits redefined by the MESSENGER team as shown in Fig. 2.

2.3 *Defining the Symbology*

Choosing the symbology may seem just a secondary problem that affects only the map appearance. However, many features use ornamental ticks to indicate a specific direction (e.g. crater rim inner scarp, fault vergence). If ornamental ticks are added in a later stage of the work, the final layout may present a randomly oriented feature symbology and it would require a thorough revision of the work. Moreover, in-depth analysis of symbology communicates a significant amount of science information, thus the mappers must agree on that in order to avoid working at cross-purposes.

The symbology used our mapping project is mainly based on the Federal Geographic Data Committee (FGDC) Digital Cartographic Standard for Geologic

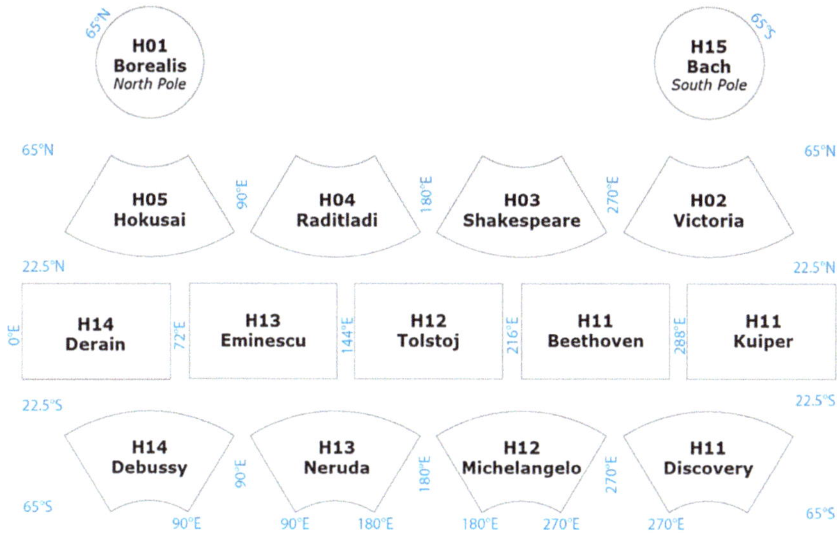

Fig. 2 Quadrangles of Mercury with new names and boundaries defined by the MESSENGER team. Equatorial: cylindrical projection; Mid-Latitude: Lambert conformal conic projection; Polar: stereographic projection

Map Symbolization prepared by the USGS. However, we adapted the symbols to some specific needs based on Mercury peculiarities. For example, *hollows*, which are a feature peculiar to Mercury (Blewett et al. 2011; Thomas et al. 2014; Blewett 2015), do not have a corresponding symbol in the FGDC symbol chart, thus we decided to represent them with a compatible 'pattern' symbol (see Table 1). Moreover, we decided to represent *lobate scarps* and *high-relief ridges* with an appropriate *fault* symbology for reasons discussed in the following paragraph.

3 Surface Features Classification Methods

Interpretation is subjective, and it cannot be expected that all mappers would interpret the same features in exactly the same way; however, this can be prevented by defining some *a priori* feature classification methods. Depending on the quality and the availability of different data sets, several kinds of classifications can be chosen, starting from purely morphologic, to field-specific classifications. In order to avoid most problems, we tried to define some classification guidelines for the mappers. Two main fields need a dedicated discussion: structures and stratigraphy.

Table 1 Some examples of the adopted FGDC symbology

Type	Used symbol	FGDC code	FGDC name
Geological contacts			
Contact, certain	————	25.1	Contact, planetary—Location accurate
Contact, approximate	— — — — — —	25.2	Contact, planetary—Location approximate
Linear features			
Crater rim, D ≥ 20 km		25.94	Raised rim of larger impact crater, planetary
Thrust, certain		25.21	Thrust fault, planetary—Location accurate
Wrinkle ridge	————	25.37	Wrinkle ridge, planetary
Surface features			
Hollow cluster		25.128	Airburst spot (pattern 434)[a]

[a]The original pattern for airburst spots is "434-K" (black) and we changed it into "434-C" (cyan)

3.1 Structural Features Classification

Widespread scarps called *lobate scarps*, *high-relief ridges* and *wrinkle ridges* characterize Mercury's surface and are interpreted as contractional faults (i.e. thrusts, Strom et al. 1975; Dzurisin 1978; Melosh and McKinnon 1988; Schultz 2000; Watters and Nimmo 2010; Byrne et al. 2014; Massironi et al. 2015; Massironi and Byrne 2015). Figure 3 shows several ways of classifying these structures in a small area of Mercury.

Using a 'level 0' approach, or *pure-morphologic* classification (i.e. 'what you see is what you draw'), every scarp is drawn with a simple scarp symbol (Fig. 3b). Although this is a seldom-used minimalist way of drawing features, it can represent a good quick-look analysis of the surface. On the other hand, a 'level 1' *morpho-structural* approach is based on a closer examination of the scarp morphometry, so that it can be mapped as, for example, a *lobate scarp*, or a *high-relief ridge* (Fig. 3c). This kind of classification could be useful for thematic morphological maps, although the vast literature concerning these structures (and the fact that they are faults, e.g. Watters and Nimmo 2010, and references therein) easily allows automatically skipping to the next type of classification. Finally, a 'level 2' *pure-structural* approach consists in completely characterizing the nature of a feature, according to its structural interpretation. Therefore, both a *lobate scarp* and a *high-relief ridge* will be interpreted and mapped as thrust faults, with the slip vergence highlighted by ornamental saw-teeth (Fig. 3d).

We opted for a pure-structural classification of the linear features of Mercury in order to provide an exhaustive framework on the geology and tectonics of Mercury.

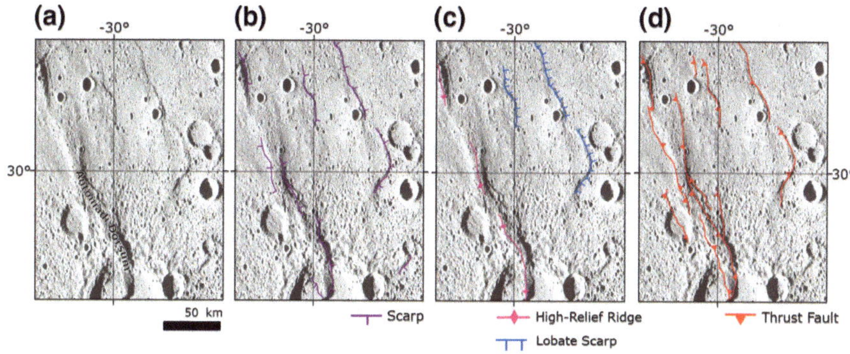

Fig. 3 Structural features classification examples. **a** 166 m/pixel BDR basemap of the Antoniadi Dorsum area of Mercury. **b** Pure-morphologic classification of scarps. Symbology shows hachures pointing down-scarp. **c** Morpho-structural classification of scarps. Symbology shows lobate scarp hachures pointing down-scarp, and mid-line diamonds on top of high-relief ridges. **d** Pure-structural classification of scarps. Symbology shows saw-teeth on the hanging-wall of the faults. In this representation, many more features are mapped and interpreted as faults, even if the fault scarp is not as prominent as the main lobate scarps or ridges

However, we kept *wrinkle ridges* as an exception and mapped them using a morpho-structural approach rather than a pure-structural approach. This choice was justified by the fact that they are a feature peculiar to the most recent volcanic plains of Mercury and it is important that they stand out on a global output. Nonetheless, we recommend and look forward to adopting a pure-structural approach even for *wrinkle ridges* for local-scale independent maps, since their nature is not much different from *lobate scarps* (e.g. Schultz et al. 2000).

3.2 *Stratigraphic Interpretation*

Because a planetary geologic map is built exclusively (or mainly) on remotely sensed data, the drawn features and geologic contacts are often based on a morphologic interpretation of the surface, sometimes referred to as *photo-interpretation*. Trask and Guest (1975) assess that '*on Mercury, surface morphology reflects the age, composition, lithology, and mode of formation of the underlying rock unit*'. This statement could as well be applied to any terrestrial body that has avoided resurfacing by wind or water.

In the Mariner 10 1:3M-scale map series (Spudis and Guest 1988, and references therein), numerous units were distinguished and mapped differently from one quadrangle to another, and this resulted in mismatches between unit extent and names along the quadrangle boundaries due to inconsistent scientific interpretation of the basemaps (see Frigeri et al. 2009). In order to minimize this kind of mismatch, we established three main unit types in advance (i.e. *smooth plains*, *intermediate plains*, *intercrater plains*) based on literature, local observations and crater counting (e.g. Galluzzi et al. 2016). Some extra units were mapped depending on the quadrangle location (e.g. *Caloris group*, Mancinelli et al. 2016; Guzzetta et al. 2017), though Rothery et al. (2017) subsequently demonstrated that the smooth plains member of the Caloris group (the Caloris interior plains) is not clearly distinct from the smooth plains outside the basin.

Craters, in turn, define impact units that can help better constrain the terrain unit stratigraphy. Crater materials are often classified based on their degradation state, but no official classification exists that can be used as a standard method. On Mercury, Kinczyk et al. (2016) classified craters larger than 40 km into five different crater degradation stages, which they correlated with the five time-stratigraphic systems of the planet, updating the previous M10 classification system of McCauley et al. (1981). However, in our case, we mapped the materials of all craters larger than 20 km. This choice increased the likelihood of size-dependent bias in evaluating the crater degradation stage. In fact, it can been observed that smaller craters may sometimes appear more degraded than larger craters of the same age, because of their less prominent morphology that tends to degrade faster (see also Galluzzi et al. 2016; Kinczyk et al. 2016). Moreover, impact secondaries work as a fast ageing process for craters located in the proximity of the impact. In order to reduce the bias in the crater morpho-stratigraphical classification, we reduced the degradation classes from five to three providing degradation class-type morphologies (see Galluzzi et al. 2016).

4 Towards a Global Map

A complete quadrangle geological map already represents a pleasing goal for a mapper; each line and polygon is the result of a series of interpretations and more or less frequent change-of-mind. A quadrangle of Mercury may contain thousands of craters and hundreds of faults and contacts, covering up to 8% of the planet's surface (e.g. Fig. 4).

However, a global map series requires paying specific attention to the quadrangle boundaries, where mappers usually discover that their 'neighbours' took different decisions in, e.g. mapping the ejecta extent or assigning the degradation class of a crater. Thus, the effectiveness of the pre-defined guidelines is usually revealed when two quadrangles need to be merged and the mappers compare their shared boundary. An example of boundary mismatch is given in Fig. 5.

In order to facilitate the quadrangle merging phase, we decided to extend the mapping linework to five degrees overlap with the adjacent quadrangles. The overlap is useful for mapping all those features that fall between two quadrangles and are cut by their shared boundary. Doing so, the mappers are able to reach a common decision by simply comparing and adjusting the overlapping features. Our group communicates via three different ways: (a) personal online communications, (b) annual group meetings and (c) face-to-face brainstorming. Hitherto, we proved

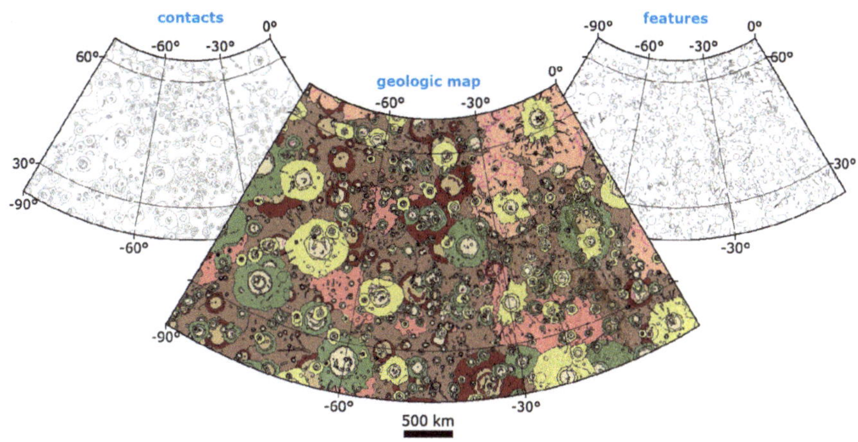

Fig. 4 Geologic map of the Victoria quadrangle covering a 5 million square km area of Mercury in Lambert conformal conic projection (Galluzzi et al. 2016), scale is true at 30°N and 58°N. The geologic map output is composed of geologic contacts that define unit and crater material boundaries, and of linear and surface features

that meeting face-to-face and working together in front of the same monitor is the best way to reach quickly and confidently a common agreement on the changes to apply at the quadrangle boundaries.

As already mentioned, a series of 1:3M-scale quadrangle maps cannot be merged into a single physical 1:3M-scale global map. However, the global merged output can be used as a digital full-scale product, which will permit more detailed global or regional analyses of Mercury's surface than the 1:15M geologic map of Prockter et al. (2016).

5 Conclusion

The Mercury 1:3M-scale global mapping project was born from the necessity of providing an exhaustive geological map of the planet derived from MESSENGER data at the best resolution available. The results of this work will be useful for the selection and prioritization of scientific targets of the ESA/JAXA BepiColombo mission. Our goal requires the production of a series of geological maps of Mercury's quadrangles, which will be merged as a digital global product.

A quadrangle map may be an individual result; however, the results of many mappers have to evolve into a coherent global output. In order to reduce the time spent in correcting the linework, we follow a main workflow that consists in defining: (a) a common mapping scale range; (b) a standard symbology; (c) structural and stratigraphic classification and interpretation guidelines; (d) the unit-type locations; (e) the crater-type morphologies; (f) a quadrangle overlap extent.

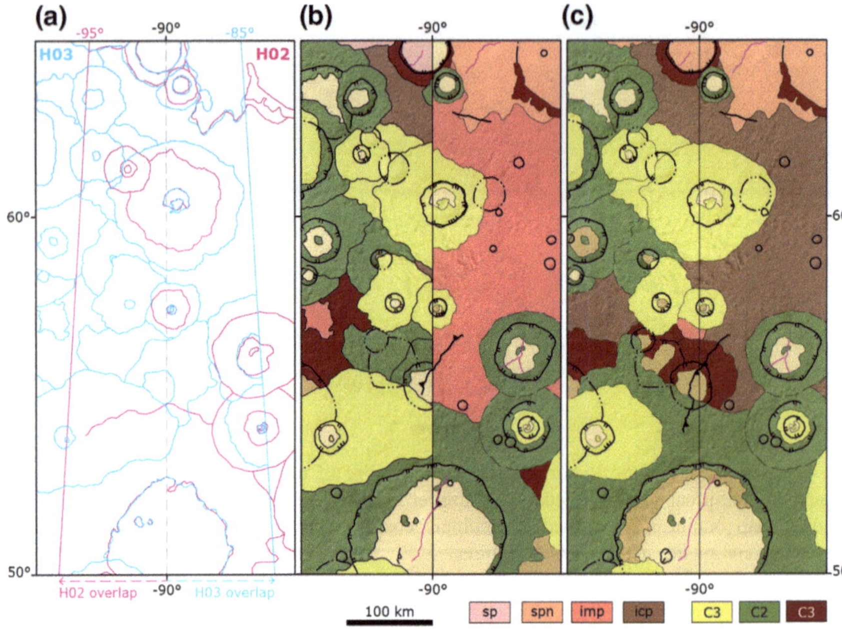

Fig. 5 This area of Mercury shows the boundary between the Shakespeare quadrangle (H03, Guzzetta et al. 2017) and the Victoria quadrangle (H02, Galluzzi et al. 2016). **a** The geologic contacts drawn by the mappers are shown in blue for H03 and in pink for H02; the linework on both sides extends beyond the quadrangle boundary (grey dashed line) by 5° in longitude (blue and pink meridians, respectively). Here several contact mismatches are visible. **b** When the linework is clipped to the actual quadrangle boundary, the mismatches are highlighted also by a different classification of the mapped geological units and a different interpretation of crater ejecta extent. **c** The final correct output of the map is obtained by a detailed review of the boundary made by the involved mappers. Projection: Lambert Conformal Conic with standard parallels at 30°N and 68°N and central meridian at 90°W. Legend: sp, smooth plains; spn, northern smooth plains; imp, intermediate plains; icp, intercrater plains; C3, fresh crater; C2, moderately degraded crater; C3, heavily degraded crater

The guidelines described in this review were born from a series of mapping trials-and-errors that allowed us to gather enough experience to facilitate the work of our other teammates. By defining a clear workflow and dealing with the main interpretation conflicts, the mappers tend to finish the work much quicker, and the quadrangle boundaries are merged more easily. The experience gained and lessons learned from this work are not to be considered peculiar to the Mercury case alone, but could be applied to most planetary surfaces that need a global geologic survey involving many mappers.

Acknowledgements The mapping review presented in this chapter was possible thanks to the effort of many people that are contributing to the 1:3M-scale geologic mapping of Mercury, in particular: Lorenza Giacomini, Laura Guzzetta, Alexander M. Lewang, Christopher C. Malliband, Paolo Mancinelli, Alessandro Mosca, David Pegg, Jack Wright, Luigi Ferranti, Harald Hiesinger,

Matteo Massironi, Cristina Pauselli, Pasquale Palumbo, and David A. Rothery, who also revised this chapter. I thank Dr. R. Aileen Yingst, who gave precious suggestions for improving the chapter content. I gratefully acknowledge funding from the Italian Space Agency (ASI) under ASI-INAF agreement 2017-47-H.0.

References

Becker KJ., Robinson MS, Becker TL, Weller LA, Edmundson KL, Neumann GA, Perry, ME, Solomon SC (2016) First global digital elevation model of mercury. In: Lunar and planetary science conference, vol 47, p 2959

Benkhoff J, van Casteren J, Benkhoff J, Hayakawa H, Fujimoto M, Laakso H, Novara M, Ferri P, Middletond HR, Ziethe R (2010) BepiColombo—comprehensive exploration of mercury: mission overview and science goals. Lunar Planet Space Sci 58:2–20. https://doi.org/10.1016/j.pss.2009.09.020

Blewett DT (2015) Hollows (Mercury). In: Hargitai H, Kereszturi Á (eds) Encyclopedia of planetary landforms. Springer Science+Business Media, New York, pp 935–937

Blewett DT, Chabot NL, Denevi BW, Ernst CM, Head JW, Izenberg NR, Murchie SL, Solomon SC, Nittler LR, McCoy TJ, Xiao Z, Baker DMH, Fassett CI, Braden SE, Oberst J, Scholten F, Preusker F, Hurwitz DM (2011) Hollows on mercury: MESSENGER evidence for geologically recent volatile-related activity. Science 333:1856–1859. https://doi.org/10.1126/science.1211681

Byrne PK, Klimczak C, Celâl Sengör AM, Solomon SC, Watters TR, Hauck SA II (2014) Mercury's global contraction much greater than earlier estimates. Nat Geosci 7:301–307. https://doi.org/10.1038/NGEO2097

Davies ME, Dornik SE, Gault DE, Strom RG (1978) Atlas of mercury. NASA Special Publication, p 423

Dzurisin D (1978) The tectonic and volcanic history of mercury as inferred from studies of scarps, ridges, troughs, and other lineaments. J Geophys Res 83:4883–4906. https://doi.org/10.1029/JB083iB10p04883

Frigeri A, Federico C, Pauselli C, Coradini A (2009) Fostering digital geologic maps: the digital geologic map of mercury from the USGS atlas of mercury, geologic series. In: Lunar and planetary science conference, vol 40, p 2417

Galluzzi V, Guzzetta L, Ferranti L, Di Achille G, Rothery DA, Palumbo P (2016) Geology of the victoria quadrangle (H02), mercury. J Maps 12:227–238. https://doi.org/10.1080/17445647.2016.1193777

Galluzzi V, Guzzetta L, Mancinelli P, Giacomini L, Lewang AM, Malliband CC, Mosca A, Pegg D, Wright J, Ferranti L, Hiesinger H, Massironi M, Pauselli C, Rothery DA, Palumbo P (2018) The making of the 1:3M geological map series of Mercury: status and updates. LPI Cont. 2047:6075

Guzzetta L, Galluzzi V, Ferranti L, Palumbo P (2017) Geology of the Shakespeare quadrangle (H03), mercury. J Maps 13:227–238. https://doi.org/10.1080/17445647.2017.1290556

Kinczyk MJ, Prockter LM, Chapman CR, Susorney HCM (2016) A morphological evaluation of crater degradation on mercury: revisiting crater classification using MESSENGER data. In: Lunar and planetary science conference, vol 47, p 1573

Mancinelli P, Minelli F, Pauselli C, Federico C (2016) Geology of the raditladi quadrangle, mercury (H04). J Maps 12:190–202. https://doi.org/10.1080/17445647.2016.1191384

Massironi M, Byrne PK (2015) High-relief ridge. In: Hargitai H, Kereszturi Á (eds) Encyclopedia of planetary landforms. Springer Science+Business Media, New York, pp 932–934. https://doi.org/10.1007/978-1-4614-3134-3

Massironi M, Byrne PK, van der Bogert CH (2015) Lobate Scarp. In: Hargitai H, Kereszturi Á (eds) Encyclopedia of planetary landforms. Springer Science+Business Media, New York, pp 1255–1262

McCauley JF, Guest JE, Schaber GG, Trask NJ, Greeley R (1981) Stratigraphy of the caloris basin. Mercury Icarus 47:184–202. https://doi.org/10.1016/0019-1035(81)90166-4

Melosh JH, McKinnon WB (1988) The tectonics of mercury. In: Vilas F, Chapman CR, Matthews MS (eds) Mercury. University of Arizona Press, Tucson, Ariz, pp 374–400

Mest SC, Crown DA, Yingst RA, Berman DC, Williams DA, Buczkowski, DL, Scully, JEC, Platz, T, Jaumann, R, Roatsch T, Preusker F, Nathues A, Raymond CA, Russel CT (2016) Update on the global geologic map of ceres from NASA's dawn mission, GSA, paper #110-07

Murray BC, Belton MJS, Danielson GE, Davies ME, Gault DE, Hapke B, O'Leary B, Strom RG, Suomi V, Trask N (1974) Mercury's surface: preliminary description and interpretation from Mariner 10 pictures. Science 185(4146):169–179. https://doi.org/10.1126/science.185.4146.169

Rothery DA, Mancinelli P, Guzzetta L, Wright J (2017) Mercury's Caloris basin: continuity between the interior and exterior plains. J Geophys Res: Planets 122(3):560–576

Schultz RA (2000) Localization of bedding plane slip and backthrust faults above blind thrust faults: keys to wrinkle ridge structure. J Geophys Res 105:12035–12052. https://doi.org/10.1029/1999JE001212

Spudis PD, Guest JE (1988) Stratigraphy and geologic history of mercury. In: Vilas F, Chapman CR, Matthews MS (eds) Mercury, pp 118–164, University of Arizona Press. ISBN 0816510857

Strom RG, Trask NJ, Guest JE (1975) Tectonism and volcanism on mercury. J Geophys Res 80:2478–2507. https://doi.org/10.1029/JB080i017p02478

Tanaka, KL, Skinner JA Jr, Hare TM (2011) Planetary geologic mappers handbook. USGS Astrogeology Science Center

Thomas RJ, Rothery DA, Conway SJ, Anand M (2014) Hollows on mercury: materials and mechanisms involved in their formation. Icarus 229:221–235. https://doi.org/10.1016/j.icarus.2013.11.018

Trask NJ, Guest JE (1975) Preliminary geologic terrain map of Mercury. J Geophys Res 80:2461–2477. https://doi.org/10.1029/JB080i017p02461

Tobler W (1987) Measuring spatial resolution. In: proceedings of the land resources information systems conference, Beijing, pp 12–16

Watters TR, Nimmo F (2010) The tectonics of mercury. In: Watters TR, Schultz RA (eds) Planetary tectonics. Cambridge Univ Press, New York, pp 15–80

Williams DA, Yingst RA, Garry WB (2014) Introduction: the geologic mapping of Vesta. Icarus 244:1–12. https://doi.org/10.1016/j.icarus.2014.03.001

Williams DA, Buczkowski DL, Mest SC, Scully JEC, Jaumann R, Raymond CA, Russell CT (2016) Geologic mapping campaign for ceres from NASA dawn mission, LPSC, Abstract #1515

Yingst RA, Mest SC, Berman DC, Garry WB, Williams DA, Buczkowski D, Jaumann R, Pieters CM, De Sanctis MC, Frigeri A, LeCorre L, Preusker F, Raymond CA, Reddy V, Russell CT, Roatsch T, Schenk PM (2014) Geologic mapping of vesta. Planet Space Sci 103:2–23. https://doi.org/10.1016/j.pss.2013.12.014

Zuber MT, Smith DE, Phillips RJ, Solomon SC, Neumann GA, Hauck SA, Peale SJ, Barnouin OS, Head JW, Johnson CL, Lemoine FG, Mazarico E, Sun X, Torrence MH, Freed AM, Klimczak C, Margot J-L, Oberst J, Perry ME, McNutt Jr. RL, Balcerski JA, Michel N, Talpe MJ, Yang D (2012) Topography of the northern hemisphere of mercury from MESSENGER laser altimetry. Science 336(6078):217–220. https://doi.org/10.1126/science.1218805

Planetary Map Design: The Chang'E-1 Topographic Atlas of the Moon

Lingli Mu, Jianjun Liu and Longfei Liu

Abstract This chapter presents how the Chang'E-1 Topographic Atlas of the Moon (Li et al. in The Chang'E-1 topographic atlas of the Moon. Sinomaps Press, Beijing China, 2013; Li et al. in The Chang'E-1 topographic atlas of the Moon. Springer, Berlin, 2016) was designed and produced covering map designing, data processing, atlas editing, and publishing subtasks. Under this mapping framework, China's Lunar Exploration Program has released several maps and atlases with high efficiency and will carry out mapping of Mars in the future. The tasks described here can be used for any planetary map, from map design to map publishing.

Keywords Planetary map design · Lunar mapping · Framework · Chang'E · Topographic Atlas of the Moon

1 History of Lunar Mapping

Lunar mapping has a long and intriguing history (Schimerman et al. 1973). About 400 years ago, William Gilbert believed that the dark-toned regions on the Moon are continents and the light-toned ones are seas. In his book, *De Mundo Nostro Sublunari Philosophia Nova,* he included a lunar map, created from naked eye observation before the invention of the telescope. In several years, Galileo Galilei portrayed the Moon using his telescopic observations (Fig. 1), published in his book *Sidereus Nuncius* in 1610.

L. Mu (✉)
Key Laboratory of Space Utilization, Technology and Engineering Center for Space Utilization, Chinese Academy of Sciences, Beijing 100094, China
e-mail: mulingli@csu.ac.cn

L. Mu · J. Liu
Key Laboratory of Lunar and Deep Space Exploration, National Astronomical Observatories, Chinese Academy of Sciences, Beijing 100101, China

L. Liu
National Disaster Reduction Center of China, Beijing 100124, China

© Springer Nature Switzerland AG 2019

H. Hargitai (ed.), *Planetary Cartography and GIS*, Lecture Notes in Geoinformation and Cartography, https://doi.org/10.1007/978-3-319-62849-3_10

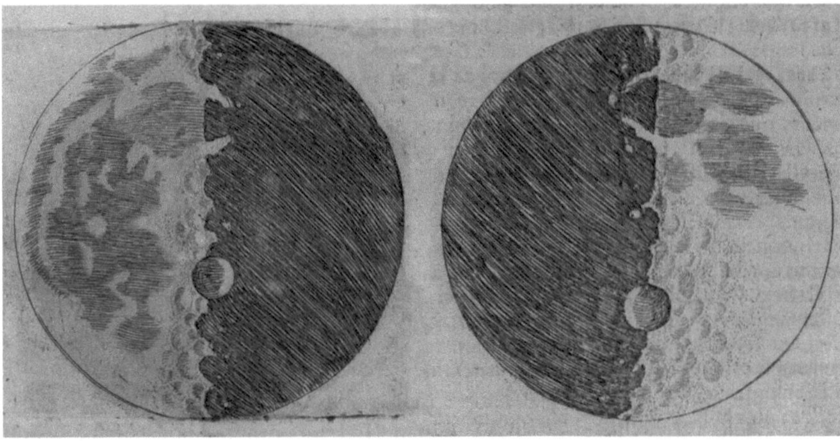

Fig. 1 The first telescopic lunar map, drawn by Galileo Galilei, 1610 (Siderius Nuncius, Venice 1610)

In the following 300 years, many engineers, mathematicians, and astronomers invented more powerful telescopes to observe the near side of the Moon and produced progressively larger-scale maps and then photographs.

Lunar remote sensing arrived in the years 1958–1976. The "Moon race" between the Soviet Union and the USA accelerated, and a series lunar exploration programs were launched, such as Luna, Pioneer, Ranger, Cosmos, Lunar Orbiter, Surveyor, Zond, and Apollo. In 1959, the first photographs of the far side of the Moon were taken by the Soviet probe Luna 3. In the 1960s, the US Air Force ACIC released the *Lunar Astronautical Chart* (LAC) sheets, as the first lunar subdivision map, which divided the Moon into 144 sheets at a scale of 1:1, 000, 000. In 1971, Bowker and Hughes produced the Lunar Orbiter Photographic Atlas of the Moon with a global coverage. For the Apollo program, high-resolution images were acquired and utilized to make topographic maps with the photogrammetric method. During this period, the topographic maps used contour line, shaded relief, and photo to portray the lunar topography.

In the 90's of last century, the Japanese Hiten, and the American Clementine and Lunar Prospector spacecraft visited the Moon and obtained a vast amount of remote sensing images and elevation data. In 1999, Eliason et al. produced the *UVVIS 750 nm Basemap* using the Clementine 750 nm images. In 2002, the 1:10, 000, 000 *Clementine Color-Coded Topography and Shaded Relief Maps* (Rosiek et al. 2002) were produced. In 2004, Bussey and Spudis published the *Clementine Atlas of the Moon* with images and corresponding shaded relief sheets based on the LAC series.

At the beginning of the twenty-first century, ESA, NASA, JAXA, China, and India launched a series of lunar exploration programs, such as SMART-1, LRO and GRAIL, SELENE, Chang'E, and Chandrayaan-1, respectively. During this period, global high-resolution images and laser altimeter data were obtained (Liu et al. 2015), which were used to derive topographic data for a large-scale lunar global

Table 1 Lunar global mapping products

Data	Mission	Sensor	Resolution	Coverage
ULCN2005 DEM	Clementine (USA, 1994.01)	UV-VIS camera and LIDAR	1. 8 km	Nearly 100%
SLN LALT-DEM	SELENE (Japan, 2007.09)	LALT	1. 8 km	100%
SLN DEM	SELENE (Japan, 2007.09)	Terrain camera	10 m	100%
SLN DOM	SELENE (Japan, 2007.09)	Terrain camera	10 m	100%
CE1 DOM	Chang'E-1 (China, 2007.10)	3 line CCD	120 m	100%
CE1 DEM	Chang'E-1 (China, 2007.10)	3 line CCD	500 m	100%
CE1 LAM-DEM	Chang'E-1 (China, 2007.10)	LALT	3 km	100%
LRO WAC-DOM	LRO (USA, 2009.06)	Wide angle camera	100 m	100%
LRO WAC-DEM	LRO (USA, 2009.06)	Wide angle camera	100 m	100%
LRO LAM-DEM	LRO (USA, 2009.06)	Laser altimeter	30 m	100%
SLDEM	Selene/LRO (USA. 2015.09)	LALT and laser altimeter	60 m	100%
CE2 DOM	Chang'E-2 (China, 2010.10)	2 line CCD	7 m	100%
CE2 DEM	Chang'E-2 (China, 2010.10)	2 line CCD	7 m	100%

mapping (Table 1). Through these programs, NASA produced the LRO topographic and roughness maps of the Moon, JAXA released *The Kaguya Lunar Atlas*, and China published a series of lunar maps, such as *The Chang'E-1 Image Atlas* (Li et al. 2010a), *The Chang'E-1 Topographic Atlas* (Fig. 2), *The Chang'E Globe of the Moon* (Li et al. 2010b), *The Chang'E-2 High Resolution Image Atlas of the Moon* (Li et al. 2012), and so on.

Between 2018 and 2028, more than ten lunar missions are expected to be launched. China will send Chang'E-4/5/6. India expects to launch the Chandrayaan-2. JAXA plans lunar landing missions including OMOTENASHI and SLIM. Russia also announced to restart the Luna-Glob project. NASA will launch the EM-1/2 and LunaH-Map missions.

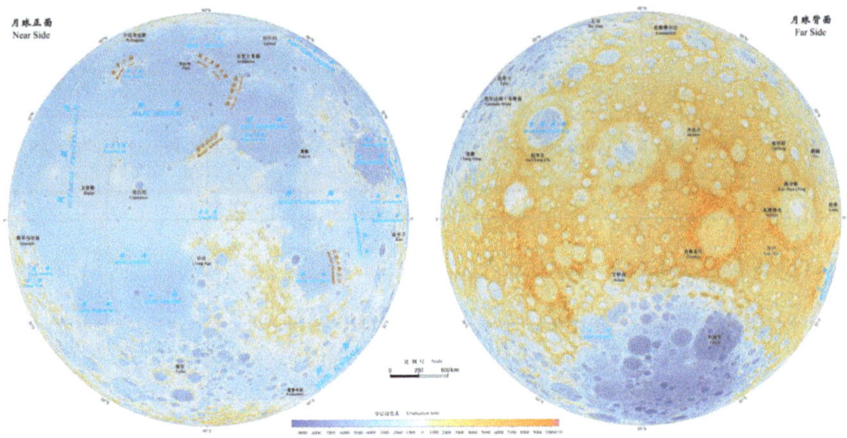

Fig. 2 Color-coded shaded topographic map of the Moon from China's first probe Chang'E-1 (Li et al. 2013, 2016)

2 Introduction to Lunar Mapping Framework

As the first Chinese unmanned lunar-orbiting spacecraft, Chang'E-1 was launched on October 24, 2007. One of the major goals for the Chang'E-1 mission was to create an accurate and high-resolution lunar 3-D map. Global stereo images of the lunar surface were obtained after a year's data acquisition. Chinese scientists released *the Chang'E-1 Topographic Atlas of the Moon*[1] (CE1TAM) in the spring of 2013. Because of the technical innovation and the high-quality topographic portrayal of the Moon, the CE1TAM won the third jury prize in the category of Atlases at the 26th International Cartographic Conference in Dresden, Germany (2013). It was the first time for this prize to be awarded to a product in planetary cartography.

To produce standard lunar map products, Mu et al. (2013) created a lunar mapping framework and map production flow (Fig. 3) with the following content:

- **Map planning**: For the purpose of the Chinese Lunar Exploration Program (CLEP), the Chang'E-1 was sent to get a global image of the Moon and DEM data for 1:2, 500, 000 mapping. Chang'E-2 obtained 7 m resolution data to support the decision on Chang'E-3's landing site. The Compiling Committee from the Ground Research and Application System (GRAS) of CLEP would make a series mapping plans for each mission.
- **Standardization**: In Earth mapping, numerous standards have been specified, most of which could not be used for lunar mapping directly. Therefore, it was proposed to set up lunar mapping standards for sharing lunar spatial data and

[1]http://159.226.88.61/CLEPWebMaps/CEAtlas/CE1/Atlas.html#page/1.

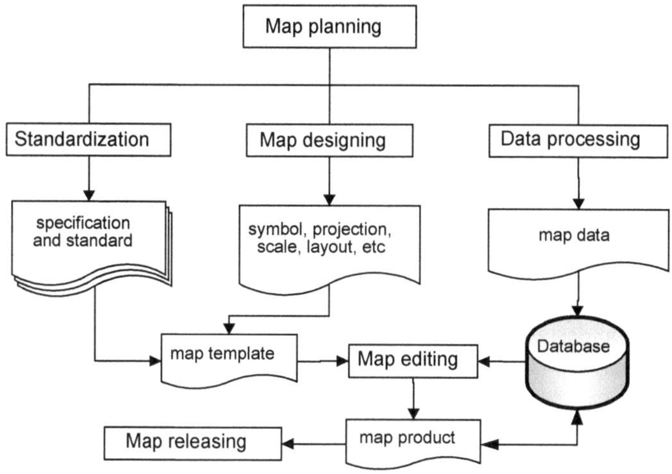

Fig. 3 Lunar map production flow chart

ensuring the quality of the lunar map products (Archinal et al. 2008). The Chinese Compiling Committee has compiled a number of Chinese national standards. Most of them are fundamental lunar specifications such as the coordinate system, map subdivision, lunar place name translations in Chinese, map control, lunar metadata, data processing, data format and storage. In recent years, some of these specifications have been released by the Standardization Administration of the People's Republic of China (SAC).

- **Map designing**: According to different map plans, the Compiling Committee defined and designed the map size, map projection, symbol, scale, layout, and so on.
- **Data processing**: Data collected from different missions have different resolutions, file formats, and projections, which makes it difficult to edit the data directly by cartographers. Consequently, all of the data would be reprocessed and stored in the database as standard map data according to the standards mentioned above (Li et al. 2010c, d).
- **Map editing**: First, the designers formed a general plan for map making and created a series of map templates following the standards of map design. Then, the editors used pre-processed map data and templates to create new maps in the digital map editing environment. Finally, the atlas or a number of maps were compiled in a relatively short time with high quality.
- **Map publishing**: After map quality check and text review, the paper map product was sent to print and release. For the Web map, all of the map data were stored in the map server, and the client could use the browser to browse and query the map information.

Leveraging the lunar mapping framework, the Compiling Committee had spent one year, six months, and one month, respectively, to produce the *CE1TAM, The Chang'E-1 Image Atlas* of the Moon[2] and *The Chang'E-2 High Resolution Image Atlas of the Moon.*[3]

3 Map Designing

Map designing includes atlas content selection, mathematical foundation design, and presentation design. Map designing is a key step, which determines the overall effect of the map. In the following sections, we present the map designing process of CE1TAM as an example.

3.1 Atlas Content Design

The key objective of the CE1TAM is to demonstrate the lunar topography using Chang'E-1 (CE-1) data. There is no creature and man-made building on the Moon, except landers and rovers sent from the Earth. Therefore, the map content focuses on the lunar landforms, which include 17 named types as defined by the IAU, including Craters, Maria, Rimae, and so on.

The CE1TAM includes three parts. The first part introduces the background of the CE1TAM with text. In this section, readers can learn about the origin of the mission and some important techniques used in the atlas. The second part is about data processing, which describes data acquisition, data preprocessing, and global topographic data processing. The last and most important part contains the subdivision maps. It is the main section of the atlas, which displays the topographic maps of the entire Moon with large scale for the first time.

The appendix of the CE1TAM is a glossary of named objects on the Moon with detailed location, geometric information, and page number.

3.2 Mathematical Foundation Design

Mathematical foundation in cartography includes the coordinate system and the projection. For the lunar coordinate system, the lunar reference ellipsoid is defined as a sphere with a radius of 1737.4 km, which was published as Chinese national standard: *The Lunar Coordinate System* (Li et al. 2014).

[2]http://159.226.88.61/CLEPWebMaps/CEAtlas/CE1_2/Atlas.html#page/1.

[3]http://159.226.88.61/CLEPWebMaps/CEAtlas/CE2/Atlas.html#page/.

Table 2 Numbering, range, and projection for 1:1,000,000 lunar subdivision map

Numbering (ID)	Latitude range	Longitude range	Projection	Central meridian	Standard parallel
A001	84°N–90°N	360°	Polar azimuthal projection	0°	90°N
B001–008	70°N–84°N	45°	Lambert conformal conic projection	Mean value of the longitude range	77°N
C001–012	56°N–70°N	30°			63°N
D001–015	42°N–56°N	24°			49°N
E001–018	28°N–42°N	20°			35°N
F001–020	14°N–28°N	18°			21°N
G001–020	0°N–14°N	18°	Mercator projection		7°N
H001–020	0°S–14°S	18°			7°S
I001–020	14°S–28°S	18°	Lambert conformal conic projection		21°S
J001–018	28°S–42°S	20°			35°S
K001–015	42°S–56°S	24°			49°S
L001–012	56°S–70°S	30°			63°S
M001–008	70°S–84°S	45°			77°S
N001	84°S–90°S	360°	Polar azimuthal projection	0°	90°S

The CE1TAM used the global maps at a global scale and subdivision maps at a regional scale to demonstrate the lunar surface. Consequently, several kinds of projections were used in the CE1TAM. The Mollweide projection and orthographic projection were adopted for the global maps. The Mollweide projection can show the global lunar map with smaller distortion compared with other projections. The orthographic projection represents the Moon in two disks. One disk is for the far side, and another shows the familiar view of the near side. Subdivision maps were created to show the topographic details at a regional scale. Because of the low resolution, the CE-1 digital elevation model (DEM) only supports small-scale mapping. According to *The Subdivision and Numbering Standard for the Lunar Primary Scale Topographic Map* (Chinese National Standard. Mu et al. 2016), the CE1TAM adopted the 1:1,000,000 subdivision scale. Table 2 shows the numbering, range, and projection for the 1:1,000,000 lunar map series, which includes 188 map sheets, and Fig. 4 shows the thumb-sized lunar subdivision topographic maps.

An index map with orthographic projection was also designed. In the index map (Fig. 5), the subdivision range, subdivision number, and corresponding page number are marked. If the coordinate of the lunar object is known, it can be found in the corresponding page easily through the index map.

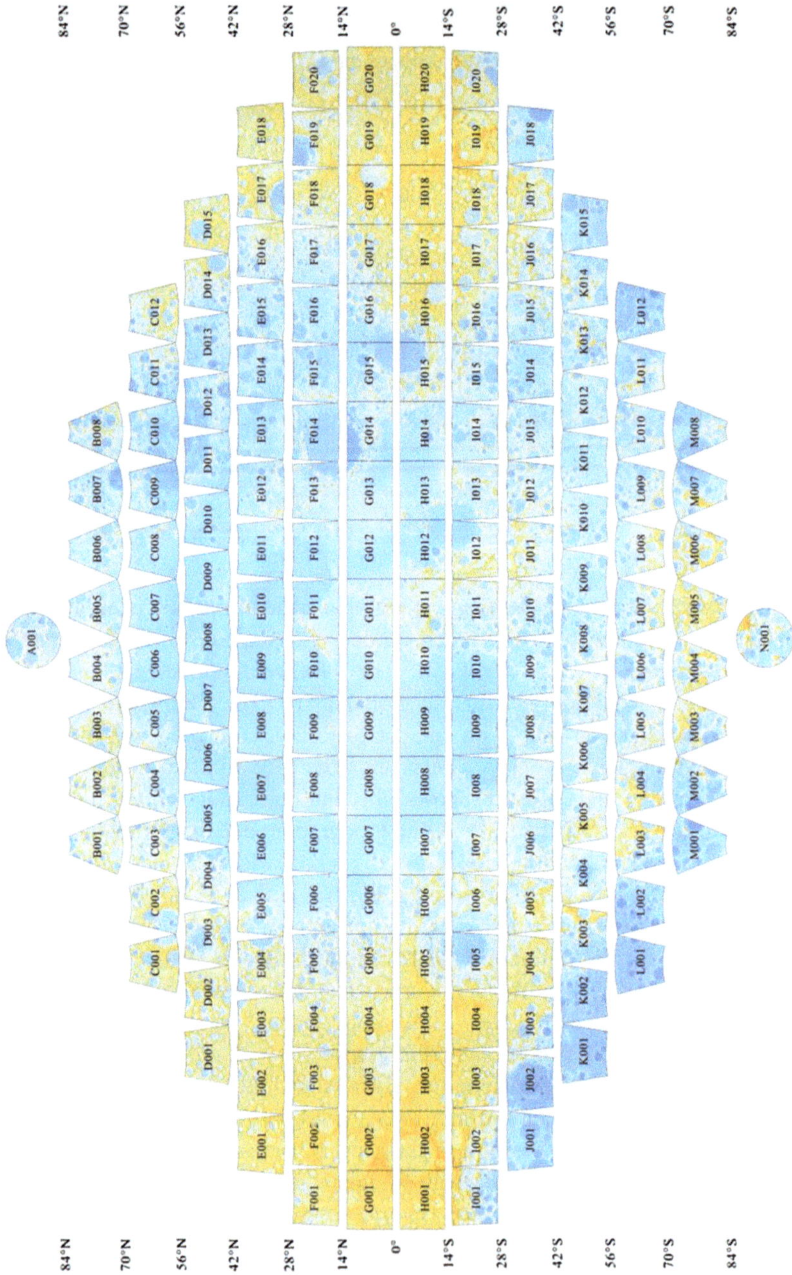

Fig. 4 Thumb lunar subdivision topographic maps

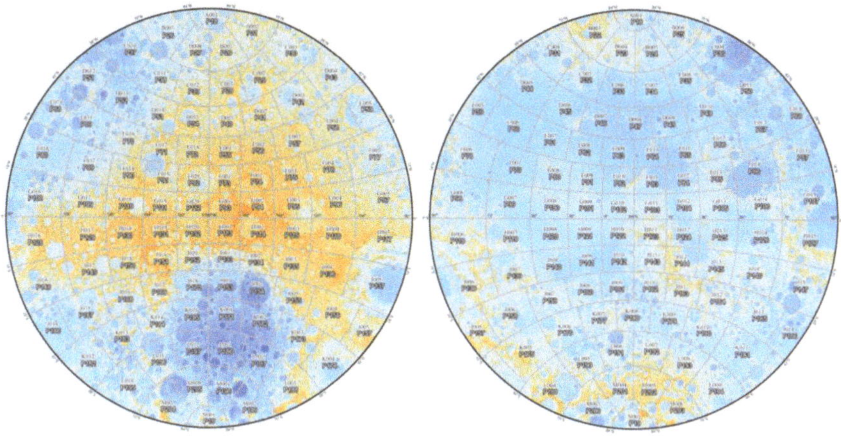

Fig. 5 Index map for the CE1TAM

3.3 *Presentation Design*

In the CE1TAM, color shaded relief and contours were utilized to present different types of lunar landforms with annotation. In the atlas, the gradient colors can reflect elevation difference on the lunar surface, and the contour provides elevation information for quantitative scientific analysis.

In the color-coded relief, the elevation information is represented by different colors (Fig. 6). From the lunar image, we can see that the tone is strong in high-altitude areas such as the highlands, but weak in low-altitude areas such as the Maria. Therefore, the warm and cold colors are used to express the altitude difference. In the map, the lowlands below the geoid are shown in blue tones and highlands in deep yellow tones. From the lowest point to the summit, the color turns blue to deep yellow gradually. In order to create shading and enhance the stereo effect, 3D modeling technique is used to render the surface by controlling the light direction and solar elevation angle.

All of the contours in the CE1TAM are extracted from the CE-1 DEM data. Taking into account the lower resolution, the contour interval is defined as 500 m. At every 5 contour intervals, there is an elevation label. In the CE1TAM, the width of the intermediate contour is defined to be 0.1 mm, and the width of the index contour is 0.15 mm.

The annotation can help the reader to locate objects. About 3698 lunar names of 17 types were selected according to size and importance. These annotations of

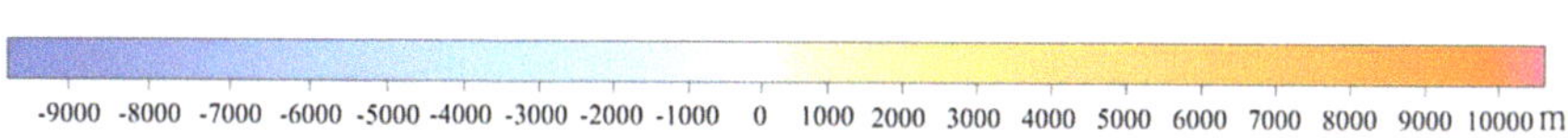

Fig. 6 Basic graduation tints for the color-coded relief

Mare, Rima, Sinus, Lacus, and Palus features are labeled in blue. Those annotations of the Montes and Mons features are labeled in deep brown. Other annotations are labeled in black. The text font, size, inclination angle, and direction are also defined for each type of annotation, respectively. For example, the font of Montes annotation is Times New Roman, the font size is 9 pt, and the editor places it along the ridgeline with 15° inclination.

3.4 Layout Design

According to the lunar subdivision dimension and the reading habit for atlases, the size of the first version of CE1TAM was defined as 370 mm × 340 mm, and the corresponding scale was defined as 1:1, 750, 000. Three years later, the size for the second version released by Springer was scaled down.

The aim of the layout design was to balance all of the map elements. In the CE1TAM, all of the map elements were organized using the concept *layer*. One layer contained one type of map elements. The elements contained in the map border were called *main map elements,* and others called *auxiliary map elements.* The main map elements include the color shaded relief, contour, annotation, and latitude–longitude grid with label marks, which are geographically matched and superimposed from bottom to top. The annotations are bilingual (Chinese and Latin) in the first version and Latin in the second version. The priority of annotation positioning is center, top, left, right, and bottom. The color of the latitude–longitude grid is defined as light gray.

The auxiliary map elements are outside the map border, including index diagram, map code, scale, and graduation tint. The index diagram is similar to the index map. In the index diagram, located in the outside of the upper blank area, the current subdivision area is marked in dark gray and the boundary is black. The adjacent subdivision area is marked in gray with the corresponding subdivision number. The map code same as the current subdivision number is placed in the center of the upper blank area. In the second version, the index diagram and map code are on the upper left or upper right. The scale and graduation tint are placed in the lower blank area. The bar scale and the graduation tint are located in the lower blank area.

The layout design and organization of the map elements are presented in Fig. 7.

3.5 Map Templates

As a result of the above design, the map designers used test data to make a map design program and produce sample maps. The Compiling Committee evaluates the program and each sample map. After the evaluation, the designers would modify the template according to the Compiling Committee's suggestion. Generally

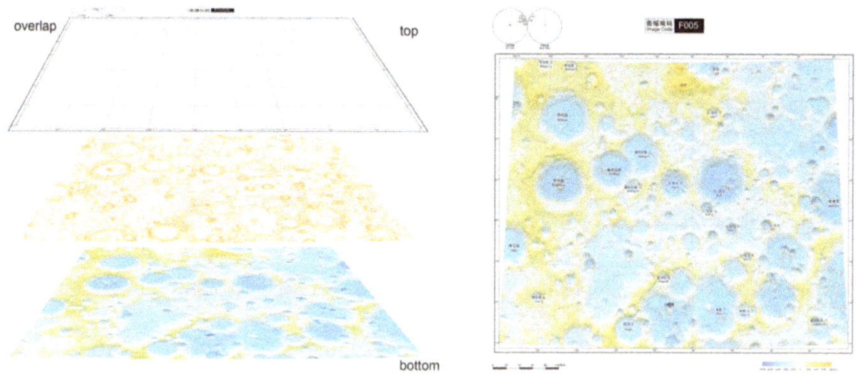

Fig. 7 Layout editing of the map elements

speaking, it is an iterative procedure. Finally, as a reference, the map templates would be used to produce new maps by the map editors.

4 Data Processing

NASA, JAXA, CLEP, and ISRO have used the stereo CCD camera and laser altimeter to obtain the lunar digital orthophoto map (DOM) and digital elevation model (DEM) data. The *photogrammetry* method was used to derive the DEM and DOM data from the stereo images obtained by CCD camera. The elevation data also can be extracted from the *location parameters* of the laser altimeter and altimetry data. The former method could result in better accuracy at a higher resolution. The above data are called *scientific data*. To compile the map, scientific data should be processed to *map data*, which is based on the standard and map design program.

4.1 Extracting Topographic Data

The Chang'E-1 (CE-1) and Chang'E-2 (CE-2) probes used stereo CCD cameras to acquire images. A three-line-array CCD push-broom camera onboard the CE-1 had three viewing angles: forward view, nadir view, and backward view. The two-line-array CCD stereo camera board on the CE-2 had two viewing angles: forward view and nadir view (Li et al. 2011). The CE-1 CCD stereo camera imaging process is shown in Fig. 8.

To get the global DOM and DEM data, the photogrammetry method was adopted. Before the processing, according to the Chinese national standard, *Subdivision and Numbering for the Lunar Primary Scale Topographic Map*,

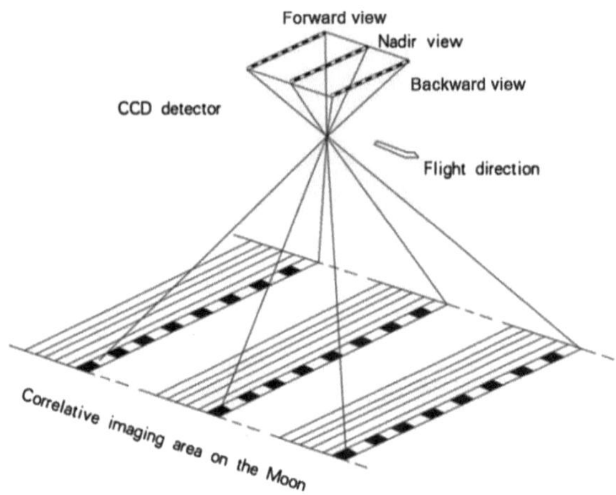

Fig. 8 CCD stereo camera imaging process (Li et al. 2016)

all the stereo images and auxiliary parameters were divided and dispatched to every subdivision block. The data processing procedures (Li et al. 2010c, 2016; Ouyang et al. 2010) are as follows:

- **Image matching**: The scale-invariant feature transform (SIFT) and least squares matching (LSM) methods were used to find the matching points in forward, nadir, and backward view images. The image matching provided sparse and dense matching points for the subdivision adjusting and DEM generation. Finally, in the three-dimensional digital photogrammetry workstation environment, the data processing technicians inspected and edited the above points one by one.
- **Block adjustment**: Three-line block independent models were used for block adjustment and global block adjustment to provide processing parameters for the absolute orientation and seamless mosaicking of lunar global topographic data.
- **DEM and DOM extracting**: In each block, the dense matching points and forward intersection methods were employed to calculate the elevation, longitude, and latitude coordinates of the matching points. Subsequently, the elevation points were interpolated into DEM data at 500 m resolution by triangulated irregular network (TIN), and the image was orthorectified to form the DOM at 120 m. Finally, in each block, the above data was mosaicked.
- **Global data mosaicking**: After quality check, all of the data were transformed and cut according to the above subdivision standard. The standard DOM and DEM data products were produced and stored in the ArcGIS Server. In this method, all of the data could be shared by data processing units.

4.2 Map Data Processing

The main map elements are all derived or transformed from scientific data, in the procedure called *map data processing*. All of the map data processing is based on the standards and design parameters. The data processing in the CE1TAM includes DEM division, contour data extracting, annotation translation, and graticule grid making.

To get the subdivision DEM data, the global lunar DEM was divided into 188 sheets according to the 1:1,000,000 subdivision, and each sheet was transformed with the corresponding projection. The contour extraction was based on the subdivision DEM data. In the CE1TAM, objects smaller than 2 mm × 2 mm were dropped and the remaining ones were generalized.

The first version of the CE1TAM was released in China, so the text and annotation had to be in Chinese. However, all of the lunar place names released by IAU are in Latin. The first step on the annotation processing was to create a translation standard and translate the names (Hargitai et al. 2014). Because of the layout size, all of the names could not be marked in the subdivision map. The second step was to select place names. In the CE1TAM, the key problem of name selection was to decide the standard. In each subdivision, the number of annotations was not more than 20. So some important place names and a part of crater names were retained.

The latitude–longitude grid can help the reader to locate and measure the distance between objects. In low latitudes, the interval was defined to be 2°. In higher latitudes, the longitude interval was defined to be more than 2°.

ArcMap provides all of the functions to support map data processing. All of the main map elements that resulted from the map data processing were stored in the ArcGIS Server with projection, which could provide the georeferenced match in the digital map editing environment.

5 Map Editing

The atlas editing was carried out in a digital map editing environment. The cartographers put the map data into the template and edited the map elements according to the design program.

5.1 Editing Environment

The editing environment was composed of professional software, database, computer, printer, and network elements. In the data module of ArcMap, some data tools were used to process the map data, such as the data dividing, mosaicking, projection transformation, contour extraction, and generalization. In the layout module of ArcMap, the cartographers edited the scale, color, font, line style, and other properties of the map elements. The software *Surfer* was used to produce color shaded relief. The cartographers utilized *Adobe Illustrator* to modify the layout of the map elements again and *Adobe In Design* to make the page layout of the text, picture, and subdivision for the atlas. All of the map data were stored in the database ArcServer. The above professional software and the database were set up on the workstation and data server. The printer provided the function to print the paper map samples. Based on the samples, the designers and the editors could check the quality of the map elements. All of the devices were connected through the network.

5.2 Map Editing

According to the design program, the cartographers used ArcMap to set the map scale to 1:1,750,00 and modify the color, font, line style of the vector map elements. Each type of the edited map element was exported as a file in Adobe Illustrator data format. For the color shaded relief, the DEM data was rendered according to the graduation tint using the 3D surface function in Surfer. The light direction was defined in the north and the solar elevation angle between 60–70°.

5.3 Layout Editing

5.3.1 Subdivision Map Typesetting

Putting all of the map elements into Adobe Illustrator, the cartographers placed the elements again according to the layout design to form digital standard subdivision map products. The following is the procedure:

- Update the auxiliary map elements, and locate them according to the map template.
- Put the map border in the center of the subdivision.
- Overlap the color shaded relief, contours, annotation, and longitude–latitude grid in turns.

Figure 7 shows the layout editing of the map elements.

5.3.2 Atlas Typesetting

The typesetting of the CE1TAM included text, picture, and subdivision map. In Adobe InDesign, the designers made a page layout template. The template defined the content area and the page pattern, such as the position, font, size of the page header and number. In a new page, the editors inserted the text and picture or linked the subdivision map in the content area and made some modification according to the page template.

6 Map Publishing

When the atlas editing was finished, all of the sheets would be printed as samples. The cartographers would made quality check according to the standards and design program. If some mistake occurred in the sheets, those sheets would be re-edited for correction. When there was no mistake, the atlas would be sent to the publishing house. The first version of the *Chang'E-1 Topographic Atlas of the Moon* was published by SinoMap Press in 2013, and the second one was published by Springer in 2016.

7 Conclusion

In the past, there was no systematic method or framework to support planetary mapping in China. In this chapter, the lunar mapping framework defined by Chinese cartographers was presented. The cartographers have used the framework to publish the *Chang'E-1 Topographic Atlas of the Moon* and other maps in relatively short time, which proved that the framework was very efficient.

Lunar mapping is a direction of planetary mapping. The current standards and programs mentioned above cannot fully meet all requirements of lunar mapping. In the next step, new standards and presentation methods on thematic maps would be considered. In the future, other planetary mapping efforts will be carried out in China, such as the Martian mapping in China's Mars Exploration Program 2020.

References

Archinal B, The Lunar Geodesy and Cartography Working Group (2008) Lunar mapping standards and the NASA LPRP lunar geodesy and cartography working group, Scientific Event B01, The Moon: Science, New Results, Ongoing Missions, Future Robotic and Human Exploration. 37th COSPAR Scientific Assembly, 13–20 July, Montreal, Canada, Abstract no. B01-0050-08

Bowker D, Hughes JK (1971) Lunar orbiter photographic atlas of the Moon. Lunar and Planetary Institute, Houston
Bussey B, Spudis PD (2004) The clementine atlas of the Moon. Cambridge University Press, United Kingdom
Eliason E, Isbell C, Lee E, Becker T, Gaddis L, McEwen A, Robinson M (1999) The clementine UVVIS global lunar mosaic? Lunar and Planetary Institute, Houston
Hargitai H, Li C, Zhang Z, Zuo W, Mu L, Li H, Shingareva KB, Shevchenko VV (2014) Chinese and Russian language equivalents of the IAU Gazetteer of planetary nomenclature: an overview of planetary toponym localization methods. Cartographic J. http://dx.doi.org/10.1179/1743277413Y.0000000051
Li C, Liu J, Mu L, Ren X, Zuo W (2010a) The Chang'E-1 Image Atlas of the Moon. Sinomaps Press, Beijing
Li C, Liu J, Mu L, Ren X, Zuo W (2010b) The Chang'E Globe of the Moon. Sinomaps Press, Beijing
Li C, Liu J, Ren X, Mu L et al (2010c) The global image of the Moon obtained by the Chang'E-1: data processing and lunar cartography. Sci China Earth Sci 53(8):1091–1102. https://doi.org/10.1007/s11430-010-4016-x
Li C, Ren X, Liu J, Zou X, Mu L et al (2010d) Laser altimetry data of Chang'E-1 and the global lunar DEM model. Sci China Earth Sci 53(11):1582–1593. https://doi.org/10.1007/s11430-010-4020-1
Li C, Liu J, Mu L, Ren X, Zuo W (2012) The Chang'E-2 High Resolution Image Atlas of the Moon. Sinomaps Press, Beijing
Li C, Liu J, Mu L, Ren X, Zuo W (2013) The Chang'E-1 Topographic Atlas of the Moon (Chinese and English), 1st edn. Sinomaps Press, Beijing
Li C, Ren X, Mu L et al (2014) Lunar Coordinate System (GB/T 30112-2013). Standardization Administration of the People's Republic of China
Li C, Liu J, Mu L, Ren X, Zuo W (2016) The Chang'E-1 Topographic Atlas of the Moon (English only), 2nd edn. Springer, Berlin
Liu J, Ren X, Mu L et al (2015) Progress in the lunar optical remote sensing and mapping research. Bull Miner, Petrol Geochem 34(3):461–470
Mu L, Li C, Liu J, Ren X, Zou X (2013) A new mapping method for the Moon with the Chang'E-1 data. Proceedings of the 26th International Cartographic Conference, Dresden, Germany
Mu L et al (2016) Subdivision and numbering standard for the lunar primary scale topographic map (GB/T 32521-2016). Standardization Administration of the People's Republic of China
Ouyang Z, Li C, Zou Y, Mu L et al (2010) Primary scientific results of Chang'E-1 lunarmission. Sci China Earth Sci 53(11):1565–1581
Rosiek MR, Kirk R, Howington-Kraus E (2002) Color-coded topography and shaded relief maps of the lunar hemispheres. In: Lunar and Planetary Science Conference XXXIII. Lunarand Planetary Institute, Houston, Abstract no. 1792
Schimerman LA et al (1973) Lunar cartographic dossier, vol I. NASA and the Defense Mapping Agency, St. Louis

Atlas Planetary Mapping: Phobos Case

I. P. Karachevtseva, A. A. Kokhanov and Zh. Rodionova

Abstract We present a general procedure of the Phobos Atlas creation. Main principles of mapping, mathematical, and geographical basics are described and justified. Data sources for mapping are listed. Approaches in the development of legends and design are considered, and some examples of the maps are shown.

Keywords Phobos Atlas · three-axial ellipsoid · Map design · Geomorphologic mapping

1 Introduction

The Phobos Atlas (MIIGAiK 2015) is based on images collected by different spacecraft, including the ongoing European Mars Express (launch 2003), NASA's Viking Orbiters (1976–1979), and the Soviet Phobos-2 (1988–1989) missions. The Atlas covers aspects of theoretical studies and practical data analysis for Phobos and integrates scientific results obtained by Russian researchers before and after Phobos-Grunt mission (2011). The Atlas was produced by MIIGAiK Extraterrestrial Laboratory (MExLab) under the support of the Russian Science Foundation (project №14-22-00197) and has broad objectives: firstly, to collect the results of the studies of the Martian satellite performed in different years (2009–2014); to present our modern knowledge about Phobos for educational and

I. P. Karachevtseva (✉) · A. A. Kokhanov
MIIGAiK Extraterrestrial Laboratory (MExLab), Moscow State University of Geodesy and Cartography (MIIGAiK), Moscow, Russia
e-mail: i_karachevtseva@miigaik.ru

Zh.Rodionova
Sternberg State Astronomical Institute, Lomonosov Moscow State University, Moscow, Russia

© Springer Nature Switzerland AG 2019

H. Hargitai (ed.), *Planetary Cartography and GIS*, Lecture Notes in Geoinformation and Cartography, https://doi.org/10.1007/978-3-319-62849-3_11

public outreach; and finally, to provide cartographic instrument for the preparation of the future international mission Bumerang[1] planned for launch with the same scientific tasks as the unsuccessful mission Phobos-Grunt: landing and sample return to understand the origin of Phobos and the Solar System.

Russia has a long-standing tradition in studying and mapping the Martian satellite. The first Russian map of Phobos was produced jointly at MIIGAiK and several institutions from the Russian Academy of Science as early as during preparation and planning of the Phobos-1 and -2 missions (1988). This map was created using airbrush technique and images from NASA missions Mariner 9 and Viking 1. To compile the map, special projections were developed, which represented the odd-shaped Phobos body in the form of triaxial ellipsoid (Bugaevsky 1987). Later, the map (1988) was used as a basis for a globe (Bugaevsky et al. 1992) and maps of Phobos in the *Atlas of the Terrestrial Planets and Their Satellites* (Shingareva et al. 1992) as well as in the multilingual map series on celestial bodies (Shingareva et al. 2005).

The Atlas of the Terrestrial Planets and Their Satellites (Shingareva et al. 1992) as a fundamental work, which combines comparative-planetographic descriptions, history of research, and various thematic maps, was used as a basic sample of the Phobos Atlas. Another example was the *International Atlas of Mars Exploration* (Stooke 2012) that contains not only maps, but multi-page text descriptions, accompanied by annotated images, and focuses on some aspects of the history of Mars research (including Phobos) that can be presented using cartographic methods.

2 Principles and Structure of the Atlas

2.1 Concept of the Phobos Atlas

An Atlas is a systematic collection of maps, drawn up according to the general procedure as a complete product (Salishev 1982). The Phobos Atlas was created as a comprehensive project, which represents miscellaneous characteristics of surface and physical properties of one of the Martian satellites. The main idea of our Atlas is to record the knowledge and experience of Phobos research, so besides maps, the issue includes descriptions of studies and their scientific results, methods, and techniques, as well as various catalogues (images, control points network, craters). The creation of the Atlas was based on the principles of integrality and complementarity of various sources integrated into an ArcGIS geodatabase (Karachevtseva et al. 2015).

[1]https://www.laspace.ru/projects/planets/expedition-m/.

2.2 Atlas Composition

Atlas materials are divided into four chapters:

I. History of Phobos studies and mapping;
II. Control point network, shape, and gravity field;
III. Spatial analysis of Phobos' surface;
IV. Geomorphology studies of Phobos.

The Atlas is structured sequentially, and each chapter is based on the results presented in the previous ones: e.g., the first chapter combines modern knowledge about main Phobos parameters and briefly describes the history of Phobos mapping, and then, the second chapter describes the photogrammetric technique and the results of modern image processing that provide coordinates basic for further analysis. The third section includes descriptions of the implementation of GIS technologies for semi-automatic measurements of Phobos features, whereas in the fourth part the results of spatial analysis and visual interpretation of the images are summarized.

The first part is an overview of Phobos studies indicating the level of our knowledge, illustrated with pictures that show early topographic schemes, historic Russian maps and globe mentioned above, modern cartographic products created using GIS and Internet technologies like the Geoportal (Karachevtseva et al. 2014), as well as maps produced recently (Wählisch et al. 2014). The complex and irregular shape of this celestial body always raises challenges for cartographers, so the section also includes detailed discussions on the usage of conformal and quasi-conformal projections, which have been developed for Phobos mapping over time (e.g., Snyder 1985; Stooke 2012).

This chapter is accompanied by three maps, which show Phobos in various views: (1) a surface representation based on modern images using Mercator and stereographic projections (for equatorial and polar parts, respectively), as well as a 3D view; (2) a historic map (1988) in normal conformal cylindrical projection for triaxial ellipsoid developed by Bugaevsky (1987) and in the original five-sheet layout proposed by Shingareva; (3) a new topographic map in Bugaevsky projection and Shingareva's layout to maintain the continuity of cartographic heritage with previous Russian mapping of Phobos.

The second chapter presents the photogrammetric methods used for the formation of the first three-dimensional geodetic control points network (CPN) of Phobos, a global orthomosaic and DEM, and separate high-resolution orthoimages and DEMs produced at MExLab (Zubarev et al. 2012). The new MExLab CPN mainly derived from Mars Express images provides the possibility to update the Phobos three-axial ellipsoid and to establish a coordinate system for mapping, to determine the fundamental Phobos parameters such as shape and libration (Nadezhdina and Zubarev 2014), as well as to characterize gravitational field (Uchaev et al. 2013). More than ten maps derived from these studies show various physical parameters of Phobos, including attractive potential, gravitational field, and dynamic heights. Besides the global maps of physical parameters, the chapter includes large-scale topographic maps demonstrating the surrounding area of Drunlo and Stickney craters with possible large scale (1: 60,000) based on photogrammetric processing of high-resolution stereo images (at 3–10 m/pixel resolution).

The third chapter describes the results of spatial analysis of Phobos' surface, carried out using modern GIS technology. It includes the description of object catalogues (of craters, rocks, and grooves) and their application for various studies, e.g., Phobos meteoroid bombardment modeling (Dmitriev et al. 2013) that was verified using crater spatial distribution. The studies described at this section are based on Phobos information system (Karachevtseva et al. 2014) that integrates different GIS layers and provides assessment of size and spatial distribution of craters and rocks presented on the maps as well as cumulative density plots and size-frequency diagrams. Processing of multi-spectral images obtained by Mars Express HRSC camera (Jaumann et al. 2007) and their careful co-registration in GIS gives the assessment of Phobos albedo properties. Albedo parameters obtained in various spectral channels showed on the maps demonstrate the possibility to judge with sufficient confidence the presence of at least three materials with different reflecting properties on the surface of Phobos (Patsyn et al. 2012). Besides 12 single maps, the chapter includes two multi-page maps that show the spatial distribution of Phobos features and surface properties with various scales.

The fourth chapter presents the current understanding of geological composition and surface processes on Phobos. It includes the description of crater morphology (Basilevsky et al. 2014) illustrated by high-resolution images that were used for visual analysis and interpretation. Particular attention is paid to the morphology of Phobos grooves; a consequence of grooves analysis is the special zoning of the Phobos surface into some individual sections that have a different geological history (Lorenz et al. 2016). The morphometric analysis of Phobos craters is illustrated by plots showing comparison with craters on other Solar System bodies (Kokhanov et al. 2014). Estimation of the degrees of crater degradation, assessment of ejecta, and analysis of deposit distribution associated with slope processes derived from the research are presented on five geomorphologic maps.

3 Coordinate System

3.1 Geographic Basis

For Earth mapping as usual some cartographic objects such as coastline and rivers can be applied to show thematic content on the maps. In extraterrestrial cartography, orthomosaics and digital elevation models (DEM) are used as geographical basis. Our studies of Phobos are based on data produced in MExLab: orthomosaic as well as separate orthoimages with the highest resolution, and DEMs with different resolution derived from three-dimensional control point network created at the first time for Phobos (Zubarev et al. 2012). It provides the geographic basis for mapping at various levels of detail. Names of Phobos relief features are taken from the Gazetteer of Planetary Nomenclature.[2] Global mosaic and names allow navigating and matching of the cartographic images and spatial distribution of the thematic content.

As Phobos has a very irregular shape that resembles a potato (see image on Fig. 4), a three-axial ellipsoid is the most suitable for the representation of its surface. Because modern GISs do not support a reference system for irregular surfaces, only some maps were produced based on the three-axial ellipsoid. These maps (see Fig. 6) were compiled using the special tool, developed by the GIS Research Centre of the Institute of Geography of the Russian Academy of Sciences[3] (Nyrtsov et al. 2012), for transformation to Bugaevsky projection for three-axial ellipsoid (Bugaevsky 1987). Parameters of ellipsoid are derived from the Phobos control point network, created in MExLab, with axis a = 13.24 km, b = 11.49 km, c = 9.48 km (Nadezhdina and Zubarev 2014).

For all other maps, the unified planetocentric coordinate system was implemented with east positive longitude from 0° to 360° based on the sphere with radius 11.1 km as recommended by the International Astronomical Union (Archinal et al. 2011).

3.2 Scales and Cartographic Projections

All maps in the Phobos Atlas are divided into three levels of detail that are determined by their scales: global (1:200,000–1:250,000), regional (1:120,000–1:150,000), and local (1:45,000–1:75,000).

According to the scale for global maps and typographical requirements, the Atlas was printed in the size of 35 × 25 cm. This format is also convenient for the presentation of individual Phobos sites at scales selected for multi-sheeted maps (1:75,000 and 150,000).

For global maps, a set of cartographic projections was defined according to features of the mapping area and acceptable distortions. For the characterization of image resolution and control point errors, as well as for maps showing distribution of surface parameters for entire body, the simple cylindrical projection was used (Fig. 1). This projection is often applied to represent the results of processing the of initial data, so it is also most often used for mapping of planetary bodies, for example, in the Astrogeology Science Center,[4] since it does not require to re-arrange maps into other projections.

For topographic maps and maps of relief features, conformal Mercator and polar stereographic projections (Fig. 2) were selected to show the objects in both equatorial and polar areas with the same details.

[2]http://planetarynames.wr.usgs.gov/Page/PHOBOS/target.

[3]http://geocnt.geonet.ru/en/3_axial.

[4]https://astrogeology.usgs.gov/maps.

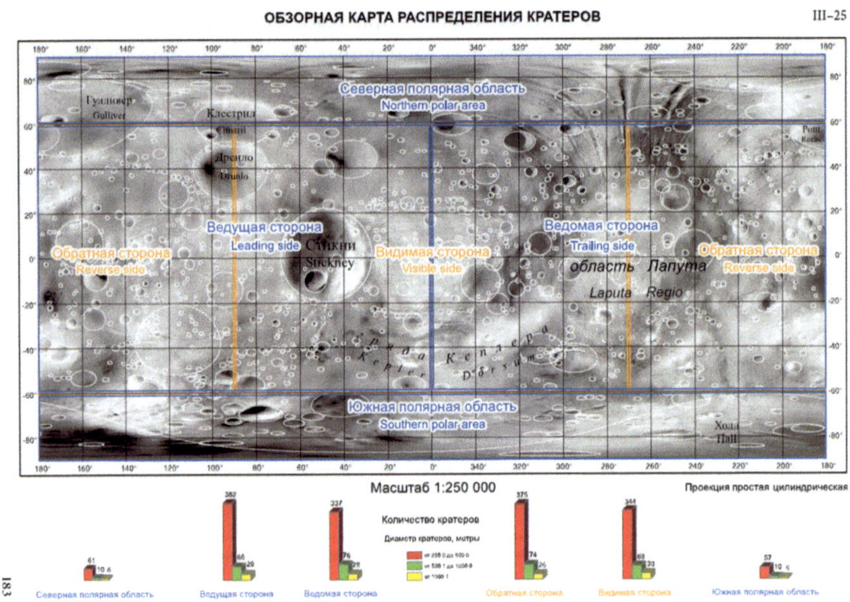

Fig. 1 Index of fifth sheet map of crater distribution (Phobos Atlas, MIIGAiK, 2015)

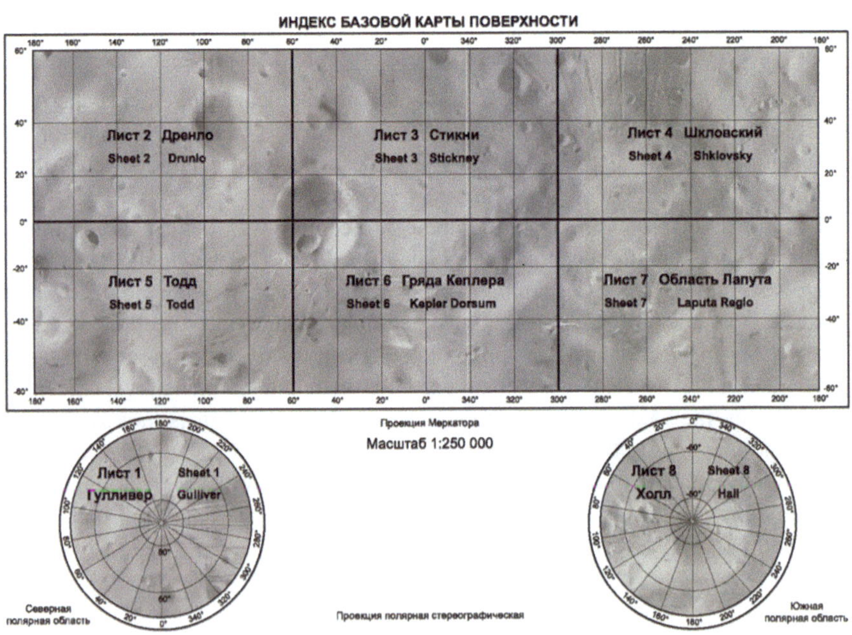

Fig. 2 Index of Phobos base map sheets

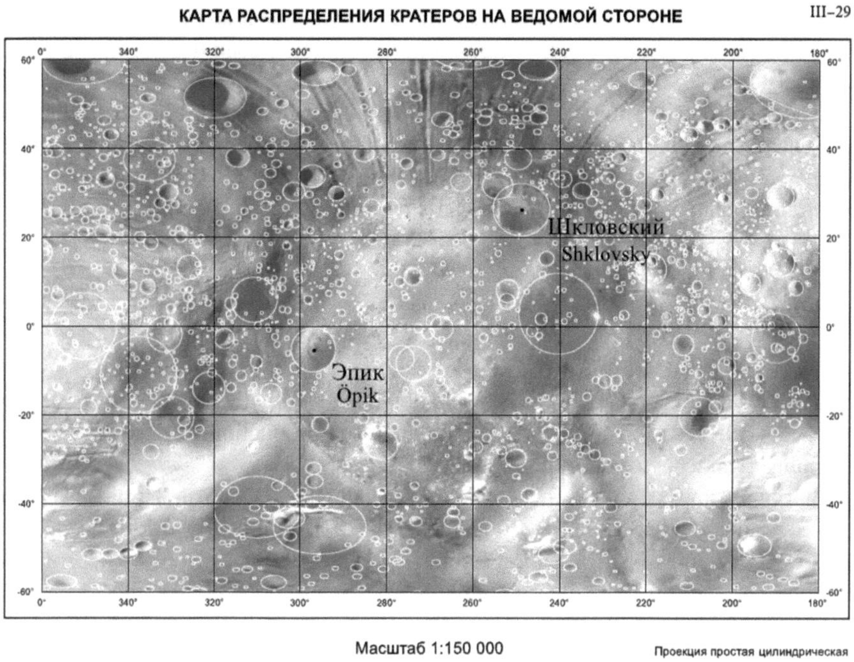

Fig. 3 Map of crater distribution, fourth sheet, trailing side of Phobos (see also Fig. 1)

Mollweide projection was chosen to demonstrate albedo properties (central meridian 90°) as well as crater density (central meridian 0°). In the first case, the choice is due to the fact that the multi-spectral data was available only for a certain Phobos area, whereas in the second case a *density* map requires an *equal area* projection.

A base map as usual depicts background reference information such as landforms and landmarks. The Phobos base map is a map derived from an orthomosaic to show relief features with possible large scale (1:75,000) for the entire body; it also includes contour lines and elevation marks added for outstanding landforms and locational reference. This map follows the eight-sheet layout of Greeley and Batson (1990) (Fig. 2).

Each of the sheets between −60° and 60° parallels was transformed into Mercator projection with the main parallel and main meridian in the center of the map. For polar areas, a stereographic projection was used.

To show crater distribution derived from manual crater detection, a five-sheet map (1:150,000) was prepared. Four sheets represent an area between −60 and 60 parallels in Mercator projection with the main parallel and main meridian in the center of each sheet. The fifth sheet shows the polar areas, for which the same stereographic projection, as described for global maps, is used. Craters outlined on the map (Fig. 3) were digitized semi-automatically using Crater Tools (Kneissl et al. 2011). Our crater catalogue that includes about 5500 objects was produced using the MExLab global mosaic and refined by available images of high resolution (4–10 m/pixel) obtained under various solar illumination conditions (Karachevtseva et al. 2012).

The most detailed map (1:45,000) shows the distribution of boulders, which is presented on a high-resolution image (1.5 m/pixel), obtained by Mars Orbiter Camera onboard Mars Global Surveyor (Malin et al. 2010). Boulders were outlined as circle objects with Crater Tools. The smallest boulder has diameter of 1.7 m.

4 Data Source and GIS

The stand-alone geodatabase has been developed with the proprietary software ArcGIS (ESRI™) for data storage, data analysis, production of derived data and mapping (Karachevtseva et al. 2014).

Various data sets—the Super Resolution Channel (SRC) of the High Resolution Stereo Camera (HRSC) images (Oberst et al. 2008) onboard the European Mars Express, as well as NASA Viking Orbiter-1 data, and images from Soviet mission Phobos 2—were used for photogrammetric image processing, performed in MExLab (Zubarev et al. 2012). They provided the basic layers: a geodetic control point network (CPN), a global orthomosaic, and a DEM. Our technique is based on proprietary photogrammetric software PHOTOMOD (Rakurs™) that was specially adopted in MExLab for planetary data processing.

The MExLab CPN includes 813 points measured 9738 times. This means that the coordinates of each point were measured accurately, in average on 12 images. Altogether, 191 images were used: 165 SRC images, and additional 16 Viking Orbiter 1 images and 10 Phobos 2 images to fill gaps in Mars Express coverage. The SRC images with a resolution ranging from 2.5 to 20 m per pixel cover 91% of the Phobos surface; the remaining 9% are covered with Viking Orbiter 1 images with an analogous resolution. The Phobos 2 images made it possible to supplement the network with new measurements, enhancing its rigidity, since the Phobos 2 orbit was different from orbits of other missions. Thus, frame images of three Phobos missions with a resolution up to 80 m/pixel were jointly processed. The accuracy of the CPN ranges from 4.5 to 67.0 m, and the mean uncertainty of three-dimensional point location is 13.7 m (Nadezhdina and Zubarev 2014).

Global Phobos orthomosaic and DEM (with resolution 20 and 200 m/pixel, respectively) were referenced to the CPN, whose coordinates have been re-analyzed recently (Oberst et al. 2014) due to new SRC images obtained by ongoing Mars Express mission since the first release of our CPN (2012). The DEM derived from automatic stereo measurements of elevation points on image pairs was augmented manually by including 3D shape structural lines along terrain features (rims of craters and grooves). As a result, a digital terrain model (DTM) was created that accurately reflects the position of relief objects. It is very important for further analysis, for example, for morphometric measurements such as crater form or depth, because sometimes the position of objects picked on orthomosaic and then measured on DEM diverges due to different processing techniques.

Based on the DTM, a number of secondary products (slope, shaded relief) were generated using the standard spatial tools of ArcGIS. Phobos roughness that shows variations of heights was calculated in many ways (Karachevtseva et al. 2012). Topographic roughness is an important geomorphological index that depends on the scale of the territory and studied forms of relief, as well as the type and resolution of input data (Florinsky 2016). To define roughness parameters of Phobos, we applied various statistical methods: area ratio, standard deviations of elevation, slope, and profile curvature (Grohmann et al. 2010) as well as Laplacian as interquartile range of the second derivative of heights (Kokhanov et al. 2013). It should be noted that none of the considered roughness indexes gives a satisfactory result using the existing global Phobos DEM. There are two reasons for this. Firstly, the actual amount of topographic information is still too small, and the ratio of body size to the resolution of the Phobos DEM is not large enough. Secondly, the quality of the original image set (resolution and accuracy) that was used for DEM formation is too heterogeneous because it is derived from various missions at different times. Therefore, a reliable computation of surface roughness will be possible in the future, when homogeneous, high-resolution topographic data will be available. However, the method of area ratios is independent of scale (Grohmann et al. 2010) and shows stable results regardless of the resolution of the original DEM, so a map of Phobos roughness was created based on this topographic index. Various roughness parameters have been calculated using specially developed tools embedded in ArcGIS (Kokhanov et al. 2013).

Efforts were made to develop the morphometric catalogue of Phobos craters—5485 features in total—(Karachevtseva et al. 2012) and inventory of grooves—862 features in total (Lorenz et al. 2016). Features included in these catalogues were measured with special morphometric tools integrated into ArcGIS (Kokhanov et al. 2016b) and used for compiling the geomorphologic maps in the Atlas (Kokhanov et al. 2016a).

5 Legends

The *Digital Cartographic Standard for Geologic Map Symbolization of Federal Geographic Data Committee* (FGDC 2006, Chapter 25: Planetary Geology Features) describes the cartographic symbols for planetary mapping.[5] A set of the symbols adopted for implementation in ArcGIS is presented in Nass et al. (2010). Since a common basis for the symbolization is offered, for the Atlas maps we proposed our own legend, which, although is based on the standard in general, includes original symbols to reflect the distinctive features of Phobos (Fig. 4).

[5]https://ngmdb.usgs.gov/fgdc_gds/geolsymstd/fgdc-geolsym-sec25.pdf.

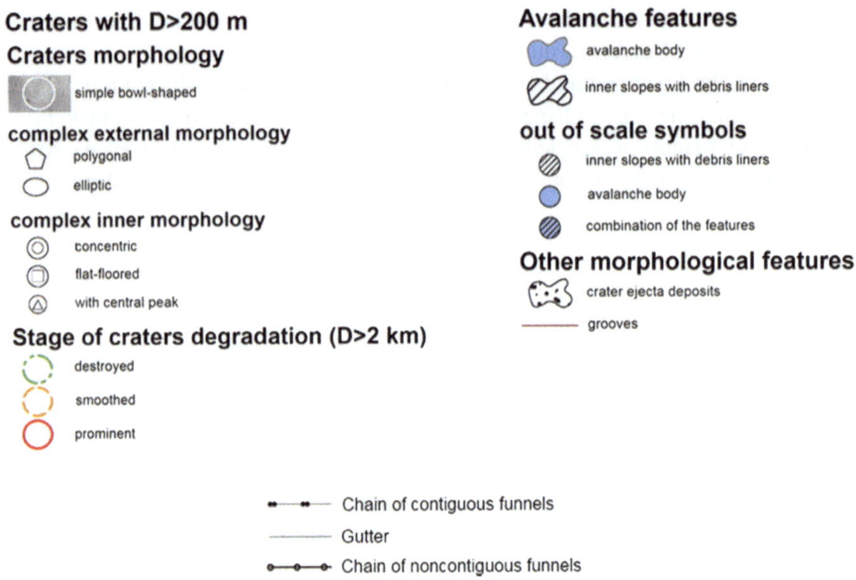

Fig. 4 Legends developed for geomorphological maps of Phobos

Although Phobos, similarly to other rocky bodies of the Solar System, is intensively cratered, a presence of several sets of grooves, crossing each other and forming a dense network, is a unique feature of the Martian satellite. Objects like Phobos are not known elsewhere among the small rocky worlds. The origin of the grooves remains unclear, and their detailed measurements and analysis carried out in our research contribute to understanding of their nature.

The uniqueness of Phobos was the main motivation for the creation of the new symbols as well as the lack of point-located cartographic symbols for morphologic objects. The developed symbols are derived from the suggested geologic classification of grooves: *gutters* (simple line depressions), *chains of contiguous funnels*, and *chains of noncontiguous funnels* (Lorenz et al. 2016). Specialty of presentation of morphological types of impact craters is chosen considering the Atlas as *an instrument for planning future missions*, including selection of landing sites, where fresh craters are of great danger.

Since symbols of craters morphology represent inner and outer geometric peculiarities, we use simple geometric off-scale signs for their classification. Areal signs are used for color-coded visualization of crater degradation stages, as well as for avalanche features and ejecta deposits visible at defined scale. Grooves are shown in linear signs.

Fonts were used to code the types of relief features: craters (normal serif), dorsae (ridges) (italic serif), regions (italic sans serif), and planitiae (plains) (normal sans serif) (Fig. 5).

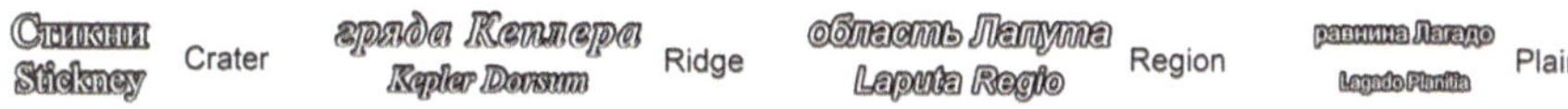

Fig. 5 Implementation of fonts for coding of relief feature types

While the text descriptions and map legends in the Atlas are prepared in Russian, feature names on the maps are presented in bilingual form (Cyrillic and Latin), because the Atlas is intended to be used by the international community, including scientists, students, and anyone interested in astronomy and planetary sciences.

6 Design

The color ramps in the Phobos hypsometric and topographic maps have been created specifically for the Atlas (Fig. 6). In accordance with the traditions of Russian cartography, various color hypsometric ramps based on perceptual approach should be chosen for different celestial bodies as it presented in the *Atlas of the Terrestrial Planets and Their Satellites* (Shingareva et al. 1992). It is taking into account the laws of perception of graphic information, for planets—these are images obtained by space missions. For the Phobos Atlas, we created the design of hypsometric ramps using an integrated perceptual analytical approach suggested by Vereshchaka and Kovaleva (2016). The analytical approach is based on the use of quantitative color parameters of ramp steps and patterns of their variation in different color models. Its advantage is the objectivity, mathematically conditioned by the characteristics of color, and the possibility for use in computer technologies.

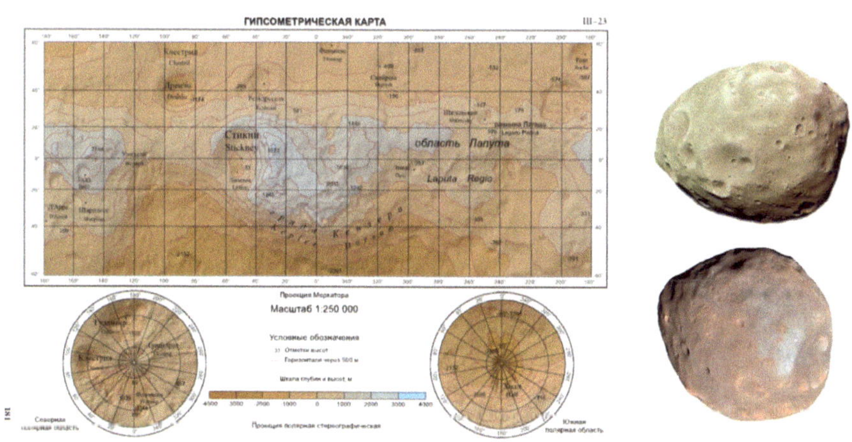

Fig. 6 Hypsometric color-coded map from Phobos Atlas (left) and color-synthesized images of Phobos obtained by Mars Express HRSC camera (right, top), credit (ESA/DLR/FU Berlin, HRSC, G. Neukum) and Mars Reconnaissance Orbiter HiRISE camera (right, bottom), credit (NASA/JPL-Caltech/University of Arizona)

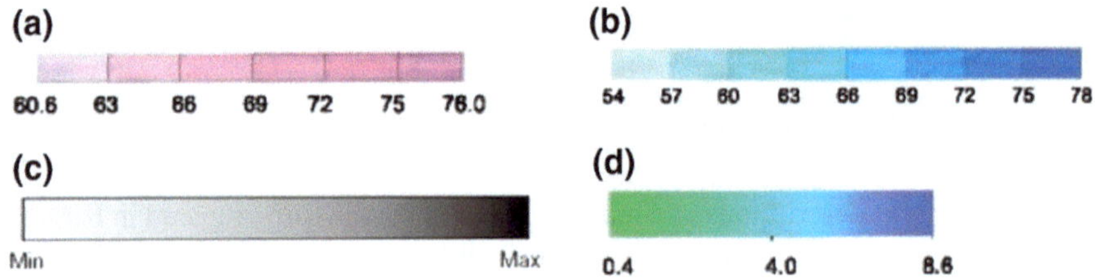

Fig. 7 Types of color scales: **a** stepped, changing in lightness for map of gravitational potential, **b** stepped, changing in lightness for map of attractive potential, **c** gradient grayscale for map of roughness, **d** gradient spectral for map of crater density

Having applied the perceptual analytical method, the boundaries of the color ramp of the Phobos hypsometric map (Fig. 6, left) are chosen in accordance with the visual perception of the Martian satellite surface on color-synthesized images (Fig. 6, right). Then, within the specified color gradient, the quantitative color parameters of the chosen ramp steps are programmed analytically using the ArcMap coloring tools.

For the other maps, we developed color ramps according to various types of data (qualitative/quantitative, absolute/relative) and different surface parameters: stepped, changing in lightness for gravitational characteristics (Fig. 7a, b), gradient grayscale for roughness, and gradient spectral scale for crater density (Fig. 7c, d). The choice of colors was determined both by cartographic traditions (e.g., for topographic maps usually red-brown ramp is used) as well as according to the color solutions implemented earlier in the *Atlas of the Terrestrial Planets and Their Satellites* (Shingareva et al. 1992), e.g., purple color for geophysical maps.

Preparation, layout compiling, design, and correction of maps were carried out by ArcMap tools. Then maps were converted into PDF format for pre-press and further publication in printing house.

7 Conclusion

The production of the Phobos Atlas—such a versatile cartographic product—requires the involvement of specialists and consultants of different specialties, whose work had to be coordinated by an editorial board. Being a comprehensive monograph, the Atlas attracted a large group of researchers from various organizations—geodesists, celestial mechanicians, geomorphologists, geologists, and cartographers. They were united by various projects: *"Geodesy, cartography and study of planets and satellites"* (Ministry of Education of Russian Federation, № 11. G34.31.0021, 2010–2012), *"Geodesy, cartography and study of Phobos and Deimos"* (Russian Foundation for Basic research, №11-05-91323, 2011–2013), *"Research of fundamental geodetic parameters and relief of planets and satellites"* (Russian Science Foundation, № 14-22-00197, 2014–2016).

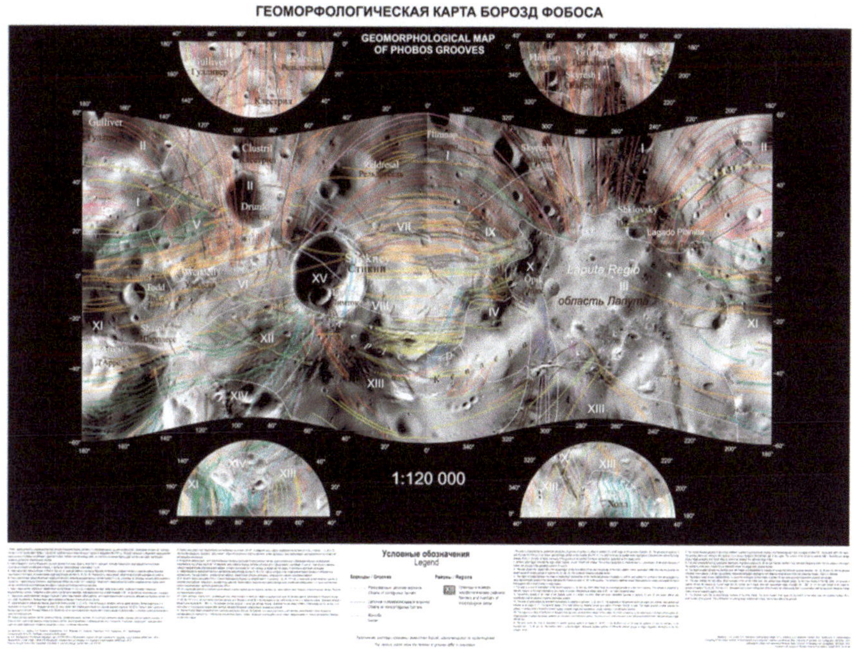

Fig. 8 Geomorphological wall map of Phobos grooves (MIIGAiK, 2016). Projections for three-axial ellipsoid developed by Bugaevsky (1987): normal conformal cylindrical (for central belt); azimuthal equidistant (for sub-polar areas)

The main results from these recent projects, which provided a coordinate-cartographic basis, were published in various scientific journals. Therefore, the Phobos Atlas was developed accumulating intensive research on the Martian satellite, such as the creation of a geodetic control point network (Zubarev et al. 2012) and the determination of shape parameters (Nadezhdina and Zubarev 2014), modeling and study of gravity field (Uchaev et al. 2013), surface compositional studies using HRSC color-channel data (Patsyn et al. 2012), the preparation of craters and grooves catalogues, the creation of a Phobos Information System (Karachevtseva et al. 2012, 2014), statistics of crater size-frequency distributions based on multi-fractal approach (Uchaev et al. 2012), and visual geologic analysis of Phobos features (Basilevsky et al. 2014). The geomorphological study of the grooves was extended (Lorenz et al. 2016) and mapped in three-axial ellipsoid that better demonstrates the irregular Phobos shape (Fig. 8).

As a collective work, students also contributed to the Atlas. Mainly geodesists and cartographers, at least ten MIIGAiK students took part at various stages of the work (collecting and pre-processing images, data co-registration, digitizing of craters and grooves, crater measurements using GIS). Having joined at the very

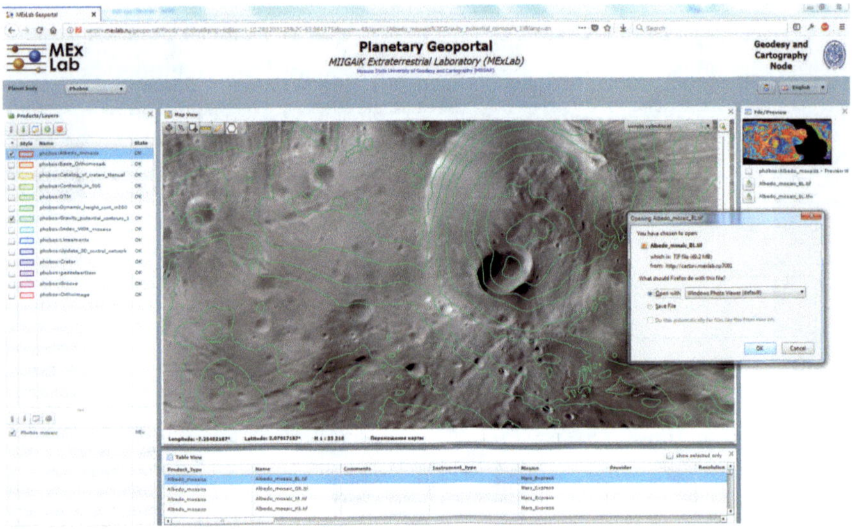

Fig. 9 Phobos layers in MExLab Geoportal: User interface for data set uploading

beginning of Phobos studies during educational practice, some of the students continued the research and defended their diploma theses in the frame of the projects and research mentioned above. Finally, maps created especially for the Atlas became a part of Ph.D. dissertation related to the cartographic support for the planning of future missions.

The Phobos Atlas is a practical instrument for landing site selection and planning mission operation at the surface, useful for the International Phobos/Deimos Landing Site Working Group,[6] organized to maintain focus on future international projects such as *Bumerang* and *Martian Moons Exploration* (MMX) missions (Kuramoto et al. 2017) to provide Phobos sample return.

Spatial data products that were applied for Phobos cartography are integrated into a stand-alone ArcGIS geodatabase, and freely accessible via the MExLab Geoportal[7] in GIS-ready format: vector layers can be downloaded as shape files, and raster data sets as georeferenced images (geotiff) (Fig. 9).

[6]https://www.lpi.usra.edu/sbag/meetings/jan2017/presentations/Duxbury.pdf.

[7]http://cartsrv.mexlab.ru/geoportal/.

References

Archinal BA, A'Hearn MF, Bowell E, Conrad A, Consolmagno GJ, Courtin R, Fukushima T, Hestroffer D, Hilton JL, Krasinsky GA, Neumann G, Oberst J, Seidelmann PK, Stooke P, Tholen DJ, Thomas PC, Williams IP (2011) Report of the IAU working group on cartographic coordinates and rotational elements: 2009. Celest Mech Dyn Astron 109:101–135

Astrogeology Science Center. https://astrogeology.usgs.gov/maps. Accessed 08 Dec 2017

Basilevsky AT, Lorenz CA, Shingareva TV, Head JW, Ramsley KR, Zubarev AE (2014) Surface geology and geomorphology of Phobos. Planet Space Sci 102:95–118. https://doi.org/10.1016/j.pss.2014.04.013

Bugaevsky LM (1987) On the problem of elaborating of isometric coordinates and equiangular cylindric projection of the triaxial ellipsoid. Izvestiya vuzov «Geodeziya i aerofotosyemka» («Geodesy and Aerophotosurveying») 4:79–90 (in Russian)

Bugaevsky LM, Krasnopevtseva BV, Shingareva KB (1992) Phobos map and Phobos globe. Adv Space Res 12(9):17–21. https://doi.org/10.1016/0273-1177(92)90314-N

Cartographical Projections of Triaxial Ellipsoid. http://geocnt.geonet.ru/en/3_axial. Accessed 25 May 2017

Dmitriev V, Lupovka V, Christou A, Oberst J (2013) Meteoroid population near Mars. In: 2013, European planetary science congress, London, EPSC 2013-629-1

FGDC (2006) Digital cartographic standard for geologic map symbolization. Federal Geographic Data Committee prepared by the U.S. Geological Survey (FGDC-STD-013-2006)

Florinsky IV (2016) Digital terrain analysis in soil science and geology, 2nd edn. Academic Press, Amsterdam

Gazetteer of nomenclature. http://planetarynames.wr.usgs.gov/Page/PHOBOS/target. Accessed 15 May 2017

Geoportal of Planetary data. Geodezy and Cartography Node. http://cartsrv.mexlab.ru/geoportal/

Greeley R, Batson G (1990) Planetary mapping. Cambridge University Press, Cambridge

Grohmann CH, Smith MJ, Riccomini C (2010) Multiscale analysis of topographic surface roughness in the Midland Valley, Scotland. IEEE Trans Geosci Remote Sens. PP(99):1–14

International Phobos/Deimos Landing Site Working Group. https://www.lpi.usra.edu/sbag/meetings/jan2017/presentations/Duxbury.pdf. Accessed 08 Dec 2017

Jaumann R, Neukum G, Behnke T, Duxbury TC, Eichentopf K, Flohrer J, Gasselt SV, Giese B, Gwinner K, Hauber E, Hoffmann H, Hoffmeister A, Kohler U, Matz K-D, McCord TB, Mertens V, Oberst J, Pischel R, Reiss D, Ress E, Roatsch T, Saiger P, Scholten F, Schwarz G, Stephan K, Wahlisch M, the HRSC Co-Investigator Team (2007) The high-resolution stereo camera (HRSC) experiment on Mars Express: instrument aspects and experiment conduct from interplanetary cruise through the nominal mission. Planet Space Sci 55:928–952

Karachevtseva IP, Shingareva KB, Konopikhin AA, Mukabenova BV, Nadezhdina IE, Zubarev AE (2012) GIS mapping of Phobos on the results of data processing of remote sensing satellite Mars Express. Mod Prob Remote Sens Earth Space 9(4):304–311 (in Russian)

Karachevtseva IP, Oberst J, Zubarev AE, Nadezhdina IE, Kokhanov AA, Garov AS, Uchaev DV, Uchaev DV, Malinnikov VA, Klimkin ND (2014) The Phobos information system. Planet Space Sci 102:74–85. https://doi.org/10.1016/j.pss.2014.11.024

Karachevtseva I, Kokhanov A, Rodionova J, Konopikhin A, Zubarev A, Nadezhdina I, Mitrokhina L, Patratiy V, Oberst J (2015) Development of a new Phobos atlas based on Mars Express image data. Planet Space Sci 108:24–30. https://doi.org/10.1016/j.pss.2014.11.024

Kneissl T, Gasselt S, Neukum G (2011) Map-projection-independent crater size-frequency determination in GIS environments—New software tool for ArcGIS. Planet Space Sci 59: 1243–1254. https://doi.org/10.1016/j.pss.2010.03.015

Kokhanov A, Lorenz C, Karachevtseva I (2016a) GIS-study and new Geomorphologic Mapping of Phobos. In: Geophysical research abstracts, vol. 18. EGU2016-1726, European Geosciences Union General Assembly, Vienna, 17–22 April 2016

Kokhanov AA, Bystrov AY, Kreslavsky MA, Matveev EV, Karachevtseva IP (2016b) Automation of morphometric measurements for planetary surface analysis and cartography. Int. Arch. Photogramm. Remote Sens Spatial Inf Sci XLI-B4:431–433. https://doi.org/10.5194/isprs-archives-xli-b4-431-2016

Kokhanov AA, Kreslavsky MA, Basilevsky AT, Karachevtseva IP, Zubarev AE (2014) Morphometry of large craters on Phobos and comparison with other bodies. In: 45th LPSC, abs. #1084

Kokhanov AA, Kreslavskiy MA, Karachevtseva IP, Matveev EN (2013) Mapping of the statistical characteristics of the lunar relief on the basis of the global digital elevation model GLD-100. Mod Probl Remote Sens Earth Space 10(4):136–153 (in Russian)

Kuramoto K, Kawakatsu Y, Fujimoto M, MMX Study Team (2017) Martian moons exploration (MMX) conceptual study results. In: XLVIII lunar and planetary science conference (LPSC), Abstract #2086

Lavochkina Association, POSKOCMOS. Bumerang project. https://www.laspace.ru/projects/planets/expedition-m/. Accessed 08 Dec 2017 (in Russian)

Lorenz CA, Kokhanov AA, Karachevtseva IP (2016) Morphological study of Phobos surface and mapping of the grooves. In: 47th LPSC, abs. #1831

Malin MC, Edgett KS, Cantor BA, Caplinger MA, Danielson GE, Jensen EH, Ravine MA, Sandoval JL, Supulver (2010) An overview of the 1985–2006 Mars Orbiter Camera science investigation. Mars—Int J Mars Sci Explor 5:1–60 Mars Informatics Inc.

Nadezhdina IE, Zubarev AE (2014) Formation of a reference coordinate network as a basis for studying the physical parameters of Phobos. Sol Syst Res 48(4):269–278. https://doi.org/10.1134/S003809461404008X

Nass A, van Gasselt S, Jaumann R, Asche H (2010) Implementation of cartographic symbols for planetary mapping in geographic information systems. Planet Space Sci 59(11–12):1255–1264. https://doi.org/10.1016/j.pss.2010.08.022

Nyrtsov MV, Fleis ME, Borisov MM (2012) Mapping asteroid 433 Eros with equidistant along meridians cylindrical and azimuthal projections of triaxial ellipsoid. Izvestiya vuzov «Geodeziya i aerofotosyemka» («Geodesy and Aerophotosurveying») 1:54–56 (in Russian)

Oberst J, Schwarz G, Behnke T, Hoffmann H, Matz K-D, Flohrer J, Hirsch H, Roatsch T, Scholten F, Hauber E, Brinkmann B, Jaumann R, Williams D, Kirk R, Duxbury T, Leu C, Neukum G (2008) The imaging performance of the SRC on Mars Express. Planet Space Sci 56 (3–4):473–491

Oberst J, Zubarev A, Nadezhdina I, Shishkina L, Rambaux N (2014) The Phobos geodetic control point network and rotation model. Planet Space Sci 102:45–50. https://doi.org/10.1016/j.pss.2014.03.006

Patsyn VS, Malinnikov VA, Grechishev AV (2012) Research of spectrometric characteristics of the surface of Phobos on the HRSC data from the Mars Express spacecraft. Mod Probl Remote Earth Sens Space 9(4):312–318 Space Research Institute, Moscow (in Russian)

Phobos from 6,800 Kilometers (Color): NASA/JPL-Caltech/University of Arizona/PIA10368. https://photojournal.jpl.nasa.gov/catalog/PIA10368/. Accessed 08 Dec 2017

Phobos: a colored view made from images taken from two different flybys in March 2010: ESA/DLR/FU Berlin (G. Neukum). https://www.universetoday.com/82302/mars-express-set-for-phobos-flyby/. Accessed 08 Dec 2017

Salishev KA (1982) Kartovedeniye (Cartography). 3rd edn, 400 p (in Russian)

Shingareva KB, Krasnopevtseva BV, Leonenko SM, Fleis ME, Buchroithner MF, Waelder O, Stooke P (2005) Phobos and Deimos—a new map in the series of multilingual relief maps of terrestrial planets and their moons. In: Proceedings of 22nd international cartographic conference, A Coruña, Spain, 9–16 July 2005

Shingareva KB, Krasnopevtseva BV, Rodionova JF et al (1992) Atlas of the terrestrial planets and their satellites. MIIGAiK, Moscow, p 208 (in Russian)

Snyder JP (1985) Conformal mapping of the triaxial ellipsoid. Survey Rev 28(217):130–148

Stooke PJ (2012) The international atlas of mars exploration. the first five decades, volume 1, 1953 to 2003. Cambridge University Press, Cambridge, p 376

Uchaev DV, Uchaev DmV, Malinnikov VA, Oberst J (2012) Multifractal model for Phobos crater size-frequency distribution. In: EPSC abstracts 7, EPSC2012-705-3

Uchaev DmV, Uchaev DV, Prutov I (2013) Multiscale representation of gravitational fields of small celestial bodies. Izvestiya vuzov «Geodeziya i aerofotosyemka» («Geodesy and Aerophotosurveying») 4:3–8 (in Russian)

Vereshchaka TV, Kovaleva OV (2016) Presentation of the relief on the maps: theory and methods (design aspect). The Scientific World, Moscow, 181 p. ISBN 978-5-91522-427-7 (in Russian)

Wählisch M, Stooke PJ, Karachevtseva IP, Kirk R, Oberst J, Willner K, Nadejdina IA, Zubarev AE, Konopikhin AA, Shingareva KB (2014) Phobos and Deimos cartography. Planet Space Sci 102:60–73. https://doi.org/10.1016/j.pss.2013.05.012

Zubarev AE, Nadezhdina IE, Konopikhin AA (2012) Problems of processing of remote sensing data for modeling shapes of small bodies in the Solar system. Mod Probl Remote Sens Earth Space 9(4):277–285 (in Russian)

The Role of Maps During Long-Term Analog Planetary Missions and Future Mars Missions

Anna Losiak, Izabela Gołębiowska, Nina Sejkora and Gernot Groemer

Abstract We discuss the process of designing, using, and evaluating maps during planetary analog missions. We present the evolution of the mapping strategy within the Austrian Space Forum (OeWF) that took place between 2006 and 2018 in response to increasing complexity and fidelity level of the field missions. We also provide suggestions for the efficient preparation of maps to be used during future planetary analog missions.

Keywords Analog missions · Human mars missions · Planetary cartography

1 Introduction

The first planetary exploration program associated with a large-scale mapping was the Apollo program. Five years before the first landing, global to regional scale maps were prepared based on the photographs of the lunar surface taken by Ranger, Surveyor, and Lunar Orbiter missions. This allowed building a perceptive model of the geological evolution of our satellite (Wilhelms et al. 1987), formulating hypotheses to be tested, and identifying the most scientifically interesting places (e.g., Muehlberger 1981). A couple of months before each landing, detailed mapping was performed within the pre-selected regions in order to plan extra vehicular activity (EVA) traverses. Mapping (geological and topographical based on stereo-pairs) was in large part based on images obtained by the earlier Apollo

A. Losiak (✉)
wildFIRE Lab, University of Exeter, Exeter, UK
e-mail: a.i.losiak@exeter.ac.uk

A. Losiak · N. Sejkora · G. Groemer
Austrian Space Forum, Innsbruck, Austria

A. Losiak
Institute of Geological Sciences, Polish Academy of Sciences, Warsaw, Poland

I. Gołębiowska
Faculty of Geography and Regional Studies, University of Warsaw, Warsaw, Poland

© Springer Nature Switzerland AG 2019

H. Hargitai (ed.), *Planetary Cartography and GIS*, Lecture Notes in Geoinformation and Cartography, https://doi.org/10.1007/978-3-319-62849-3_12

missions from the Apollo Command/Service Module orbiting the Moon (e.g., maps of the Apollo 17 traverses were prepared based on Apollo 15 photographs: Manned Spacecraft Center 1972). Crew safety during landing and exploration was the main objective of selecting specific locations, and science was, in fact, a secondary objective. All surface activities during every Apollo mission were planned in great detail in terms of selection of locations to be visited by astronauts, traverses, as well as time allowed for different activities (Manned Spacecraft Center 1972), so the exploration flexibility during a single mission was limited. However, because there was at least a few months' lag time between every lunar landing, it was possible to implement improvements to the next missions based on the previous mission's experiences.

For Mars missions, due to the 10–20 min signal travel time between Mars and Earth, all surface activities will either need a significantly more time, or more decision-making autonomy for the crew. The first of those approaches is currently taken by the rover missions, but future crewed missions will have to be performed in a semi-autonomous mode. They are likely to involve astronauts spending up to 17 months on the surface of the Red Planet (e.g., Drake 2009), during which communication between the Earth and the Mars crew will be performed with a time delay. Designing a mission architecture, including the design of useful cartographic products, mapping tools and work-flows, that will optimize the efficiency of the science operations under these conditions, is a complex task. This complexity requires extensive analog testing under terrestrial conditions (e.g., Eppler et al. 2013).

This paper reflects on the evolution of the role of maps during the Mars analog missions organized by the Austrian Space Forum between 2006 and 2018 and provides suggestions for future similar projects.

2 Evolution of Maps Usage at the Austrian Space Forum Analog Mars Missions

The Austrian Space Forum (OeWF, Österreichisches Weltraum Forum) is a non-profit research organization that has conducted 12 Mars analog field campaigns since 2006. Missions took place in diverse locations: (1) representative of average current Mars conditions such as the Mars Desert Research Station (MDRS) in Utah in 2006 (Groemer et al. 2007) and Northern Sahara near Erfoud, Morocco in 2013 (Groemer et al. 2014), the desert in the Dhofar region, Oman in 2018; (2) the early and wet Mars analog site of Rio Tinto, Spain in 2011 (Orgel et al. 2013); (3) subsurface simulations at Dachstein Ice Caves, Austria in 2012 (Groemer et al. 2012); and (4) (rock) glacier analogs, resembling features present currently in the northern regions of Mars, at the Kaunertal glacier, Austria in 2015 (Groemer et al. 2016). These OeWF field missions included >70 experiments and engineering tests in the fields of astrobiology, robotics, human factors, geoscience, and spacesuit operations. Most of these experiments were performed in cooperation with external partners and scientists from >30 nations. Major recurring assets include two advanced spacesuit simulators (Aouda. X and Aouda. S; Groemer et al. 2011).

Between 2006 and 2018, analog missions of the OeWF became gradually more complex, which was reflected in the evolution of the cartographical support required by the mission. The first missions in 2006 (Groemer et al. 2007) and 2010 were focused on technical tests and development of operational procedures, and because of that they did not require extensive cartographical support. The only map-related activity was related to marking GPS-acquired coordinates of sampling locations on a map (at first on a paper map, then also in Google Earth).

The first OeWF mission that was using a more complex approach toward mapping was MARS2013. The Remote Science Support's (RSS—a team responsible for managing science-related activities) and Flight Plan's (FP—a team planning all actions of the Field Team) activities were centered around a single-file, easy-to-use, spatially referenced database that included all basic information about the conditions at the site of study, as well as previous and planned activities (Fig. 1). The database was prepared in Google Earth to minimize the barriers to entry for all team members as RSS and FP teams consisted mostly of people who had no background in GIS and because only limited time was available for appropriate training. Google Earth was also used by the science support team during the NASA's Desert RATS mission (Yingst et al. 2013), but detailed geological mapping for this mission was performed in ArcGIS software (Skinner and Fortezzo 2013). RSS for the MARS 2013 mission created all required maps in Google Earth, although it was a very time-consuming process and the created maps had topological shortcomings (e.g., not-fully closed polygons, overlapping polygons, and alignment errors). Google Earth lacked advanced analytical functionalities of GIS software, such as calculation of the total distances covered by every astronaut or calculation of slope inclination. This more advanced analysis was performed after the mission, and Google Earth KMZ files were transferred to popular GIS software products and formats (Losiak et al. 2014).

During an OeWF mission in Oman in 2018, a slightly modified approach was taken: Maintaining a mapping system was moved from RSS to FP team and QGIS software was used instead of Google Earth. This approach enabled automated assignment of the formating style for every element on the map, improved quality of exported maps, and simplified GIS analysis. However, only a couple of people in the mission support center were able to operate it.

3 Suggestions for Maps Preparation During the Future Analog Missions

3.1 Topographic Data

Up-to-date, high-resolution topographic maps are vital for an effective: (1) planning of technical tests (e.g., checking if rover can move through a slope of a certain

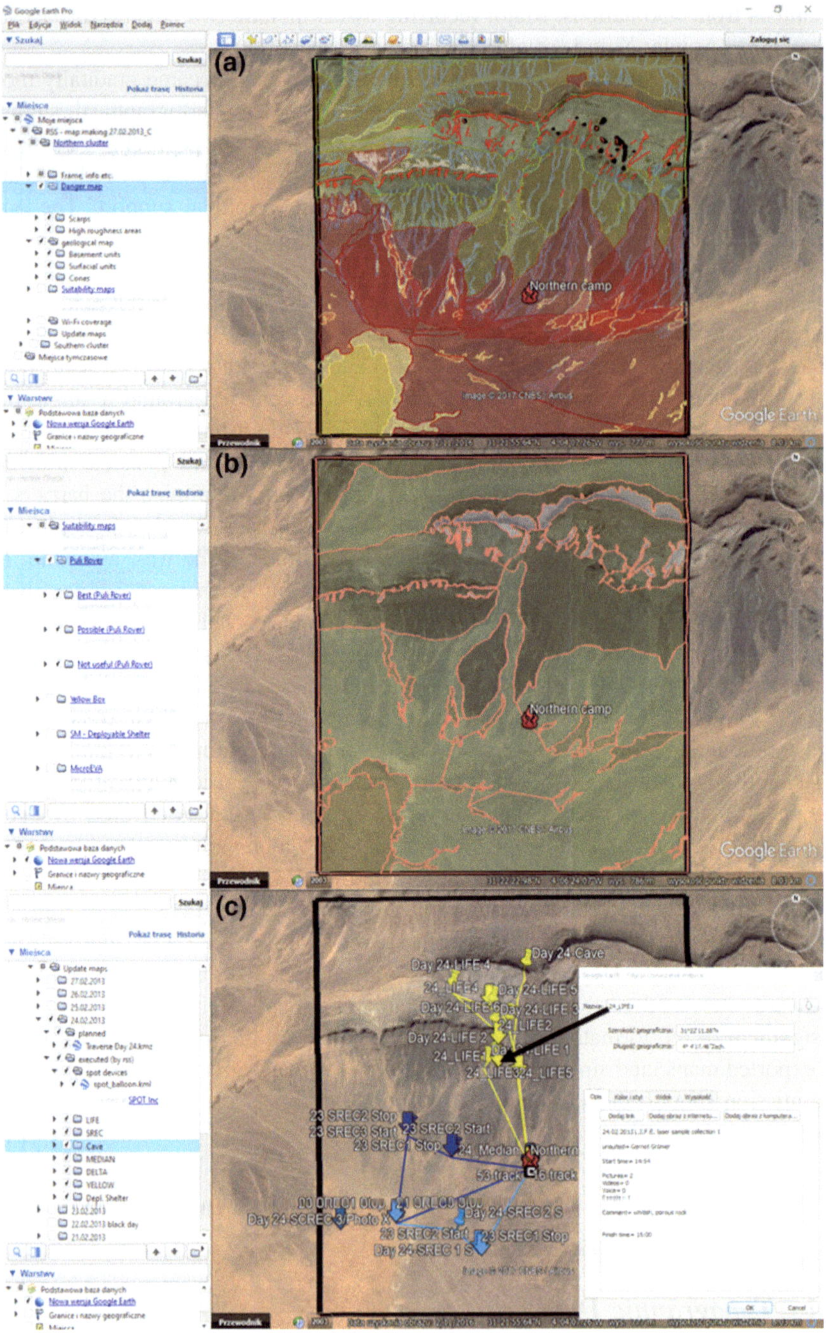
(a)
Northern camp
(b)
Northern camp
(c)
Day 24 Cave
Day 24 LIFE 4
Day 24 LIFE 2
Google Earth

◂**Fig. 1** Screen shots from the spatially referenced database that was developed in Google Earth as part of the MARS2013 analogue mission. The structure of the database is visible in the panel on the left side. Detailed maps were prepared for the area 5 km × 5 km around the camp (context maps presenting area 90 km × 90 km, not shown here). **a** A "Geological map" with a "Danger map" (red lines and areas) overlaid on it. Both of them were based solely on analysis of satellite images and were prepared following the suggestions of Skinner and Fortezzo (2013). **b** A set of suitability maps, one for every scientific experiment, depicted the suitability of the area for performing a specific experiment. It divided the area into three categories "best" (dark green), "possible" (light green), and "not useful" (white). This classification was made based on the experiment requirements combined with the analysis of the satellite images and the digital elevation model. For every map the criteria of the suitability were clearly described within the database. The FP team was responsible for designing traverses through areas in the "best" and/or "possible" categories, whereas "not useful" areas were to be avoided. Moreover, points indicated by FP (as experiment locations) needed to be placed within the areas categorized as having "best" suitability. This system enabled to speed up the traverse planning, but was time-consuming to set up. The image is showing a suitability map for a Puli Rover (PULI) which was a rover designed to explore rough terrains under extreme conditions. Based on its technical properties and previous tests, we knew that it was able to transect all areas except those with: 1. high average inclination $\geq 60\%$ OR 2. high probability of encountering high local inclination above $\geq 100\%$ (e.g. scarps) OR 3. with numerous large rocks (>20 cm in diameter) laying on the surface (e.g., close to the high inclination slopes) which were marked as "not useful". **c** All planned and executed activities were included on the update maps. Each device/experiment was color-coded in the same way on both planned and executed maps (e.g., LIFE experiment is marked with yellow lines and signs). Planned and executed maps differ by a shape of point signs indicating sites of interest ("pins" for planned, "arrows" for executed). Information about all the actions undertaken at a given place are provided in a properties file of the point on the map (e.g., descriptions of the area provided by analog astronauts are included in the text box, images taken at the site were named and their location on the server was provided). The full Google Earth database can be obtained upon request from the Austrian Space Forum, or corresponding author of this paper

inclination), (2) identifying high-risk areas (so that traverse will not be planned through an area that may be too hazardous), (3) traverse timing including the projected astronaut physiological workloads, (4) increasing the quality of geological mapping and (5) providing a context for virtual reality experiments for a post-EVA revisit of a test site by the RSS.

The optimal time span between the date of topographical data acquisition and when field activities performed based on those data are executed, strongly depends on the level of lability of the environment. In some environments (e.g., most of stony-deserts) data acquired several (~ 10) years earlier is sufficient for analog planetary missions mapping activities. However, in areas that are changing dynamically (e.g., sandy-deserts with active dunes, areas near active lava flows, glaciers, or operational mines) data should be obtained more recently (less than 1–2 years earlier). In the most extreme case (e.g., an active surface of a glacier), even data obtained weeks before the mission is already obsolete because this surface changes every day (e.g., mission at the Kaunertal glacier, Austria in 2015; Groemer et al. 2016). This means that a traverse that was optimal on one day, on the next day can be impassable. In this case, if it is impossible to set up a system that is able to process a new 3D model every day (e.g., based on drone-based photography), it is best to assume that topographic data for the surface of the glacier are of limited reliability and modify field procedures accordingly. Another option would be to use drone or baloon images that are updated daily, and examined by a person experienced in the analysis of aerial photographs for possible changes in topography. An optimal resolution of the data should be <1 m.

Georeferenced topographical data with a resolution of $\sim$30 m, covering 99% of Earth, is available from the advanced spaceborne thermal emission and reflection radiometer (ASTER)—these data are also included in the Google Earth software. For some regions, higher-resolution data (e.g., <1 m) are available; however, it may be hard to find and acquire it. It is caused by the fact that this data may be curated by many different institutions; some databases incorporate an entire country (e.g., ESTONIA), are restricted to municipality (e.g., WROCLAW) or are available based on land ownership (e.g., AUSTRIA). The problem with access to good quality topographical data at small or no cost can be solved by involving in the analog mission scientists from the region where field work takes place, who have the best knowledge about local circumstances.

For some locations (especially in Europe and North America), topographical maps with spirit levels marked every $\sim$5 m (for scales $\sim$1:25 000) or every 1–2 m (for scales 1:10 000) are available. This source of data needs to be treated with caution because measurements that were the basis of this map may be decades old, even if map was published quite recently.

3.2 *Aerial Photography*

The most important source of information for the analog planetary missions is aerial photographs. They are used to (1) determine surface properties of the area in order to estimate traversability (e.g., sandy area, rocky area, etc.) and to identify potentially dangerous places (e.g., prone to rock avalanches), (2) perform geological mapping of the area and identify the most interesting sites, (3) recognize inclination patterns (if no high-resolution DEM is available, an experienced geographer or geologist can determine the extend of relative changes in elevation). In combination with the topographical data, aerial photo analysis identifies the most and the least suitable areas for exploration during the mission (Losiak et al. 2014) what is crucial during the traverse planning.

Aerial images should have a resolution of <1 m (optimally <0.5 m) which means that photographs with sufficient resolution are available for most of the areas on Earth for free, through Google Earth. Additionally, for some areas even higher resolution images are available from similar types of data sources as DEMs discussed above (e.g., ESTONIA, WROCLAW, AUSTRIA). A very high-resolution images (<5 cm) can be obtained with a drone and automatically stitched together to form a photo mosaic; but the area covered with this kind of imaginery is usually limited.

3.3 *Software*

Software selection for an analog planetary mission depends on the level of mission complexity and users' needs and skills. For the low fidelity level (e.g., technology tests), Google Earth is sufficient.

For more complex, longer missions, especially if they include geological or astrobiological exploration, it may be better to use an advanced GIS software such as ArcGIS or QGS. This allows to keep data in a geo-referenced database and analyze them on the spot. However, such programs are usually complicated and hard to use without a proper training. So, only a small group of team members will be able to use the system. This solution may be the most efficient one, but a simplified subsystem enabling other people to interact with the maps (e.g., by reading traverses) needs to be introduced into procedures. On the other hand, previous large scale, high-fidelity missions such as MARS2013 (Losiak et al. 2014) and D-RATS (Yingst et al. 2013) successfully used Google Earth for most of their mapping activities.

3.4 *Cartographic Training*

Mapping experts tend to overestimate map interpretation skills of non-specialists (Gołębiowska 2015). It is our experience, field team members (most of whom are not geologists/geographers) experienced challenges even in locating themselves on the map (especially if GPS system is not used). It is important to introduce users (specifically the field crew) to the prepared geological, geomorphological and thematic maps that will be used during the mission, and to do it using the software that will be used by this specific team. The training should include a hands-on exercise using real data. If this step is omitted, crew members will not make full use of the prepared cartographical system (Losiak et al. 2014).

Acknowledgements Authors would like to thank all members of the Austrian Space Forum and non-associated participants of many Mars analog missions that we have performed together over the years. We are also very grateful to reviewers for their productive suggestions. We acknowledge the support of the National Science Centre Poland (Narodowe Centrum Nauki) Grant number UMO-2013/08/S/ST10/00586 and funding from the European Union's Horizon 2020 research and innovation programme under the Marie Sklodowska-Curie grant agreement No 749157: "Using charcoals as a proxy to estimate energy released from small Impact cratering processes—ImpChar."

References

ArcGIS. www.esri.com/arcgis/about-arcgis
ASTER. http://asterweb.jpl.nasa.gov/gdem.asp

AUSTRIA: Digital Elevation Model of the State of Tyrol, Kaunertal region, under a research license by the Government of Tyrol, TIRIS (Tiroler Rauminformationssystem), using tirisMAPS under a creative commons license, 2015. The source DEM model was obtained through: Jarvis A., H.I. Reuter, A. Nelson, E. Guevara, 2008, Hole-filled seamless SRTM data V4, International Centre for Tropical Agriculture (CIAT). http://srtm.csi.cgiar.org

Drake BG (ed) (2009) Human Exploration of Mars Design Reference Architecture 5.0, NASA-SP-2009-566, NASA Center for AeroSpace Information, Hanover, MD

Eppler D, Adams B, Archer D, Baiden G, Brown A, Carey W, Cohen B, Condit C, Evans C, Fortezzo C, Garry B, Graff T, Gruener J, Heldmann J, Hodges K, Hörz F, Hurtado J, Hynek B, Isaacson P, uranek C, Klaus K, Kring D, Lanza N, Lederer S, Lofgren G, Marinova M, May L, Meyer J, Ming D, Monteleone B, Morisset C, Noble S, Rampe E, Rice J, Schutt J, Skinner J, Tewksbury-Christle CM, Tewksbury BJ, Vaughan A, Yingst A, Young K (2013) Desert Research and Technology Studies (DRATS) 2010 science operations: operational approaches and lessons learned for managing science during human planetary surface missions. Acta Astronaut 90:224–241

ESTONIA. http://geoportaal.maaamet.ee/eng/Maps-and-Data/Topographic-Data/Elevation-data-p%20308.html

Gołębiowska I (2015) Legend layouts for thematic maps: a case study integra-ting usability metrics with think aloud method. Cartographic J. 52(1):28–40. https://doi.org/10.1179/1743277413Y.0000000045

Groemer G (ed) (2012) Dachstein Mission Report public, 05 August 2012. http://oewf.org/en/portfolio/austria-dachstein-2012/

Groemer G, Frischauf N, Soucek A, Sattler B (2007) AustroMars—a simulated high-fidelity human Mars analogue mission, Mars2030—AustroMars Science Workshop. In: Groemer G (ed) Proceedings of the Conference held in at the University of Salzburg, 24–26 September 2006. Austrian Space Forum, pp 4–12

Groemer G, Hauth S, Luger U, Bickert K, Sattler B, Hauth E, Föger D, Schildhammer D, Agerer C, Ragonig C, Sams S, Kaineder F, Knoflach M (2012) The Aouda. X Space suit simulator and its applications to astrobiology. Astrobiology 12(2):125–134. https://doi.org/10.1089/ast.2011.067

Groemer G, Soucek A, Frischauf N, Stumptner W, Ragonig C, Sams S, Bartenstein T, Häuplik-Meusburger S, Petrova P, Evetts S, Sivenesan C, Bothe C, Boyd A, Dinkelaker A, Dissertori M, Fasching D, Fischer M, Föger D, Foresta L, Fritsch L, Fuchs H, Gautsch C, Gerard S, Goetzloff L, Gołebiowska I, Gorur P, Groemer G, Groll P, Haider C, Haider O, Hauth E, Hauth S, Hettrich S, Jais W, Jones N, Taj-Eddine K, Karl A, Kauerhoff T, Khan MS, Kjeldsen A, Klauck J, Losiak A, Luger M, Luger T, Luger U, McArthur J, Moser L, Neuner J, Orgel C, Ori GG, Paternesi R, Peschier J, Pfeil I, Prock S, Radinger J, Ramirez B, Ramo W, Rampey M, Sams A, Sams E, Sandu O, Sans A, Sansone P, Scheer D, Schildhammer D, Scornet Q, Sejkora N, Stadler A, Stummer F, Taraba M, Tlustos R, Toferer E, Turetschek T, Winter E, Zanella-Kux K (2014) The MARS2013 Mars analog mission. Astrobiology 14:360–376

Groemer G, Losiak A, Soucek A, Plank C, Zanardini L, Sejkora N, Sams S (2016) The AMADEE-15 Mars simulation. Acta Astronaut 129:277–290

Losiak A, Gołebiowska I, Orgel C, Moser L, MacArthur J, Boyd A, Hettrich S, Jones N, Groemer G (2014) Remote science support during MARS2013: testing a mapbased system of data processing and utilization for the future long-duration planetary missions. Astrobiology 14:417–430

Manned Spacecraft Center (1972) Apollo 17 Traverse Planning Data, 3rd edn. NASA Manned Spacecraft Center, Houston

MARS2013. oewf.org/en/portfolio/morocco-mars-2013/

Muehlberger W (1981) Apollo 16 traverse planning and field procedures. In: Ulrich GE, Hodges CA, Muehlberger WR (ed) Geology of the Apollo 16 area, Central Lunar Highlands USGS, vol 1048, pp 10–20

Orgel C, Kereszturi A, Váczia T, Groemer G, Sattler B (2013) Scientific results and lessons learned from an integrated crewed Mars exploration simulation at the Rio Tinto Mars analogue site. Acta Astronaut 94(2):736–748. https://doi.org/10.1016/j.actaastro.2013.09.014
PULI. www.pulispace.com/
QGIS. www.qgis.org
Skinner JA, Fortezzo CM (2013) The role of photogeologic mapping in traverse planning: lessons from DRATS 2010 activities. Acta Astronaut 90:242–253
Wilhelms DE, McCauley JF, Trask NJ (1987) The Geologic History of the Moon, USGS Professional Paper 1348, U.S. Geological Survey, Reston, VA
WROCLAW. http://gis.um.wroc.pl/imap/?gpmap=hipsometria
Yingst RA, Cohen BA, Ming DW, Eppler DB (2013) Comparing Apollo and Mars Exploration Rover (MER)/Phoenix operations paradigms for human exploration during NASA Desert-RATS science operations. Acta Astronaut 90:311–317

Cartography of the Soviet Lunokhods' Routes on the Moon

I. P. Karachevtseva, A. A. Kokhanov, N. A. Kozlova and Zh. F. Rodionova

Abstract Soviet missions Luna-17 (1971) and Luna-21 (1973) deployed the roving robotic vehicles Lunokhod-1 and Lunokhod-2 on the lunar surface. The Lunokhods (Moonwalkers) were the first extraterrestrial rovers that were operated remotely from Earth. Using Lunar Reconnaissance Orbiter (LRO) narrow-angle camera (NAC) (Robinson et al. 2010), the Lunokhods' routes have been reconstructed (Karachevtseva et al. 2013; 2017). Following the rover tracks that are visible on high-resolution LROC NAC images, we identified the exact rover traverses and compared them with data from archive topographic maps created during Soviet lunar missions. Derived LRO data (DEMs and orthomosaics) allowed us to analyze the topography of the Moon area at the local level and to map the Lunokhods' routes with more details.

Keywords Soviet Moon missions Luna-17 and Luna-21 · Lunar robotic vehicle (Lunokhod) · LROC NAC · DEM · Orthomosaic · Traverse reconstruction · Maps of Lunokhod-1 and Lunokhod-2 landing area

1 Introduction

The development of the theory and practice of Planetary Cartography is directly related to the progress in the field of space research beginning in the second half of the twentieth century. Russia has a long tradition in Planetary Cartography: Compilation of lunar maps began after the successful execution of the epochal space experiment to photograph the far side of the Moon by the Luna-3 mission in

I. P. Karachevtseva (✉) · A. A. Kokhanov · N. A. Kozlova
MIIGAiK Extraterrestrial Laboratory (MExLab), Moscow State University of Geodesy and Cartography (MIIGAiK), Moscow, Russia
e-mail: i_karachevtseva@miigaik.ru

Zh. F. Rodionova
Sternberg State Astronomical Institute, Lomonosov Moscow State University, Moscow, Russia

© Springer Nature Switzerland AG 2019

H. Hargitai (ed.), *Planetary Cartography and GIS*, Lecture Notes in Geoinformation and Cartography, https://doi.org/10.1007/978-3-319-62849-3_13

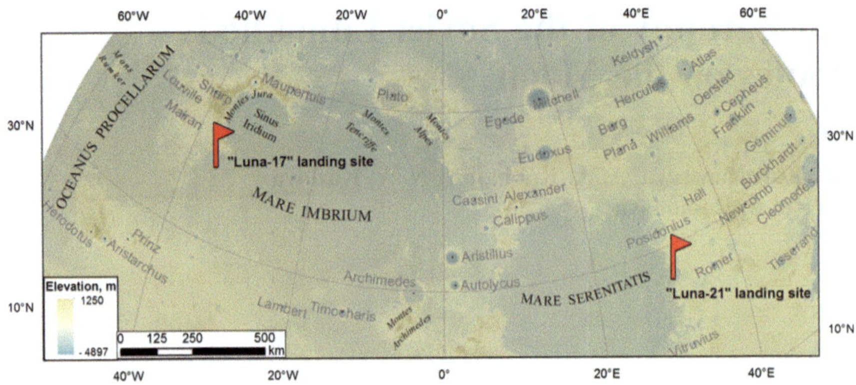

Fig. 1 Map of Lunokhods' landing sites: Luna-17 in Mare Imbrium and Luna-21 in Le Monnier crater, located on the eastern margin of the Serenitatis Basin (Mare Serenitatis)

1959 for the first time (Shevchenko et al. 2016). Later, Lunokhods, also for the first time, carried out detailed in situ exploration of large lunar surface areas remotely.

Luna-17 landed on November 17, 1970, in Mare Imbrium and Luna-21 landed on January 16, 1973, in Mare Serenitatis (Fig. 1). The goals of the missions were with the help of distantly controlled planetary vehicles to study the lunar topography, geology, and morphology (Vinogradov et al. 1971; Barsukov 1978), as well as mechanical conditions (Leonovich et al. 1971; 1978) and the chemical composition of the lunar surface (Kocharov and Viktorov 1974; Cherkasov and Shvarev 1975). The concept behind roving robotic vehicles was to support the human exploration of the Moon. It was anticipated that the rover would land and certify the site prior to human landings, examine the human lander after it landed, and transport the crew to a backup rover if necessary. Lunokhod-1 studied lunar mare (Florensky et al. 1978), and Lunokhod-2 focused on the transition area between mare and highlands (Florensky et al. 1976).

2 Lunokhods' Vehicles and Recent Remote Sensing Data

2.1 Lunokhods' Vehicles

The construction of Lunokhods was identical, but the total mass of Lunokhod-2 was 836 kg compared to 756 kg of Lunokhod-1 (Kemurdzhian et al. 1993).

Both Lunokhods carried laser retroreflectors—14 glass pyramids in one thermally insulated box—that were used for geodetic Lunar Laser Ranging (LLR) experiments in common with retroreflectors established on the lunar surface during Apollo Missions (a panel of 300 or 100 prisms). Since 1978, regular laser measurements of Lunokhod-2 are carried out providing data on the distance to the Moon with an accuracy of 25 cm (Kokurin 2003). Results from the LLR

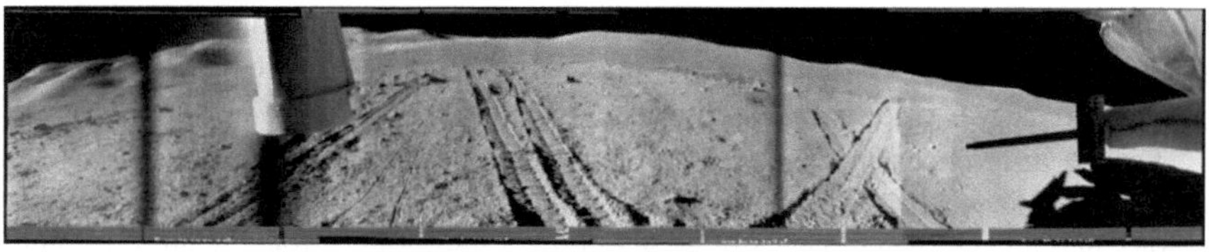

Fig. 2 New assembled archive panorama # 6-387 shows track of the rover and its ninth wheel

experiment (both Lunokhods and 3 Apollo reflectors) give us important insights concerning the dynamics and interior of the Moon (Dickey et al. 1994). In addition, with the reflector coordinates established to the cm level, the laser reflector stations mark important geodetic reference points that define the currently used lunar coordinate systems (Archinal et al. 2011).

At the time of the missions, there were no detailed maps or images of the landing sites. Therefore, it was impossible to plan the whole route beforehand; only main goals and directions were chosen. To save Lunokhod from dangerous craters and rock fragments, a special human crew was trained to control and operate the rovers from the Earth. For that purpose, Lunokhods had two television cameras for navigation (abbreviated as "MKTV"—the Russian abbreviation stands for "Small Frame Television System") mounted at a height of 950 mm above the surface. Due to the construction of cameras and considering signal delays, navigation images came to the Lunokhod driver every 6–10 s.

In order to conduct topographic and morphologic studies of the lunar surface, the rovers were also equipped with four panoramic TV scanning cameras: one horizontal and one vertical on each side (Selivanov et al. 1971). The design of the cameras allowed taking black and white panoramic images 360° × 30° (6000 lines × 500 samples) with an angular resolution of 0.06°. However as the cameras were mounted on the Lunokhod sides, horizontal cameras made panoramas with the width a bit more than 180° (Fig. 2). Therefore, stereo pairs at some selected sites were obtained by taking panoramas from two positions of Lunokhod. It allowed to determine the steepness and slopes inside craters with high level of details based on photogrammetric processing of stereo panoramas (Rodionov et al. 1971).

2.2 *LROC Data: DEMs and Orthomosaics*

As lunar landing sites, such as Luna-17 (Lunokhod-1) and Luna-21 (Lunokhod-2) working areas, have been LRO mission priority targets, stereo images with pixel size up to 0.3 m have been obtained for these regions. The orbiter is equipped with the LROCamera (LROC) system consisting of a wide-angle and two identical narrow-angle scanner cameras (NAC) for high-resolution (0.5 m/pixel) monochrome close-up views (Robinson et al. 2010). The panchromatic images are 5000 pixels wide and typically 50,000 pixels long. It provided possibilities to compare extraterrestrial in situ measurements with studies based on orbital remote sensing data. For this purpose, stereo processing of LROC NAC images for Lunokhods' activity areas was implemented (Karachevtseva et al. 2013; 2017).

Due to differences in the rovers routes, to study the Lunokhod-1 traverse only one stereo pair was used (see for details Karachevtseva et al. 2013) and about 60 stereo pairs were processed for research of Lunokhod-2 track (see for details Karachevtseva et al. 2017). 10 km^2 Lunokhod-1 area DEM with vertical accuracy ~0.1–0.5 m and 200 km^2 Lunokhod-2 area DEM with vertical accuracy 2–3 m were produced. The NAC geometric calibration and alignment information were taken from previously studied (Oberst et al. 2010) pre-launch measurements. LRO navigation data were derived from reconstructed orbit and pointing. A least-squares bundle block adjustment was carried out, following which the position and orientation of the NAC images improved. The PHOTOMOD implementation of the block adjustment is based on the RPC model (Grodecki and Dial 2003), which is characterized by an effectively reduced number of adjustment parameters (6 per image) providing a stable solution. The final 0.75 m DEM was shifted by 86 m laterally to fit the LLR-based coordinates of the Lunokhod-1 rover (Williams et al. 2013). All LRO data are provided in Mean Earth (ME) coordinate system, and LLR retroreflector array coordinates are also in the ME system (LGCWG and LRO Project 2008). Orthomosaics of the landing sites with the best visibility of the rover tracks were created using the DEMs. The results of photogrammetric image processing of LROC NAC stereo pairs—local DEMs and orthomosaics—provide a consistent coordinate base for mapping and studies of the routes, including variance of heights and slopes along the tracks.

3 Reconstruction of the Lunokhods' Routes

3.1 *Identification of Lunokhod-1 Traverse*

LRO NAC images show that the traverse of Lunokhod-1 can almost fully be recovered (~99%), except for few areas in shadow (<1%). Tracks appear dark in the images due to the disturbance of the surface by the wheels. In the available super resolution LROC NAC image (0.33 m/pixel), Lunokhod-1 tracks of left and right wheels can be seen clearly. Contrast is low on images taken at high Sun, and finding and following the rover wheel tracks are more difficult. Contrast enhancement was used to bring out image details. Analysis of surface brightness profiles across Lunokhod-1 tracks (Fig. 3) reveals that the brightness of the wheel tracks typically drops by ~1% with respect to the surrounding area. This is due to a change in the photometric properties of the surface once it has been disturbed. The brightness of the tracks is found to depend on solar elevation, but it also appears to depend on the azimuthal direction of rover motion (Figs. 4, 5, 6, and 7).

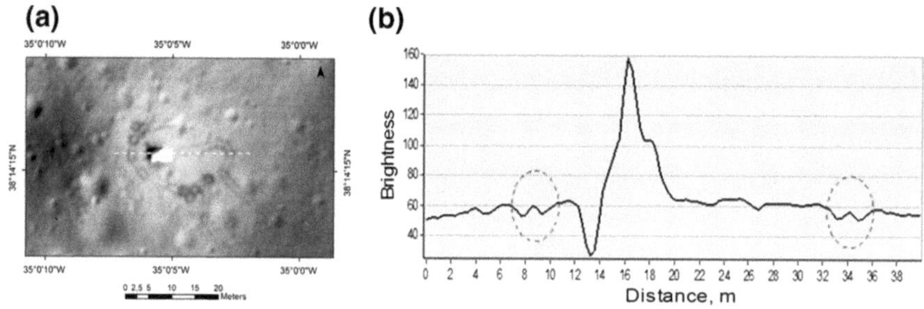

Fig. 3 Analysis of brightness of Lunokhod-1 wheel tracks: (a) LROC NAC image M175502049 with resolution 0.3 m/pixel; (b) corresponding profile of brightness along transect line with white dashed color on the image. *Note:* W-shaped pattern on the plot representing tracks of left and right rover wheels

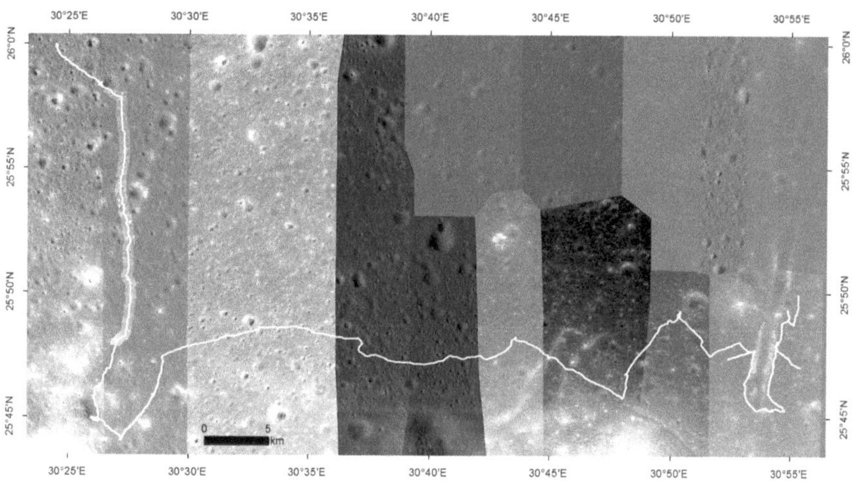

Fig. 4 Orthomosaic with high quality of visible track, produced to identify the route (equidistant cylindrical projection with center at study area—central meridian 30°40′, standard parallel 25°50′)

Poorly visible parts of the track, due to multiple crossing tracks or turning motion of the Lunokhod-1 (about ~5% of the traverse length), were identified with the help of the archive topographic maps created during Luna-17 mission as a result of panorama processing (Abramova et al. 1978). To use the large-scale maps with resolution up to cm level (see below one example on Fig. 8a), the scales of the LROC NAC images were increased to 1:500, which approximately corresponds to the scale of the historic topographic plans.

The identified track was digitized and measured using ArcGIS software. Following identification of the track, the total length of the traverse is measured in GIS using equidistant along the meridian projection (due to meridional general orientation of traverse) as 9.93 km (Karachevtseva et al. 2013), that is shorter than

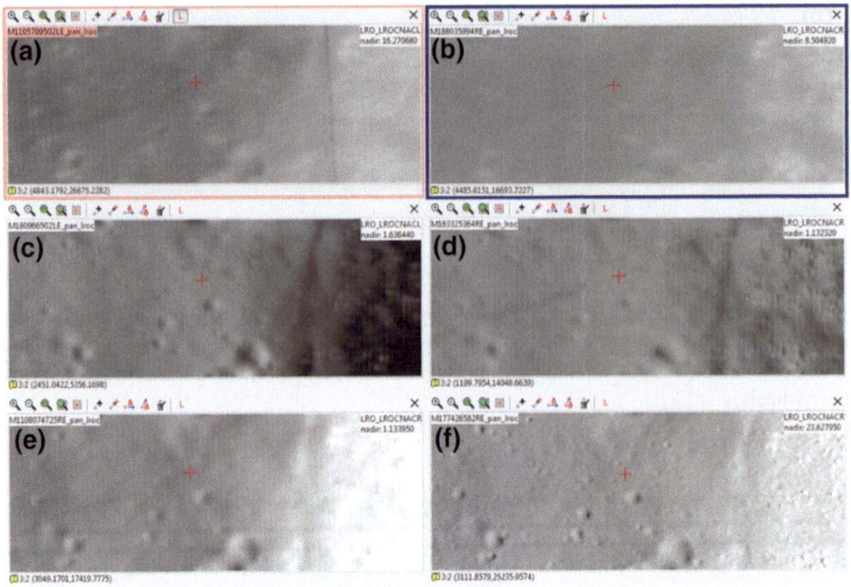

Fig. 5 Traverse visibility on different LROC NAC images: (a) M1105709502LE, (b) M188035994RE, (c) M180966502LE, (d) M183325364RE, (e) M1108074725RE, (f) M177426582RE

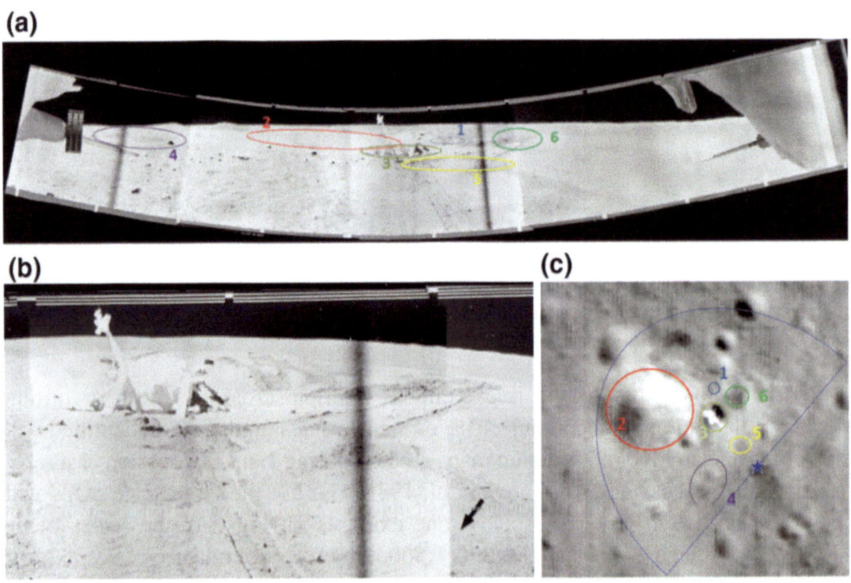

Fig. 6 Implementation of Lunokhod panoramic images for identification of the wheel tracks: (a) Lunokhod-2 panorama #6-375 with marked prominent features that are also visible in LROC NAC orthoimage; (b) zoomed in fragment of the panorama—the ninth wheel track is marked by an arrow; (c) LROC NAC orthoimage of the landing site, observation point of the panorama is marked with a star

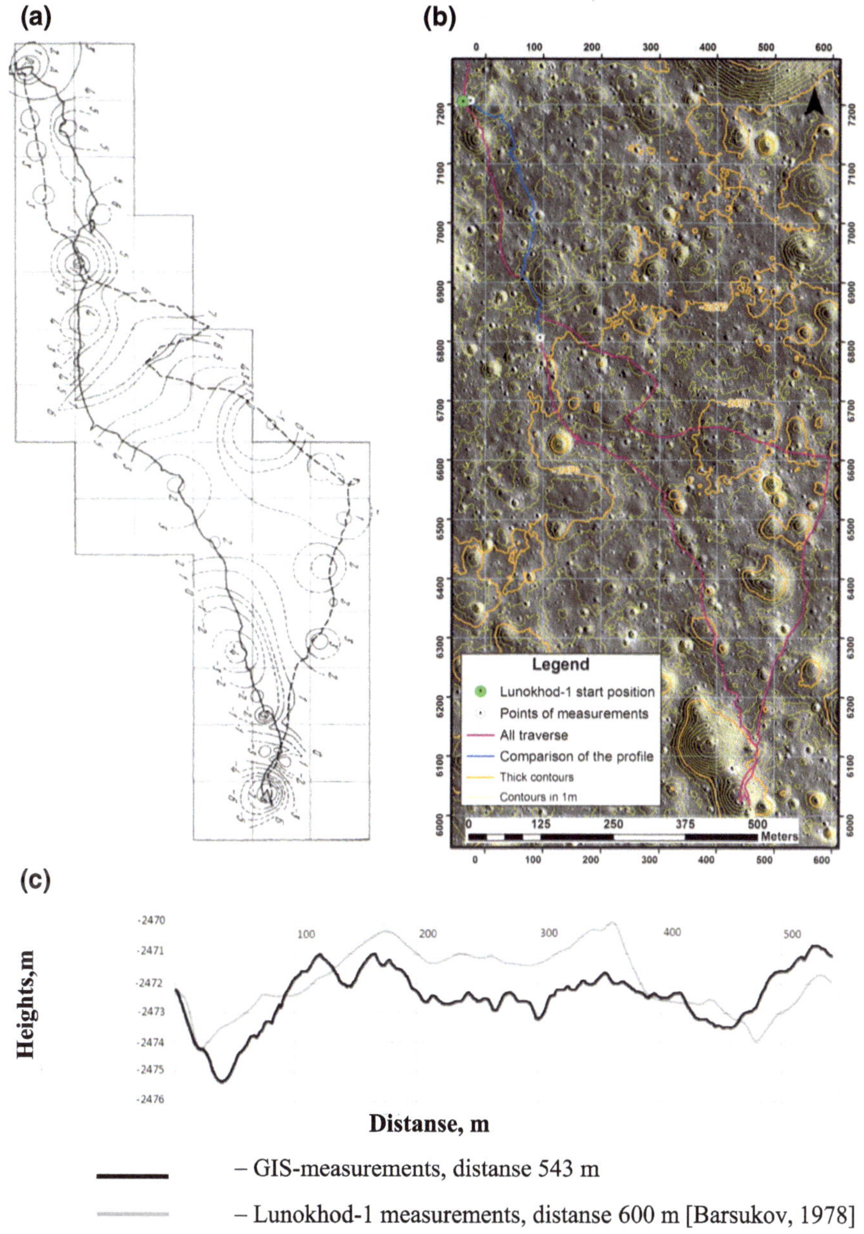

Fig. 7 Topography for the southern part of the Lunokhod 1 traverse area: (a) topographic map from Leonovich et al. (1978); (b) contour map, derived from LROC NAC data; and (c) comparison of profiles for the part of traverse (blue line at the contour map)

the previously published significantly (10%) longer path of 10.54 km (Ivanov et al. 1978). Studies of the traverse sections indicate similar differences in estimates of the track length for each lunar day. This discrepancy is probably due to wheel slippage, which would artificially increase the travelled distance. Wheel slippage is a common problem on both Lunokhods and Mars Exploration Rovers. The odometer measurements, expected to check against wheel slippage, had an accuracy of not more than 7% (Leonovich et al. 1978), which may partly explain the discrepancy.

3.2 Identification of Lunokhod-2 Traverse

The Lunokhod-2 route was primarily west to east (contrary to Lunokhod-1, which moved south to north), and Lunokhod-2 travelled a distance four times longer than Lunokhod-1 did. Hence, several sets of LROC NAC images were required to cover the Lunokhod-2 study area. These images were acquired over several different LRO mission phases and have various lighting conditions (Fig. 4).

For track identification, images taken under complementary illumination and the highest-resolution were used (pixel scales ~0.5–1.0 m). Figure 5 and Table 1 show examples of traverse visibility on different space images.

The lengths of individual parts of the Lunokhod-2 traverse from different lunar days were determined using GIS tools and equidistant projection with the main parallel crossing the most part of the route to reduce linear distortions. The total length of the traverse was measured as 39.1 km, significantly longer (~2 km) than the previously published result of 37 km, which was based on the onboard odometer measurements. The big discrepancy between historic and recent distance measurements of the Lunokhod-2 route probably has multiple causes: (1) mistakes based on odometer measurements (ninth wheel) as mentioned above for Lunokhod-1 and (2) accumulation of errors along the longer path.

3.3 Implementation of Panoramic and MKTV Images

Surface panoramas and MKTV images also were useful in local identification and refinement of the Lunokhods' traverses. For this purpose, archive analog panoramas as scanned image fragments were re-assembled in digital form (Kozlova et al. 2014) with the help of specially developed software (Zubarev et al. 2016).

Observation points for panoramas were determined on the route (Kozlova et al. 2016) using the identification of the same craters in newly processed archive panoramas (Fig. 6a) and in high-resolution LROC NAC orthoimages (Fig. 6c). One of the first panoramas taken by Lunokhod-2 captures the landing module and the beginning of the route: Wheel tracks are clearly visible, including the track of the ninth wheel (Fig. 6b).

4 Comparisons with Archive Traverse Data

Of course, the differences in recent measurements for both Lunokhods' routes also depend on two various methods implemented for estimation of the path: (1) calculation from odometer based on "in situ" data from lunar surface provided by the ninth wheel and (2) digitizing tracks using orbital images. To analyze the discrepancies, archive topographic schemas and maps produced during Luna-17 and Luna-21 missions were used.

4.1 Lunokhod-1 Archive Data

The contours at Lunokhod-1 topographic schema and profiles (Fig. 7a) published after the mission (Rodionov et al. 1971; Leonovich et al. 1978) were compared with contour map derived from DEM produced using LROC NAC stereo pairs (Fig. 7b). While the total height range in both representations is in agreement, only approximate matching of slopes and topographic trends was found. Based on available 600-m length traverse archive profile, a comparison with a traverse profile taken from the LROC DEM reveals that the basic features (e.g., point of lowest elevation) match well, while the magnitude of the topography is grossly misrepresented (Fig. 7c).

Also, comparisons between crater morphologies from the published archive topographic contour map and the LRO NAC image and DEM were made (Fig. 8). The earlier contour map has been derived by photogrammetric processing of Lunokhod-1 stereo panoramas (Abramova et al. 1978). Since the goal of mapping was to represent all recognized relief features with its characteristics, special system of cartographic signs has been developed (Shingareva and Burba 1978). Lunar relief objects shown on the archive maps were divided into three groups: (1) elements with closed outline (craters, hills, boulders); (2) linear elements (grooves, ridge, lineaments); and (3) spatial elements (fields of craters and boulders).

Considering the different nature of the archive and recent data sets, the overall agreement in size, depth, and shape of the crater is good (Fig. 8c, d).

4.2 Lunokhod-2 Archive Data

During Lunokhod-2 mission, operation route map, which original is currently held by the Lavochkin Research and Production Association Museum,[1] was compiled.

Also, various individual topographic maps of small study areas derived from operative panorama processing based on coordinate observations and navigation measurements were produced (Rodionov et al. 1973), including a topographic map

[1]http://www.laspace.ru/rus/museum.php.

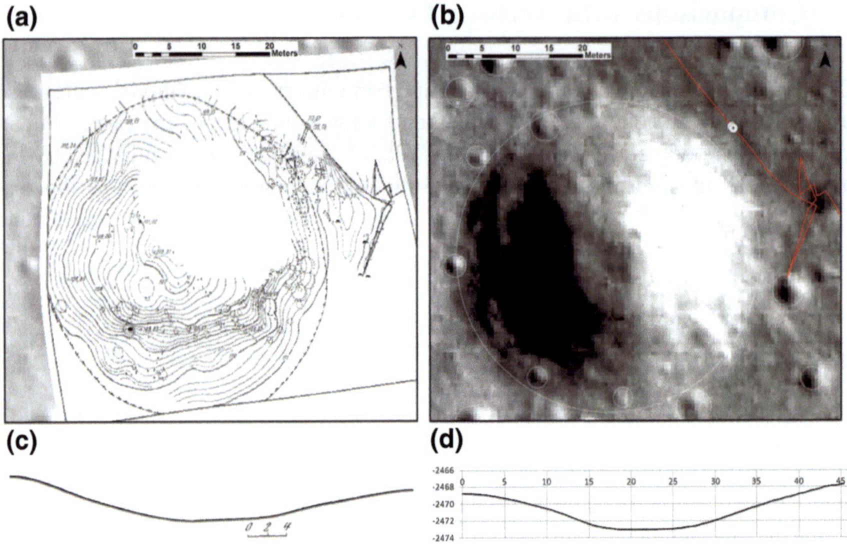

D=40.0 m; H=4.5 m D=45.3 m; H=5.5 m

Fig. 8 Comparison of archive Lunokhod-1 data and recent GIS measurements: (a) topographic map of a small crater (map# 3 from Abramova et al. 1978) referenced to the LROC NAC orthoimage; (b) the same crater identified on LROC NAC orthoimage and a part of rover track identified using the archive map; (c) height profile through the crater from the map # 3; (d) height profile obtained from high-resolution LROC NAC DEM

Table 1 Illumination conditions on LROC NAC images

M1105709502LE Resolution 1.25 Emission angle 17.57 Incidence angle 45.29 Phase angle 31.5 Solar longitude 334.3	**M188035994RE** Resolution 1.40 Emission angle 9.17 Incidence angle 29.04 Phase angle 25.06 Solar longitude 123.05
M180966502LE Resolution 1.41 Emission angle 1.78 Incidence angle 70.57 Phase angle 72.32 Solar longitude 37.28	**M183325364RE** Resolution 1.40 Emission angle 1.2 Incidence angle 46.39 Phase angle 45.29 Solar longitude 65.84
M1108074725RE Resolution 1.36 Emission angle 1.2 Incidence angle 67.48 Phase angle 66.33 Solar longitude 3.13	**M177426582RE** Resolution 0.49 Emission angle 24.13 Incidence angle 71.52 Phase angle 47.93 Solar longitude 354.88

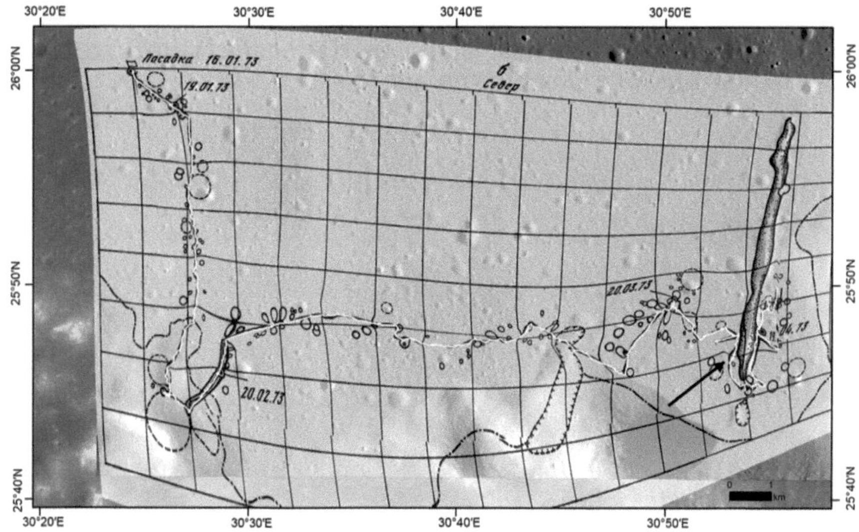

Fig. 9 Differences between Lunokhod-2 route, digitized in GIS (white), and the route from transformed historic topographic schema (black), produced as the result of analysis and mapping after Luna-21 mission (maximal discrepancy, marked with arrow, is about 80 m)

of the southwest part of the route, a relief map of an intensively studied small crater, and a topographic sketch map of the entire working area.

Archive Lunokhod-2 route maps were used for comparison with recent traverse reconstruction. For this about 50 corresponding tie points on the route from archive map and new traverse were picked. Then archive map was rubber stretched to match the new traverse recognized on LRO orthomosaic. Judging from the distortion of the map grid significant long-wavelength displacements was identified in the archive map, up to the 100-m level (Fig. 9). However, on small-scale, the historic, and the LROC NAC-based representation of traverses show excellent qualitative agreement.

5 Recent Cartography of Lunokhods' Activity Area

Lunokhods for the first time carried out detailed in situ exploration of large lunar surface areas remotely. Lunokhod-1 travelled on a typical lunar mare region, while Lunokhod-2 investigated a transition area from mare into highlands. Despite the fact that so much time has passed since the missions, these areas are still of great scientific interest, for example, for morphometric measurements (Basilevsky et al. 2014), and the archive Lunokhod data is still valuable and suitable for further research using recent technologies (Kaydash and Shkuratov 2014).

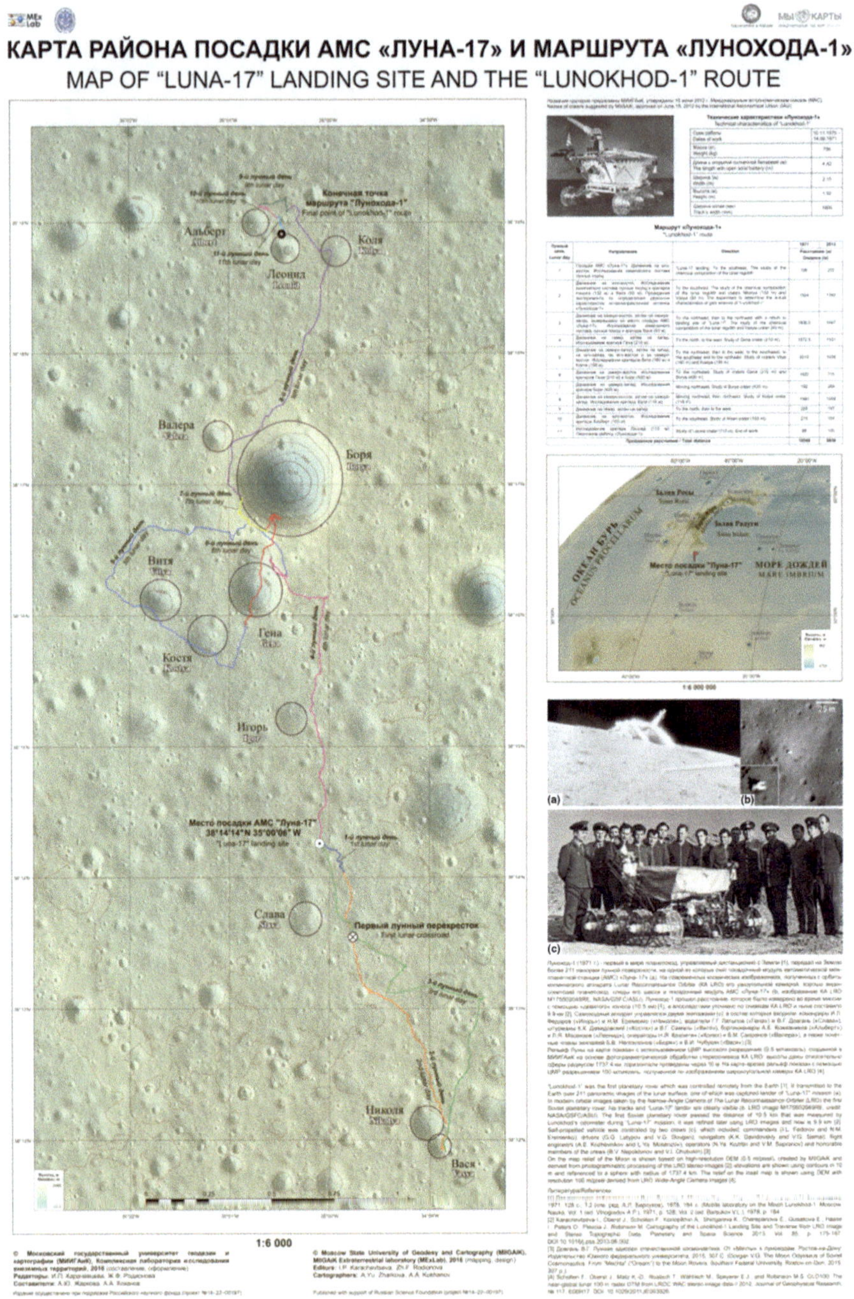

Fig. 10 Map of the Lunokhod-1 route (Karachevtseva et al. 2016). Original size of map sheet is 594 × 841 mm. Colors of route parts show distance travelled through single lunar day

To present the results of the study, Lunokhod route maps—1:60000 for the Lunokhod-1 and 1:25000 for the Lunokhod-2—were compiled (Fig. 10). The geographical basis for mapping is orthomosaic of high-resolution images, which shows the smallest craters at the defined scale. Maps are compiled in equidistant cylindrical projection, to minimize linear distortion for distance measurements. The main parallel of the projection for the Lunokhod-1 map is 35°N, and for the Lunokhod-2, it is 25.8°N (the center of mapped territory).

Since these maps constitute one series, a uniform design was developed: One gradient color scale is used for relief visualization both on the main maps and the insertion maps. Also, local relief on the main maps is shown by contours. Different contour intervals were selected depending on the minimal and maximum heights inside the mapped territory: 10 m for the Lunokhod-1 and 50 m for the Lunokhod-2 map.

The sign of traverse was drawn as a centerline between two visible tracks. As the Lunokhods worked several lunar days, each section of traverse, passed during single day, is shown in a separate color and labeled by the day's number. Both maps contain reference information, including technical characteristics of the rovers, description of routes by lunar days with comparison of old and recent distance measurements and brief historical overview. The Lunokhods' maps are presented in A1 size and give a fresh look at the history of the Soviet lunar exploration program.

GIS-supported files are available via the MIIGAiK Extraterrestrial Laboratory Geoportal.[2]

6 Summary

As lunar landing sites such as the Luna-17 (Lunokhod-1) and Luna-21 (Lunokhod-2) operation areas have been the priority targets of the LRO mission, this provided an opportunity to compare extraterrestrial in situ measurements with studies based on orbital remote sensing data. For this purpose, stereo processing of LROC NAC images for Lunokhods' activity areas was implemented. Based on LRO data and GIS, we reconstructed Lunokhods' traverses and gave new insights into Soviet lunar mission achievements.

Bringing archive data into modern spatial context provides excellent opportunities for detailed comparative analysis with new data. It provides a new view on past and recent lunar missions as, for example, studies of Lunokhod-1 and Yutu working areas (Basilevsky et al. 2015), including morphometric and geologic assessment, estimations of boulder types and densities, crater classes, and regolith structure. Moreover, the well-studied lunar regions can be used for future missions as an analog for testing and calibration of different instruments and techniques.

[2]http://cartsrv.mexlab.ru/geoportal (short URL: goo.gl/ffycQb).

For example, high-resolution DEMs and orthomosaics of the Lunokhod-1 and Lunokhod-2 areas support morphometric and safety assessments for the selection of the candidate Luna-25 (Luna-Glob) landing sites. There is a considerable amount of shadow in polar areas, and there is no opportunity to obtain good quality stereo pairs for photogrammetrically processed DEMs. Therefore, a method of estimating the distribution of slopes in portions of shaded areas measured in the images acquired at different solar incidence angles was suggested (Abdrakhimov et al. 2015). This method was calibrated on analog regions in Lunokhod-1 and Lunokhod-2 areas where we have images with various illumination conditions as well as detailed DEMs. Furthermore, the LLR coordinates of the Lunokhod-1 and Lunokhod-2 rover positions provide high absolute accuracy of the created DEMs that can be used as a reference area to control calibration of the stereo camera for planned Russian Moon projects and to perform refinement of the spacecraft trajectory during the future orbital mission Luna-26 (Luna-Resource).

References

Abdrakhimov AM, Basilevsky AT, Ivanov MA, Kokhanov AA, Karachevtseva IP, Head JW (2015) Occurrence probability of slopes on the lunar surface: estimate by the shaded area percentage in the LROC NAC images. Solar Syst Res 49(5):285–294. https://doi.org/10.1134/S0038094615050019

Abramova MV et al (1978) Topographical and navigational complex of research Lunokhod-1. In: Barsukov AP (ed) Mobile laboratory on the moon Lunokhod-1, vol 2. Nauka, Moscow, pp 67–80 (in Russian)

Archinal BA et al (2011) Report of the IAU working group on cartographic coordinates and rotational elements: 2009. Celest Mech Dyn Astron 109(2):101–135

Barsukov VL (ed) (1978) Mobile laboratory on the Moon Lunokhod-1, vol 2. Nauka, Moscow, p 183 (in Russian)

Basilevsky AT, Kreslavsky MA, Karachevtseva IP, Gusakova EN (2014) Morphometry of small impact craters in the Lunokhod-1 and Lunokhod-2 study areas. Planet Space Sci 92:77–87. https://doi.org/10.1016/j.pss.2013.12.016

Basilevsky AT, Abdrakhimov AM, Head JW, Pieters CM, Wu Y, Xiao L (2015) Geologic characteristics of the Luna 17/Lunokhod 1 and Chang'E-3/Yutu landing sites, Northwest Mare Imbrium of the Moon. Planetary and Space Sci 117:385–400

Cherkasov II, Shvarev VV (1975) Lunar soil. Nauka, Moscow, p 142 (in Russian)

Dickey JO et al (1994) Lunar laser ranging: a continuing legacy of the Apollo program. Science 265:482–490

Florensky KP, Basilevsky AT, Bobina NN, Burba GA, Grebeunik NN, Kuzmin RO, Polosukhin VP, Popovich VD, and Pronin AA (1976) Processes of the transformation of the lunar surface in the Le Monnier region studied using results of detailed investigations made by Lunokhod 2. In: Tectonics and structural geology. Planetology; international geological congress, session, 25th, Sydney, Australia, 16–25 August 1976, Reports (A78-16366 04-91). Nauka, Moscow, pp 205–234 (in Russian)

Florensky KP, Bazilevsky AT, Zezin RB, Polusuhin VP, Popova ZV (1978) Geologic and morphologic study of lunar surface. In: Barsukov VL (ed) Mobile laboratory on the Moon Lunokhod-1, vol 2. Nauka, Moscow, pp 102–135 (in Russian)

Grodecki J, Dial G (2003) Block adjustment of high-resolution satellite images described by rational polynomials. Photogram Eng Remote Sens 69(1):59–68

Ivanov OG, Ryvkina VA, Nepoklonov BV (1978) Activities of the Lunokhod-1 on the Moon. In: Barsukov VL (ed) Mobile laboratory on the Moon Lunokhod-1, vol 2. Nauka, Moscow, pp 7–20 (in Russian)

Karachevtseva I, Oberst J, Scholten F, Konopikhin A, Shingareva K, Cherepanova E, Gusakova E, Haase I, Peters O, Plescia J, Robinson M (2013) Cartography of the Lunokhod-1 landing site and traverse from LRO image and stereo topographic data. Planet Space Sci 85:175–187. https://doi.org/10.1016/j.pss.2013.06.002

Karachevtseva I, Kozlova N, Kokhanov A, Zubarev A, Patratiy V, Konopikhin A, Basilevsky A, Oberst J, Haase I, Joliff B, Plescia J, Robinson M (2017) Cartography of the Luna-21 landing site and Lunokhod-2 traverse area based on lunar reconnaissance orbiter camera images and surface archive TV-panoramas. Icarus 283:104–121. https://doi.org/10.1016/j.icarus.2016.05.021

Karachevtseva IP, Rodionova ZhF, Zharkova AYu, Kokhanov AA (2016) Map of "Luna-17" landing site and the "Lunokhod-1" route, 1:6000, MIIGAiK

Kaydash VG, Shkuratov YuG (2014) Structural disturbances of the lunar surface near the Lunokhod-1 spacecraft landing site. Sol Syst Res 48(3):167–175

Kemurdzhian AL (ed) (1993) The Rovers. Mashinostroenie, Moscow, p 400 (in Russian)

Kocharov GE, Viktorov SV (1974) The chemical composition of the lunar surface in the work region of the Lunokhod-2. Rep USSR Acad Sci 214(1):71–74 (in Russian)

Kokurin YuL (2003) Lunar laser ranging: 40 years of research. Quantum Electron 33(1):45–47 (in Russian)

Kozlova NA, Zubarev AE, Karachevtseva IP, Nadezhdina I, Kokhanov A, Patraty V, Mitrokhina L, Oberst J (2014) Some aspects of modern photogrammetric image processing of Soviet Lunokhod panoramas and their implementation for new studies of lunar surface. Int. Arch. Photogramm. Remote Sens. Spatial Inf. Sci XL-4, ISPRS Technical Commission IV Symposium, Suzhou, China, https://doi.org/10.5194/isprsarchives-xl-4-121-2014. Kozlova_Lunokhod_DOI_panoramas_full_paper.pdf., pp. 121–126. http://www.int-arch-photogramm-remote-sens-spatial-inf-sci.net/XL-4/121/2014/isprsarchives-XL-4-121-2014.pdf

Kozlova NA, Zubarev AE, Patratiy VD, Konopikhin AA, Oberst J (2016) Method of a planetary rover localization based on synthetic Lunokhod images. Int. Arch. Photogramm. Remote Sens. Spatial Inf. Sci., XLI-B4, 435–439, https://doi.org/10.5194/isprs-archives-xli-b4-435-2016

Leonovich AK, Gromov VV, Rybakov AV, Petrov VK, Pavlov PS, Cherkasov II, Shvarev VV (1971) Studies of the mechanical properties of lunar soil on self-propelled vehicle Lunokhod-1. In: Vinogradov AP (ed) Mobile laboratory on the Moon Lunokhod-1, vol 1. Nauka, Moscow, pp 78–88 (in Russian)

Leonovich AK, Ivanov OG, Pavlov PS et al (1978) Self-propelled chassis Lunokhod-1 as a tool for the study of lunar surface. In: Barsukov VL (ed) Mobile laboratory on the Moon Lunokhod-1, vol 2. Nauka, Moscow, pp 25–43 (in Russian)

LGCWG and LRO Project (2008) A Standardized Lunar Coordinate System for the Lunar Reconnaissance Orbiter and Lunar Datasets, Version 5. http://lunar.gsfc.nasa.gov/library/451-SCI-000958.pdf. Accessed 1 Oct 2008

Robinson MS, Brylow SM, Tschimmel M, Humm D, Lawrence SJ, Thomas PC, Denevi BW, Bowman-Cisneros E, Zerr J, Ravine MA, Caplinger MA, Ghaemi FT, Schaffner JA, Malin MC, Mahanti P, Bartels A, Anderson J, Tran TN, Eliason EM, McEwen AS, Turtle E, Jolliff BL, Hiesinger H (2010) The lunar reconnaissance orbiter camera (LROC) instrument overview. Space Sci Rev 150:81–124

Rodionov BN, Nepoklonov BV, Kiselev VV (1971) Topography study along the "Lunokhod-1" route. In: Vinogradov AP (ed) Mobile laboratory on the Moon Lunokhod-1, vol. 1. Nauka, Moscow, pp 55–73 (in Russian)

Rodionov BN, Nepoklonov BV, Kiselev VV, Selivanov AS, Zasetskiy VV, Ivanov OG, Babakov FI, Mikhailov RM, Reshetov MM (1973) Topographical surveys on the lunar surface from the Soviet unmanned spacecrafts. Geodesy and Cartography 10:29–41 (in Russian)

Selivanov AS, Govorov VM, Zasetskii VV, Timohin VA (1971) Construction properties and main parameters of Lunokhod-1 television systems. In: Vinogradov AP (ed) Mobile laboratory on the Moon Lunokhod-1, vol 1. Nauka, Moscow, pp 55–65 (in Russian)

Shevchenko V, Rodionova Zh, Michael G (2016) Lunar and planetary cartography in Russia. Astrophysics and Space Science Library 425. ISBN 978-3-319-21038-4. Springer International Publishing Switzerland (https://doi.org/10.1007/978-3-319-21039-1)

Shingareva KB, Burba GA (1978) The system of conventional signs for lunar topographic plans, developed on the basis of the survey materials Lunokhod-1. In: Barsukov VL (ed) Mobile laboratory on the Moon Lunokhod-1, vol. 2. Nauka, Moscow, pp 80–91 (in Russian)

Vinogradov AP (ed) (1971) Mobile laboratory on the Moon Lunokhod-1, vol 1. Nauka, Moscow, p 128 (in Russian)

Williams JG, Boggs DH, Folkner WM DE (2013) 430 Lunar Orbit, Physical Librations, and Surface Coordinates, Report No. IOM 335-JW, DB, WF-20130722-016. Jet Propulsion Laboratory

Zubarev AE, Nadezhdina IE, Kozlova NA, Brusnikin ES, Karachevtseva IP (2016) Special software for planetary image processing and research, Int. Arch. Photogramm. Remote Sens. Spatial Inf. Sci., XLI-B4, 529–536, https://doi.org/10.5194/isprs-archives-xli-b4-529-2016

Part IV
Mapping Methods

Feature Databases in Planetary Geology

Stuart J. Robbins

Abstract The study of planetary surfaces is as diverse as the surfaces themselves, yet almost all can be assisted by the existence of feature databases. Databases provide a single, standardized source that can serve as an important reference for different studies that involve the same kinds of landforms. The construction of a planetary features database involves many different considerations, each of which should be addressed before construction of the database begins. These considerations are broadly classified into three different sections in this chapter and involve pre-planning what features will be included and how measurements will be conducted, the mechanics of the database construction, and how uncertainties and biases can be minimized. This chapter also addresses an important final step for database construction, the art of sharing and leaving one's database open to future corrections.

Keywords Planetary geology · Databases · Planetary databases · Feature databases · Geology

1 Introduction

Databases of geologic features are an important component of planetary investigations. As geologic features can both affect and be effected by other features and processes on planetary surfaces and the subsurface, large databases can make investigation of the underlying processes easier and faster to carry out, and ultimately help to speed the flow of science. Ultimately, a database's utility is limited by how it was constructed, what it may include, how it is shared, and how it may be updated. This chapter explores those issues and makes recommendations based on the experience of the author.

S. J. Robbins (✉)
Southwest Research Institute, 1050 Walnut Street, Suite 300, Boulder, CO 80302, USA
e-mail: stuart@boulder.swri.edu

© Springer Nature Switzerland AG 2019

H. Hargitai (ed.), *Planetary Cartography and GIS*, Lecture Notes in Geoinformation and Cartography, https://doi.org/10.1007/978-3-319-62849-3_14

2 Planning a Planetary Features Database

Planning a database is an important first step to constructing the database. Key questions that must be answered include:

- What kind of feature(s) will be included?
- What is the limiting size of the feature(s)?
- What other data set(s) is/are required to construct this database?
- What parameters will be included for each feature?
- How will those parameters be measured, and how will objectivity and uniformity be maintained?

Feature(s): The most basic first step for planning the database is understanding what the database will be about. Will it be a database of all linear features, such that it will include both extensional and compressional? Or will it be a more limited database of just scarps? Making this decision early will speed the database construction because there will be less need to restart later if it is decided that the database was too ambitious or needs to be expanded. Making this decision should be based on the science question(s) that are desired to be answered. For example, if one wants to study Mercury's global contraction, a database of scarps is relevant, but extensional features such as graben may not be.

Limiting Size: Almost any database of planetary features could be constructed to a size only limited by the remote sensing data set(s) because features often will occur at sizes spanning several orders of magnitude. Therefore, choosing a limiting size is imperative before beginning to avoid unnecessary work effort.

Required Data: Any database of planetary surface features will require remote sensing data from which it will be constructed. As discussed in the Geologic Mapping chapter, one will often have a primary data set from which to work, and other data sets may be brought in as necessary. For example, when mapping impact craters that are kilometres in diameter and larger on Earth's Moon, one may start with the *Lunar Reconnaissance Orbiter's* (*LRO*) Wide-Angle Camera (WAC) global "morphologic" map product, which is a global, 100 m/pixel product produced at a solar incidence angle that averages 60°. While this product is an excellent starting point, many impact craters are subdued and long shadows near the poles will hide features, and so it may be supplemented with *LRO* Lunar Orbiter Laser Altimeter (LOLA) topographic data. As an active instrument, LOLA can "see" in permanent shadow and topography can give a more objective cue for finding crater rims than imagery.

Parameters: For their most basic data, any database of planetary features should include the location of the feature and its size. A database will also need some sort of identification for each parameter such that it can be referred to later. This can be as simple as an integer of the features as they were identified, or it could be based on any number of other schemes such as latitude and longitude or size. Most important is that this identification key does not change from one iteration to the next, such that it is a constant, and any reference to, *e.g.*, "feature ABC123" would always refer to the same feature in the database. Beyond these parameters, any additional parameters will be based upon the kind of feature. For example, a database of mountains might include footprint perimeter, area of the base, height of the mountain, maximum slopes, or whether it is volcanic. A database of valley networks may include total area of the network, volume of the network, average cross section/width, estimated age, and how many levels of branching. An impact crater database may also include crater elliptical properties, crater depth measurements, interior morphology, ejecta morphology, and preservation state. What parameters will be in a database should be decided early on in the project, for several might be able to be measured concurrently and save work effort.

Measurement and Objectivity: For a database of features to be useful, not only must the researcher(s) creating it have confidence that it is objective and repeatable, but so must others who may assess your science from it and/or use the database in the future for their own efforts. All morphometric parameters should be inherently objective so long as the same method of measurement is made. For example, if one were to measure crater diameter by measuring from the north rim to the south rim, then another researcher using the exact same method should achieve the same diameter measurement. If that is not the case, then some additional research may need to be conducted to understand the natural variation and variability between different researchers of the same (what should be) objective measurement; or, a better understanding of the uncertainty in such a measurement is needed (*e.g.* finite pixel size). For morphologic parameters, objectivity is more difficult because the classification is often more subjective and relies upon a researcher's judgement—classification of geologic features with remote sensing data is inherently an interpretive task. For example, if one were to try to classify an impact crater's preservation state, that requires subjective decisions about how much infilling there may be, the state of the rim, and state of the ejecta. This decision may be influenced by data type and quality—*e.g.* higher resolution images or different sun angle could change the interpretation—but ultimately, this is a judgement call by the researcher. It is always best to make the subjective decision as objective as possible (such as listing an a priori set of criteria or decision tree for any classification), but that is not always possible. Consequently, while morphologic parameters may be very useful to the overall science question(s) being asked, minimizing the number of morphologic features is advisable *if* one's goal is a more objective data set.

3 Database Construction

Constructing or compiling a database of planetary features is likely to be time-consuming (otherwise, it probably would already have been completed by someone else). Therefore, methods to reduce time spent are important. Questions that may be considered include:

- Could any portion of the database construction be *reliably* automated?
- Is it faster to identify features in a "pyramid" approach to size?
- Is it faster to search through one data set first and then another, or to do multiple searches concurrently?

Automation: Despite decades of computer vision research, the human eye–brain system is still, generally, significantly better at pattern recognition than computers for planetary features. However, some parts of the process may be automated, or computer-assisted. For example, if measuring a linear feature, it might be faster to manually identify 10 points along the feature and then for the computer to interpolate those points and find the best location for them on the feature (based perhaps on topographic data highs/lows, or change in pixel brightness from light to dark). This process takes advantage of the human's ability to recognize the feature, and it takes advantage of the computer's ability to be more precise and take what the human has given it as a starting point.

Pyramid Approach: Practically speaking, it takes time to scroll through remote sensing data and identify features. This time is extended if one needs to change the map scale for each feature due to orders-of-magnitude size differences. Instead, what may be faster is to scroll through maps and identify the largest features first, then identify the smaller features at a finer map scale. This simple approach could save a significant amount of time that would have been spent simply changing the map scale and re-recognizing the feature.

Serial or Parallel Data Sets: Similar to the pyramid approach for feature sizes, seemingly tiny amounts of time spent doing some aspect of feature recognition, parameter measurement or classification could add up to an extraordinary amount of time at the end. If using multiple data sets to identify or measure a feature, one may find that the simple function of turning on and off the visibility of the additional data set(s) will add up to a substantial amount of time. Therefore, it may be faster to exhaust one data set before moving to the next rather than identifying and characterizing all aspects of a feature at the same time. Along with this is the human brain's ability to multitask, which is not perfect, and research has shown that humans are inherently bad at multitasking, even doing only two things at once. This means that if one is identifying, measuring and classifying a single feature multiple ways at the same time, overall quality will likely drop.

4 Minimizing Random Versus Systematic Biases, and Validation

As briefly discussed above in "Measurement and Objectivity", it is highly likely that there will be some inherent variation in measurement or classification. After all, interpretation of remote sensing data is just that: interpretation. An individual's interpretation of a feature may vary from one moment in time to the next, and one individual's interpretation may vary from another person's. This is true not only for what should be objective measurements, but also for less objective classifications.

In constructing a planetary features database, one will need to keep in mind systematic biases and random biases. A single individual is much more likely to introduce a systematic bias to the data: they measure parameters and interpret features a certain way. While this systematic bias may be somewhat large compared to the average person (*e.g.* consistently hitting the left instead of the centre of a target), it should at least be relatively consistent (minimizing random biases). On the other end, numerous individuals who construct a database will introduce more of a random bias because each of them has their own sets of systematic biases. In an ideal situation, many individuals would construct the same database and this would be compared and reconciled, such that the systematic biases are cancelled out and random biases are minimized. In reality, this is almost never done. Therefore, when constructing a database, one will need to decide whether they desire to minimize systematic biases by having many people work on the same measurements for the database but in different locations on the body, or minimize the random biases by having single individuals be responsible for each task.

The comparison mentioned above is an important part of any database validation. Before a database is released, efforts should be made to understand systematic biases of anyone who was involved in constructing it. Additionally, at least small parts of the database should be validated—or checked—by independent researchers such that a measurement of the inherent disagreement between researchers can be incorporated in any release notes. While this is not a perfect solution, it is better than blindly assuming the database is a perfect representation of reality.

5 Sharing and Updating

Once a planetary features database is generated and validated, it should be released to other researchers and the general public. I will address neither the philosophical nor legal reasons behind this, but sufficive to say, the process of science generally relies upon the dissemination of information, and a features database is well within that category.

There is one final step after sharing, and there are two different approaches: update or do not update. There are good reasons for both, though the latter is mostly filled with practical considerations. For example, if one generated such a database as part of their doctoral dissertation and then chooses to leave the scientific field, there may be no incentive nor time to update the database beyond its initial release. Similarly, there may be no funding for the time required to update the database after it has been released, or there may be no new data from which one could generate a revised database (*e.g.* no new spacecraft sent to the body to take relevant data).

For the other option, there is the fact that regardless of how well one has vetted their work, there will almost always be errors. Not having some avenue by which even others may update the database once it has been released would vitiate another of the main principles of the scientific process: everything should be open to revision. Therefore, even if the person or persons who generated a planetary features database opt to not maintain it, there should always be some open, well-publicized method by which others could contribute to its updating.

6 Conclusions

Databases of planetary features are an important contribution to the study of planetary surfaces. Regardless of how extensive they may be or how much detailed information about any given type of feature they may contain, there will almost always be a use for such a product.

Databases and Metadatabases in Planetary Geology—The Mars Crater Database

Nadine G. Barlow

Abstract Impact crater databases are one of the first data products produced from images acquired by planetary spacecraft missions and are used in a variety of studies investigating the formation ages and geologic histories of surface units. Martian crater databases have been compiled since the first global images of the planet were acquired by the Mariner 9 mission and today include a variety of morphologic and morphometric data about each crater. Three of the most widely used Martian crater databases are described in this contribution.

Keywords Mars · Impact craters · Crater database

1 Introduction

The early Mars flyby missions of Mariners 4, 6, and 7 revealed the presence of impact craters on the surface of the planet (Murray et al. 1971), but the global extent of cratering was not revealed until the Mariner 9 orbital mission in 1971–1972 (Mutch et al. 1976). Mariner 9 images showed that the southern hemisphere is heavily cratered while the northern hemisphere displays a lower crater density, which is reflective of the different ages of these regions. Crater size-frequency distribution analysis of the Mariner 9 data revealed that Mars experienced an early period of high degradation rates uncoupled from the cratering history and interpreted as degradation resulting from warmer, wetter conditions on early Mars (Hartmann 1973; Chapman 1974; Jones 1974). A global database of impact craters and their interior morphologic features derived from Mariner 9 A-frame images was compiled by Brown University researchers. Craters were generally larger than about 4 km diameter and classified into four degradation classes (fresh, slightly degraded, moderately degraded, and highly degraded) based on the characteristics of the crater

N. G. Barlow (✉)
Department of Physics and Astronomy, Northern Arizona University,
Flagstaff, AZ 86011-6010, USA
e-mail: Nadine.Barlow@nau.edu

© Springer Nature Switzerland AG 2019

H. Hargitai (ed.), *Planetary Cartography and GIS*, Lecture Notes in Geoinformation and Cartography, https://doi.org/10.1007/978-3-319-62849-3_15

shape and presence of specific interior features (Mutch et al. 1976). The Mariner 9 image resolutions (ranging from 100 m/pixel to 1 km/pixel) limited the amount of detailed study of ejecta and interior morphologic features.

The arrival of the Viking 1 and 2 Orbiters in 1976 substantially improved our view of the Martian surface. Higher resolution (generally 100 m/pixel but up to 40 m/pixel in selected locations) and clearer atmospheric conditions than during the Mariner 9 mission (which arrived at the height of a global dust storm) resulted in greatly improved image quality. The images revealed a wealth of new information about crater shapes and ejecta and interior morphologies (e.g., Carr et al. 1977; Strom et al. 1992), which were compiled in both regional and global databases. The most widely used global crater database from this era was *The Catalog of Large Martian Impact Craters* (Barlow 1988).

2 Barlow's Catalog of Large Martian Impact Craters

The Catalog of Large Martian Impact Craters was compiled by Nadine Barlow from the 1:2,000,000-scale photomosaics produced from Viking Orbiter images (Barlow 1987). The surface of Mars was subdivided into 30 quadrangles by the Viking team, from MC-1 covering the north pole to MC-30 covering the south polar region. The MC-1 and MC-30 quadrangles were subdivided into 10 subquadrangles, labeled A through J. Quadrangles between 30° and 60° latitude in each hemisphere (i.e., MC-2 through MC-7 and MC-24 through MC-29) were subdivided into either four or five subquadrangles. Those in the north were labeled NW, NC, NE, SW, or SE, while those in the south were NW, NE, SW, SC, SE, representing the geographic location of each subquadrangle (northwest, north-central, northeast, southwest, south-central, or southeast). The quadrangles within the ±30° latitude zone (MC-8 through MC-23) were each subdivided into four subquadrangles (NW, NE, SW, and SE). Each subquadrangle was covered by a separate photomosaic.

Each hardcopy photomosaic map was overlain with a mylar sheet, and craters 5 km in diameter and larger were circled. Elliptical craters were drawn as ellipses where both the major and minor diameters could be measured. Ejecta and interior morphologies were noted, but generally only if both were present. Each mylar sheet was then transferred into a digitizing table, where a BASIC program was used to convert measurements off the sheet into an ASCII file containing information for each crater. Each crater was labeled with a number on the mylar sheet which correlated with an identification number for the crater in the file. The program calculated the latitude and longitude of each crater's center using a 3-point setup along the edges of each photomosaic. Each crater entry included the following information: MC subquadrangle on which the crater is found, identification number, latitude and longitude (west longitude system used at that time; values to two decimal places) of the crater center, crater diameter (in km, to one decimal place), general geologic unit on which crater was found, general preservational state of the crater (e.g., retains ejecta blanket, no ejecta blanket, highly degraded crater), ejecta morphology, one interior morphology, diameter of central pit (if applicable, to one decimal place), minor diameter if crater is elliptical (major diameter was recorded in the diameter column), azimuthal angle of orientation of elliptical craters (where north = 0°, east = 90°, etc.), and any comments such as the crater's official name (approved by the International Astronomical Union (IAU)). The resulting database contained entries for 42,283 craters, most of which were 5 km diameter and larger. The database was distributed initially in ASCII format and eventually in Excel format upon request to Barlow. In about 2000, it was converted into an ArcGIS shapefile and became publically available through the US Geological Survey's Planetary Interactive GIS on the Web Analyzable Database (PIGWAD) (https://webgis.wr.usgs.gov/pigwad/down/mars_crater_consortium.htm).

The Catalog of Large Martian Impact Craters became the primary resource for Martian crater studies through the 1990s. But the advent of new orbital missions producing higher resolution images of the Martian surface revealed that the ejecta and interior morphology classifications needed revision. In addition, a new analysis of the shape of Mars from the Mars Global Surveyor mission's Mars Orbiter Laser Altimeter (MOLA) instrument revealed a center-of-shape offset from the center of mass. The latter had been used to derive the coordinate system for Mars for the Viking datasets, but the new shape information resulted in a revision to the control network from which the coordinates of surface features were derived. This new coordinate system (MDIM 2.0; Kirk et al. 2000) resulted in the central coordinates of the craters in *The Catalog of Large Martian Impact Craters* being offset from the actual crater center and not always by the same amount or in the same direction. There also was a change in reporting longitude values from the west longitude system used during the Viking missions to an east longitude system. All of these changes led to the realization that the Martian crater database needed to be revised.

3 Robbins and Hynek Mars Crater Database

A new global database of Martian impact craters derived from Mars Odyssey Thermal Emission Imaging System (THEMIS) daytime infrared (day IR) images was published in 2012 by Stuart Robbins and Brian Hynek (Robbins and Hynek 2012). This catalog contains 384,343 craters with diameters in the range of 1 km up to ~850 km—the largest basins such as 2300-km-diameter Hellas are not included. Crater mapping was conducted on the THEMIS day IR global mosaics by drawing a polyline with vertex spacing of 500 m around each crater's rim using ArcGIS's editing tools. The data were then imported into Igor Pro software and both circle-fitting and ellipse-fitting routines were run on the data to produce crater diameters in kilometers to two decimal places. The software also calculated the latitude and longitude of the center of each crater to three decimal places. Longitudes are given in east longitude with positive values up to 180° for craters in the eastern hemisphere and negative values down to −180° for the western hemisphere. Topographic information was obtained for all craters ≥ 3 km diameter using Mars Global Surveyor Mars Orbiter Laser Altimeter (MOLA) gridded data and converted into average rim height, average surface elevation, and average floor depth. MOLA analysis was also used to derive crater diameter, which was compared to the diameters obtained from tracing the rim crest and found to agree within 1 km. Each crater is identified using the MC quadrangle number followed by a six-digit number.

Morphologic classifications of each crater were conducted using the THEMIS daytime IR global mosaic with 100 m/pixel resolution. Each crater entry includes up to three categories of interior morphologies and five categories to describe the ejecta morphology (number of ejecta layers, ejecta morphology following the classification system of Barlow et al. (2000), texture and shape of ejecta deposits, and a comments column). The database also includes measurements of ejecta layer perimeters and areas, which, with crater radius, are used to calculate the lobateness (a measure of ejecta sinuosity (Barlow 1994)) and ejecta extent (both the actual runout distance and the value normalized to the crater radius which is called the ejecta mobility ratio (Mouginis-Mark 1979)). Finally, the database includes a 4-point classification system of each crater's degradation state with 4 being a very fresh crater and 1 being a highly degraded crater, information about how confident the compilers were that a given feature is actually an impact crater, and the official IAU-approved name of the crater. The database is available for public download from Robbins' Web site (http://about.sjrdesign.net/research_mars.html) and from the PIGWAD Web site given previously.

4 Barlow's Revised Catalog of Large Martian Impact Craters

Nadine Barlow also is revising her *Catalog of Large Martian Impact Craters* using THEMIS daytime IR data (Barlow 2015). Because the global mosaics do not always utilize the best image data, the revised catalog (henceforth called *Catalog 2.0*) is using individual THEMIS frames to identify and use the best images of each crater. Diameters are measured using a chord connecting two opposite points on the crater rim in an ArcGIS program specifically designed for the project and are cited to one decimal place. A best-fit circle is drawn and the corresponding diameter is recorded. If a circle cannot be fit to the crater rim, the crater is listed as elliptical and both the major and minor diameters are measured and recorded. The program calculates the latitude and longitude of the crater center to two decimal places using the MDIM 2.0 coordinate system (Kirk et al. 2000) and the azimuthal angle of orientation of the major axis for elliptical craters. Crater ID is recorded both as the MC subquadrangle and identification number of the crater in the original *Catalog of Large Martian Impact Craters* (henceforth *Catalog 1.0*) as well as a new identification number based on the crater's latitude and longitude. This new identification number is in the format of XXXX ± YYY, where XXXX refers to the east longitude (000.0° to 360.0°E) and ±YYY is the latitude, where positive values refer to craters in the northern hemisphere (00.0° to 90.0°N) and negative values refer to craters in the southern hemisphere (00.0° to 90.0°S). Thus, the 21-km-diameter Bacolor crater, located at 32.99°N 118.58°E, has an identification number of 1185 + 329.

Catalog 2.0 includes the ejecta morphology following the standardized nomenclature recommended by the Mars Crater Consortium (Barlow et al. 2000) or "No" if the crater has no surrounding ejecta blanket, and up to two interior morphology classifications. Ejecta perimeter and area values are included along with the average and maximum ejecta mobility ratios and the lobateness of each complete ejecta layer (Barlow 1994, 2015). Central peak, peak ring, and central pit diameters are included as well as the ratio of each to the crater diameter. The qualitative preservation classification in *Catalog 1.0* has been replaced by a 7-point numeric scale with 0.5 divisions, ranging from 0.0 for a "ghost" crater up to 7.0 for a very well-preserved crater (Barlow 2004). The geologic unit on which each crater is situated, from the Tanaka et al. (2014) global Geologic Map of Mars, is included. Finally, a comments column provides space for the crater name and/or any other observations related to the crater. As of March 2017, *Catalog 2.0* is complete for MC quadrangles 1 through 22, with 26,073 craters measured and classified. A comparison of data in *Catalogs 1.0* and *2.0* reveals that while many of the ejecta and interior classifications have been revised, the overall trends previously reported from Viking data analysis are still valid (Barlow 2015). The final database will be archived on Barlow's Web site and through the USGS Planetary Data System Annex (Hare et al. 2015).

5 Conclusions

Databases containing information about impact crater diameter, location, and ejecta/interior characteristics are widely used to study the formation ages and geologic/degradation histories of planetary surfaces. Martian crater databases have been popular ever since the Mariner 9 mission provided our first global views of the planet. As image resolution has improved, crater databases have undergone revision by different researchers, either globally for general use or regionally to address specific questions. While location and diameter information have become more consistent between different datasets, there is still considerable variation in the classification of morphologic and preservational data (Barlow 2017), and researchers are encouraged to confirm such data before using it for their own studies.

References

Barlow NG (1987) Relative ages and the geologic evolution of Martian terrain units. PhD Dissertation, University of Arizona, Tucson, AZ, p 124

Barlow NG (1988) Crater size-frequency distributions and a revised Martian relative chronology. Icarus 75:285–305

Barlow NG (1994) Sinuosity of Martian rampart ejecta deposits. J Geophys Res 99:10927–10935

Barlow NG (2004) Martian subsurface volatile concentrations as a function of time: clues from layered ejecta craters. Geophys Res Lett. 31, Cite ED L05703. https://doi.org/10.1029/2003gl019075

Barlow NG (2015) Characteristics of impact craters in the northern hemisphere of Mars. In: Osinski GR, Kring DA (eds) Large meteorite impacts and planetary evolution V, Geological Society of America Special Paper, vol 518, pp 31–63

Barlow NG (2017) Revision of the "Catalog of Large Martian Impact Craters" and comparison to other Martian crater databases. In: Lunar and planetary science conference XLVIII, Abstract #1562

Barlow NG, Boyce JM, Costard FM, Craddock RA, Garvin JB, Sakimoto SEH, Kuzmin RO, Roddy DJ, Soderblom LA (2000) Standardizing the nomenclature of Martian impact crater ejecta morphologies. J Geophys Res 105:26733–26738

Carr MH, Crumpler LS, Cutts JA, Greeley R, Guest JE, Masursky H (1977) Martian impact craters and emplacement of ejecta by surface flow. J Geophys Res 82:4055–4065

Chapman CR (1974) Cratering on Mars: I. Cratering and obliteration history. Icarus 22:272–291

Hare TM, Gaddis LR, Bailen M, LaVoie SK, Padams J (2015) PDS ANNEX: a PDS imaging node repository for geospatial planetary research products. In: Planetary data workshop 2, Abstract #7060

Hartmann WK (1973) Martian cratering 4. Mariner 9 initial analysis of cratering chronology. J Geophys Res 78:4096–4116

Jones KL (1974) Evidence for an episode of crater obliteration intermediate in Martian history. J Geophys Res 79:3917–3931

Kirk RL, Lee EM, Sucharski RM, Richie J, Grecu A, Castro SK (2000) MDIM 2.0: a revised global digital image mosaic of Mars. In: Lunar and planetary science conference XXXI, Abstract #2011

Mouginis-Mark P (1979) Martian fluidized ejecta morphology: variations with crater size, latitude, altitude, and target material. J Geophys Res 84:8011–8022

Murray BC, Soderblom LA, Sharp RP, Cutts JA (1971) The surface of Mars. 1. Cratered terrains. J Geophys Res 76:313–330

Mutch TA, Arvidson RE, Head JW, Jones KL, Saunders RS (1976) The geology of Mars. Princeton University Press, Princeton, p 400

Robbins SJ, Hynek BM (2012) A new global database of Mars impact craters ≥ 1 km: 1. Database creation, properties, and parameters. J Geophys Res 117, Cite ID E05004. https://doi.org/10.1029/2011je003966

Strom RG, Croft SK, Barlow NG (1992) The Martian impact cratering record. In: Kieffer HH, Jakosky BM, Snyder CW, Matthews MS (eds) Mars. University of Arizona Press, Tucson, pp 383–423

Tanaka KL, Skinner JA, Dohm JM, Irwin RP, Kolb EJ, Fortezzo CM, Platz T, Michael GG, Hare, TM (2014) Geologic map of Mars. 1:20,000,000 scale, USGS Scientific Investigations Map 3292

Grid-Mapping: Quantifying the Distribution of Landforms

Martin Voelker and Jason D. Ramsdale

Abstract Grid-mapping is a valuable method to quantify the distribution and geography of landforms using remote sensing data. It can be applied at many mapping scales; from small local study areas to global examinations. Therefore, it is a particularly convenient approach to planetary mapping. By separating the study area into tick boxes of a predefined size, it is possible to determine if a landform occurs in one box or not. The result is an extensive geospatial data set in the form of a spreadsheet which can be used for both simple and advanced statistical calculations and evaluations. Hence, it is possible to study the geographic behaviour of landforms within their planetary environment. And as grid-mapping is not necessarily dependent on high-resolution data sets, it can be applied to almost any celestial body in the Solar System that has been photographed by a space probe.

Keywords Grid-mapping · Geostatistics · Geography

1 Introduction

Traditional geologic mapping is based on the delineation of continuous landforms, providing detailed maps with high information content. But as each planetary mapper may interpret geologic contacts differently, these data sets have limited access to reliable statistical evaluations. For geoscientific subjects requiring both intersubjectivity and quantitative analyses, we suggest a supplemental numeric approach described as grid-mapping. However, the downscaling of mapping to a fixed grid size is also related to a certain loss of information. This fact has to be

M. Voelker (✉)
European Space Astronomy Center ESAC, European Space Agency ESA,
Camino Bajo del Castillo s/n, 28692 Villanueva de la Cañada, Madrid, Spain
e-mail: martin.voelker@esa.int

J. D. Ramsdale
Department Physical Sciences, Open University, Walton Hall,
Milton Keynes MK7 6AA, UK

© Springer Nature Switzerland AG 2019
H. Hargitai (ed.), *Planetary Cartography and GIS*, Lecture Notes in Geoinformation and Cartography, https://doi.org/10.1007/978-3-319-62849-3_16

considered before starting and depends upon the scientific problem being analysed. As grid-mapping generalizes the distribution of landforms it might be not adequate for detailed analyses, e.g., when focusing on one specific landform.

Like classic mapping, grid-mapping is also conducted within a GIS-environment, but requires only basic GIS skills. Hence, grid-mapping can be executed in cooperation with both unskilled mappers (e.g., for crowd-sourced mapping) and specialists scanning the research area for its own specific topic or landform.

The potential application of grid-mapping is diverse; (1) analysing the distribution and geographical behaviour of landforms, (2) recognizing relations between several landforms, (3) when finished, mapped landforms can be reclassified for extracting even more information (e.g., summarizing all landforms requiring ice or liquid water).

2 Approach and Data Basis

Before starting grid-mapping careful preparations are highly recommended. Later changes and modifications are only possible under time-consuming efforts. The objectives of the work need to be clearly defined; (1) Which surface features/parameters are needed to be mapped? (2) What is the required resolution of grids? (3) Is there sufficient data coverage available? (4) Delineation of the research area. A detailed flow chart of the grid-mapping process is presented in Fig. 1.

The pre-selection of landforms is vital for the work. If landforms are added during the mapping process, all grids that already have been scanned need to be examined again. It is also helpful to test if the resolution of the imagery is sufficient in order to recognize the landforms.

Within the GIS-environment the grids themselves are polygonal features containing all the information of the mapping. The length of their edges defines the resolution of the grids. The smaller the grids, the higher is the resolution and information content for later statistical analyses. Grid-mapping work on Mars with a resolution of 20 $\times$ 20 km using CTX data sets (Context Camera onboard Mars Reconnaissance Orbiter; Malin et al. 2007; 6 m/pixel) is adequate (Ramsdale et al. 2017; Voelker et al. 2017). If using an even higher resolution in small research areas fitting, e.g. into HiRISE imagery with a resolution of up to 0.3 m/pixel (McEwen et al. 2007), smaller grids are helpful. For more small-scale and global analyses the HRSC experiment (Neukum et al. 2017; Jaumann et al. 2015) provides a good coverage with a nominal resolution of up to 10 m/pixel. Moreover, even old data sets can be used if no better resolution is available, e.g. Voyager imagery (Smith et al. 1977) for the outer Solar System. For planetary bodies covered by a dense atmosphere (especially Venus and Titan), it is also possible to grid-map their surfaces on basis of topographical and thermal infrared data sets.

When image processing is completed, create a polygonal shapefile in GIS, e.g., by using the *Repeating Shapes* tool by Jenness (2012) for ArcGIS, with the requested mesh size of the grids. The attribute table of the compiled shapefile can

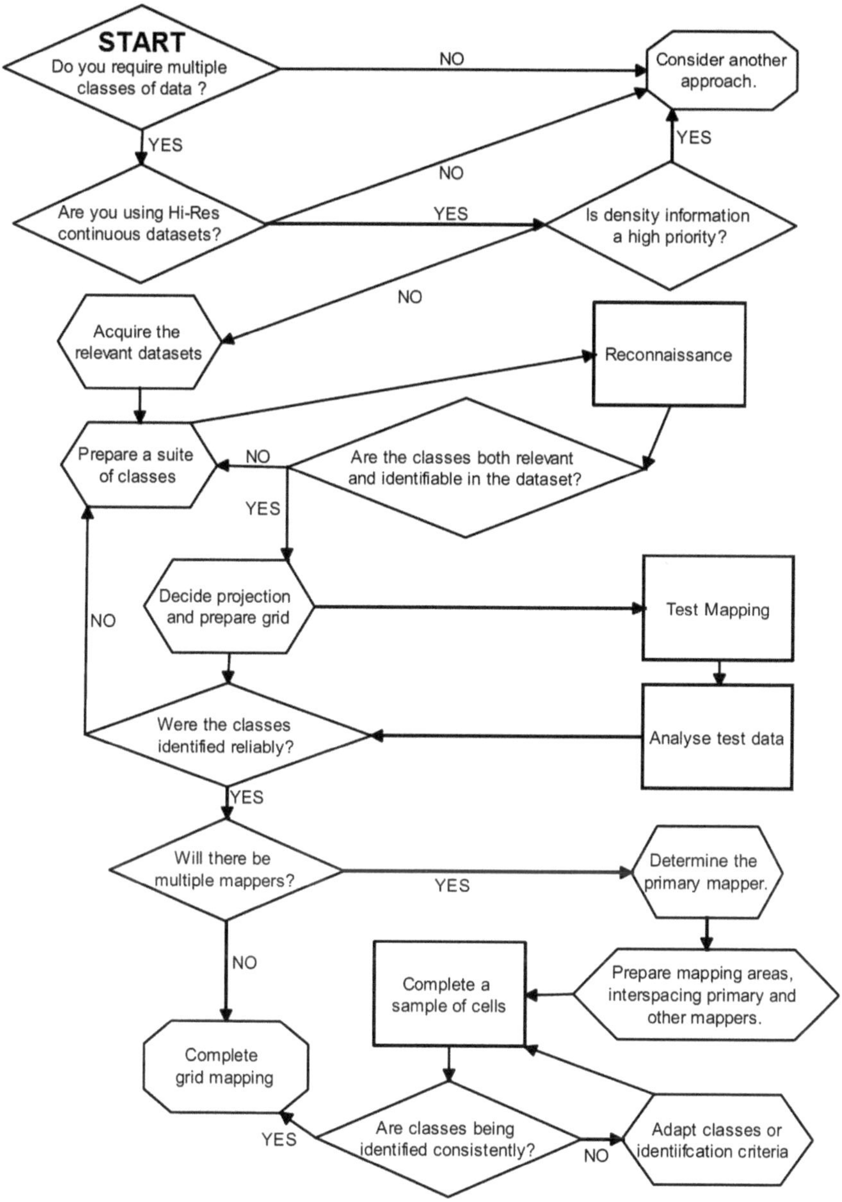

Fig. 1 Flow chart demonstrating an idealised work process for grid-based mapping. Diamonds represent decisions to be made by the mappers, hexagons for preparation steps, quadrangles for mapping stages and octagons for endpoints (Ramsdale et al. 2017)

be extended by a new column for each landform (see Table 2). Other parameters that might be gathered are data coverage and/or the quality of the images used.

3 Mapping

The actual mapping is to inspect each grid for each pre-selected landform in detail. It is very important to map in a consistent mapping scale (e.g., 1:30,000 for CTX images)! The results are being entered in the related cell of the attribute table of the shapefile according to the mapping key presented in Table 1.

As grid-mapping may be time-consuming it is also effective to map only every second grid for very large research areas. The remaining grids can be interpolated later by using a spreadsheet environment or other statistical software (e.g., R). In order to ensure a reliable interpolation, we suggest a weighted interpolating key, considering a wider spatial frame (Fig. 2).

If further parameters are desired (e.g., elevation, slope, aspect) they can be calculated automatically if topographical raster data sets (DEMs) are available. Thus, the distribution and geographical behaviour of each landform related to its elevation or so can be analysed, e.g., in line diagrams.

Tipp: If quick results are required, start by making an observation in each nth-grid. The preliminarily skipped grids can be scanned and added later.

4 Visualisation

When collection of data has been completed, the results of grid-mapping can be represented in two ways: graphically in maps and statistically in numbers and graphs. In order to increase the reliability of the results, we recommend to consider

Table 1 Recommended mapping key as used in Table 2

Code	Prominence	Image quality	Image coverage
0	Absent	N/a	0%
1	Possible	Bad	<33%
2	Present/yes	Medium	66%
3	Dominant	Good	>99%
4	No data		

Prominence: if a surface feature does not occur in a grid it is categorized as "absent". "Possible" landforms are features that cannot be clearly distinguished. Landforms are indicated as "present" when they appear in a grid without being dominant. If a landform prevails the area they are described as "dominant". Image quality: classifies the subjective quality of the image. Image coverage: provides information how much of the grid is covered by images, and hence, provides information about the reliability

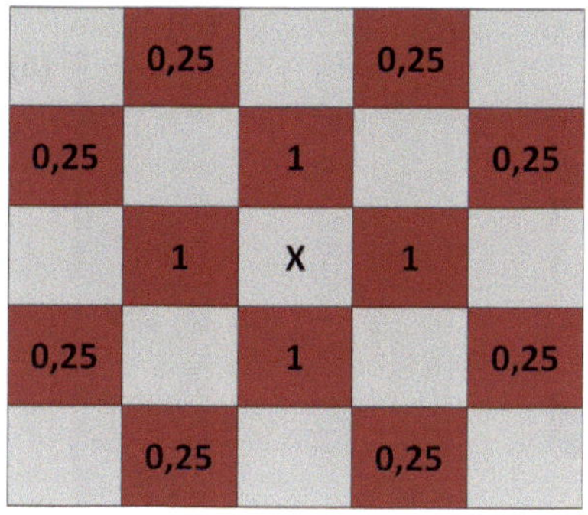

Fig. 2 Suggested weighting factor for interpolation of non-mapped grids (bright), based on the results of the mapped grids (dark). The numbers shown are the weighting factors, decreasing with increasing distance from the grid (x) being interpolated (Voelker et al. 2017)

only grids where landforms clearly exist ("present" and "dominant"). Surface features classified as "possible" may contain significant errors, so they should be excluded.

Depending on the amount of landforms mapped, it is helpful not to show more than six different landforms in one map. If there are more than six landforms, it is convenient to classify landforms; for example, into one class containing all fluvial landforms (Fig. 3) and a second one containing aeolian landforms.

In order to keep the maps comprehensible and simple, it is important to use an easy symbol key. Each of the landforms shown should have its own symbol, e.g. colours, lines or icon. It is highly recommended to use common symbol keys, for example blue colours for possible fluvial landforms or horizontal lines for layered and light-toned deposits. Besides the landforms themselves, it is also possible to express their significance as presented in Table 2, by using different brightness values. Thus, landforms classified as "dominant" appear as dark-coloured (e.g. dark blue for a dominant channel), and landform that are "present" but not prominent appear bright-colored (e.g. light blue for a small and narrow channel within one grid).

For better understanding, grid-maps can be combined with other geospatial datasets, e.g. with maps presenting topography, slope inclination, aspect or thermal inertia of the study area. Existing geological maps can be added too (Ramsdale et al. 2017).

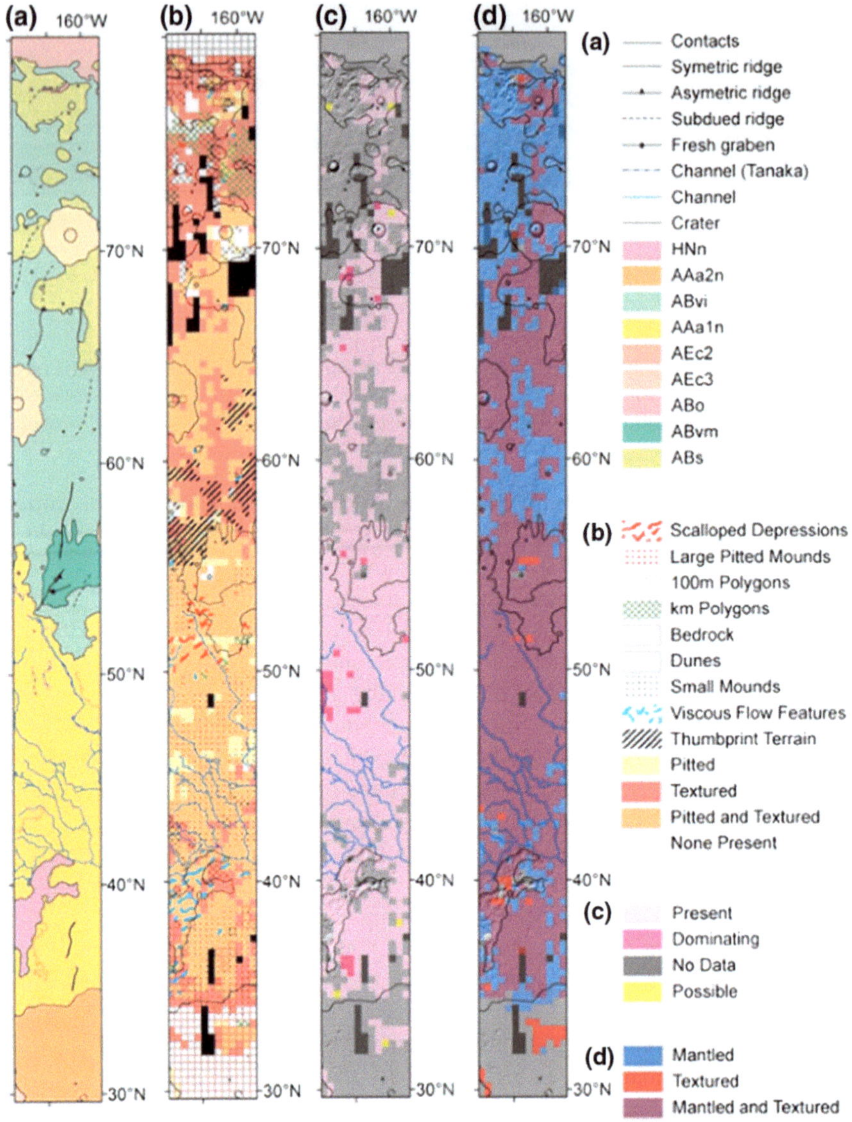

Fig. 3 Grid-map of Arcadia Planitia, Mars (Ramsdale et al. 2017). **a** geological reconnaissance map (Tanaka et al. 2005). **b** geomorphologic map of selected cold-climate landforms. **c** grid-map presenting the distribution of pitted terrain. **d** map showing both pitted and textured terrain. Basemap of b, c, and d is a MOLA hillshade. For further information, the reader is referred to Ramsdale et al. (2017)

Table 2 Model table for four landforms (gullies, channels, yardangs and dunes)

A	B	C	D	E	F	G	H	I
FID	Gullies	Channels	Yardangs	Dunes	Image quality	Image coverage	Comments	Elevation [m]
0	0	1	1	3	3	1	Barchans	1,115
1	0	0	2	2	3	3	–	1,120
2	1	2	1	0	2	3	–	1,179
3	3	2	0	0	2	2	–	1,184

See Table 1 for explanation of mapping key. A: FID (feature identifier) provides a consecutive and unique number for each record, B–E: selected landforms for this study, F–G: quality and coverage of the imagery used (e.g., CTX), H: comments for extra remarks; I: further column for statistical evaluations; within a GIS-environment this field can be calculated automatically

5 Statistics

Statistic evaluation methods are diverse. They range from simple to complex calculations, depending on the examined scientific questions and the mapper's skill level in geostatistics and programming. Because of the vast range of methods, we present only basic statistical analysis.

One of the most important factors to investigate is relation between the distribution of landforms and topographic parameters (e.g. elevation, slopes inclination, aspect and latitude). Below, we are focusing on these three parameters for the glacier-like viscous-flow features on Mars according to Voelker et al. (2017). However, the method can be easily applied to other parameters too, like thermal inertia, temperatures and dust cover. Hence, it is possible to relate e.g. thermal inertia to certain landforms, for instance, dunes.

Before starting the calculation it is necessary to compute the mean elevation and slope values for each grid, based on the elevation and slope maps. This process can be automatized in a GIS-environment. The results for the average elevation and slope values are inserted into two new columns within the attribute table of the grid-map layer. For the aspect values, we recommend not to use the mean values, as mathematical distortions may occur when a northern-exposed grid contains values around 360° and 0°, resulting in a "wrong" mean value of 180°. Instead, we suggest to use the majority function for each grid.

After the attribute table has been completed, the table can be exported in order to be used within a spreadsheet or programming environment.

Independent of which method the user chooses, it is highly recommended to classify the additional topographical parameters. As the topography of most study areas is very diverse and irregularly distributed (e.g. there are more low elevated than high-elevated grid cells in the study area), it is helpful to normalize the distribution of the landforms (by each elevation bin) in order to remove this sampling bias. This statistical correction is necessary for compensating both under- and over-represented elevation classes. It is conducive to classify, e.g. the elevation into

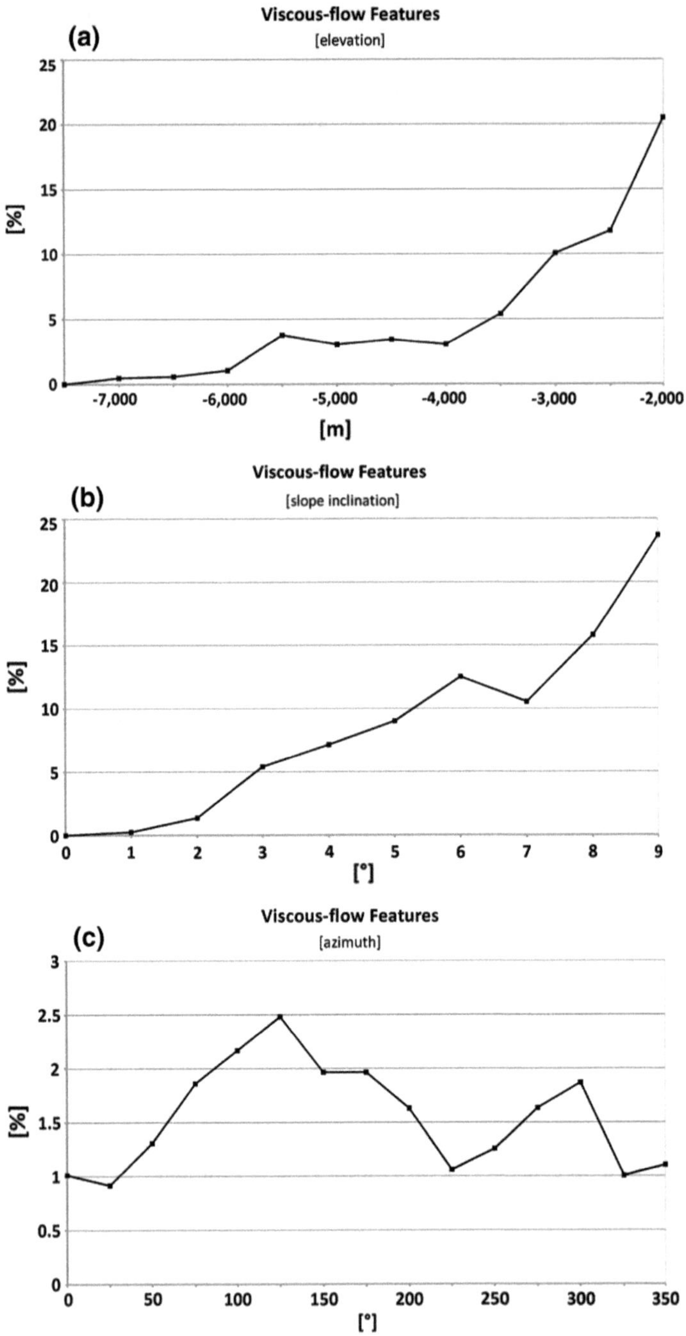

Viscous-flow Features
(a)
[elevation]
25
20
15
10
5
0
[%]
-7,000
-6,000
-5,000
-4,000
-3,000
-2,000
[m]
Viscous-flow Features
(b)
[slope inclination]
25
20
15
10
5
0
[%]
0
1
2
3
4
5
6
7
8
9
[°]
Viscous-flow Features
(c)
[azimuth]
3
2.5
2
1.5
1
0.5
0
[%]
0
50
100
150
200
250
300
350
[°]

◂**Fig. 4 a–c** Frequency of occurrence at certain bins (normalized distribution) of viscous-flow features in Hellas Planitia, Mars from grid-mapping, according to their elevation (a), slope inclination (b), and azimuth (c); e.g., 10% of all grids at an elevation of 3,000 m below the Martian datum contain viscous-flow features (Voelker et al. 2017). a: most viscous-flow features in Hellas Planitia appear at elevations higher than −4000 meters. b: as viscous-flow features are gravity-driven morphologies. They mainly occur on steeper slopes. c: their azimuth (orientation) is preferentially towards east and west. Their dependence on certain aspect values supports theories that they consist of a material prone to sublimation caused by solar insolation

classes (for example 0 m, 100 m, 200 m, 300 m and so on). Thus, it is possible to determine the density, and hence, distribution of each landform regarding its elevation.

The results can be plotted in line graphs (Fig. 4a–c) or box-plots. Box-plots also provide information about several descriptive statistical values (median, quartiles and interquartile range). Dependent on the mapper's skill level further statistical analyses like correlation, regression or three-dimensional geospatial analyses are viable.

The results can be used for interpretation of each landform's topographical and geographical behaviour, like the examples of viscous-flow features shown in Fig. 4. Viscous-flow features predominantly occur on higher elevations, steeper slope inclinations and are mainly found on east- and west-exposed slopes. Their appearance on steep slopes supports theories that these features are composed of a viscoplastic material, e.g. a mixture of ice and debris. Moreover, grid-mapping has revealed that their position on east and west-facing walls unveils a relation to solar insolation, and hence, an ice-related origin too. Further applications of grid-mapping can be found in Ramsdale et al. (2017) and Voelker et al. (2017).

6 Conclusions

Grid-mapping is a convenient method to quantify the distribution of landforms. In contrast to conventional methods it also provides an intersubjective, and hence, reproducible mapping method. However, the benefits of this method strongly depend on each scientific task. As grid-mapping is closely associated with a reduction of geospatial resolution, it also means a certain loss of geographical information. So its application needs to be carefully assessed in advance.

This approach is especially helpful in order to gain insight into the regional and global relations of landforms and the topographic, geological and environmental conditions that formed them; relations which only can be discovered from a wider perspective.

Furthermore, grid-mapping is a method requiring only basic skills. It is easy to learn and can produce quick results if requested. However, the larger the study area or the smaller the grids are, the more time is needed for a completion of grid-mapping. Once all data has been raised it is easy to calculate simple statistic parameter helpful for understanding the geographic behaviour of the surface features mapped. With advanced skills in geostatistics, even more information can be derived from the, often huge, data sets.

References

Jaumann R et al (2015) Quantifying geological processes on Mars—results of the high resolution stereo camera (HRSC) on Mars express. Planet Space Sci 112:53–97. https://doi.org/10.1016/j.pss.2014.11.029

Jenness J (2012) Repeating shapes for ArcGIS. Jenness Enterprises. http://www.jennessent.com/arcgis/repeat_shapes.htm

Malin MC et al (2007) Context camera investigation on board the Mars Reconnaissance Orbiter. J Geophys Res **112**:E05S04. https://doi.org/10.1029/2006je002808

McEwen AS et al (2007) Mars reconnaissance orbiters high resolution imaging science experiment HiRISE. J Geophys Res Planets **112**(E5). https://doi.org/10.1029/2005je002605

Neukum G, Jaumann R (2004) The HRSC co-investigator and experiment team HRSC: the high resolution stereo camera of Mars Express. ESA Spec Publ 1240:1–19

Ramsdale JD et al (2017) Grid-based mapping: a method for rapidly determining the spatial distributions of small features over very large areas. Planet Space Sci 140:49–61. https://doi.org/10.1016/j.pss.2017.04.002

Smith BA et al (1977) Voyager imaging experiment. Space Sci Rev 21(103):103–127

Tanaka KL, Skinner JA, Hare TM (2005) Geologic Map of the Northern Plains of Mars, Scientific Investigations Map 2888

Voelker M, Hauber E, Schulzeck F, Jaumann R (2017) Grid-mapping Hellas Planitia, Mars – insights into distribution, evolution and geomorphology of (Peri)-glacial, fluvial and lacustrine landforms in Mars' deepest basin. Planet Space Sci 145:49–70. https://doi.org/10.1016/j.pss.2017.07.012

Distribution Pattern Analysis in Planetary Mapping

Petr Brož and Mátyás Gede

Abstract The rapid development of geospatial methods in recent decades offers new ways how to investigate the evolution of the surfaces of the terrestrial bodies of the solar system. New methods of spatial distribution and alignment analyses can be used to find spatial and genetic relations between surface objects of our interest, their distribution patterns and structural alignments. Such relations can reveal the causal physical process behind their formation and hence improve our knowledge of the geological processes responsible for modifications of surfaces of these celestial bodies.

Keywords Cluster analysis · Nearest neighbour technique · Structural alignment · Two-point azimuth technique

1 Introduction

Our knowledge of the surface processes on solid bodies of the Solar system is mainly derived from remote sensing data as these bodies are not easily accessible for human explorers. Therefore, we are facing the difficulties to clearly interpret our observations as the cause of the origin of observed objects is out of the reach of our instruments (e.g. deeply in the subsurface) and as we also lack the possibility to confirm our hypothesis by basic field works. To partly overcome this problem, we can analyse the patterns in distribution of objects of our interest by applying *spatial distribution and alignment analyses.* Such approach offers the possibility to find spatial and genetic relations between various sets of surface features sharing a

P. Brož (✉)
Institute of Geophysics, Czech Academy of Science, Boční II/1401,
141 31, Prague, Czech Republic
e-mail: Petr.broz@ig.cas.cz

M. Gede
Eötvös Loránd University, Pázmány P St 1/A, Budapest 1088, Hungary

© Springer Nature Switzerland AG 2019
H. Hargitai (ed.), *Planetary Cartography and GIS*, Lecture Notes in Geoinformation and Cartography, https://doi.org/10.1007/978-3-319-62849-3_17

common formation mechanism and hence provides insight into the link between their distribution patterns and causal physical process behind them.

In order to apply these methods, the investigated objects need to be localized first (Fig. 1a). Such localization is often done manually or semi-automatically in geographic information system (GIS) environment on the basis of satellite imageries showing the surfaces in different wavelengths, illumination conditions or shadings. Images taken in visible range of light are used for recognition of surface features exhibiting a different appearance from the surrounding units, e.g. impact craters, faults or conical edifices (Fig. 1b). On the other hand, the surface thermal anomalies associated with interior processes, for instance, higher surface temperature caused by volcanic activity on Io (Lopes-Gautier et al. 1999) or with variations in primary composition, weathering or grain size, for instance, with compositional differences in lava flows on Venus (Smrekar et al. 2010), can be revealed from infrared and near infrared images (e.g. Lopes-Gautier et al. 1999; Christensen et al. 2004; Mueller et al. 2008). Similarly, the topographic data are used to recognize features that have been significantly modified by other processes over time, like impact basins infilled by sedimentary rocks.

1.1 Feature Identification

The output of such analysis is a simplified *map* in which the positions of objects of our interest are marked (Fig. 1c). Only those objects which we assume to be sharing similar mechanism of their formation should be mapped. Their position can be then used for further analysis. It enables us to search for patterns in their distribution or for correlations between their position and variables of our choice (like latitude, longitude, topography, leading and trailing hemispheres, sub-planet and the anti-planet facing hemispheres, etc.). For example, the high gully density in the northern hemisphere of Mars (e.g. Kneissl et al. 2010), similarly as the distribution of the Martian paleo-valley networks (Bouley et al. 2016), can be linked with specific latitudes.

An important aspect of pattern analysis is the choice of the size of investigated area which should be closely related to the scientific question which we are trying to answer (Illian et al. 2008). For example, to reveal the mechanism of heat dissipation on Jupiter moon Io or the level of the global contraction of Mercury, the global distribution of active volcanic centres or thrust-fault-related landforms has to be considered and mapped (Lopes-Gautier et al. 1999; Byrne et al. 2014). On the other hand, to reveal the source of magma which supplied a field of Martian volcanic cones, only the spatial distribution within that field has to be analysed (Bleacher et al. 2009). Therefore, before the analysis, we should carefully consider the extension of the studied phenomenon.

An important prerequisite for application of the spatial distribution and alignment analyses is simplification of studied phenomena from the surface features with the complex geometry, to features with zero dimensionality represented by points

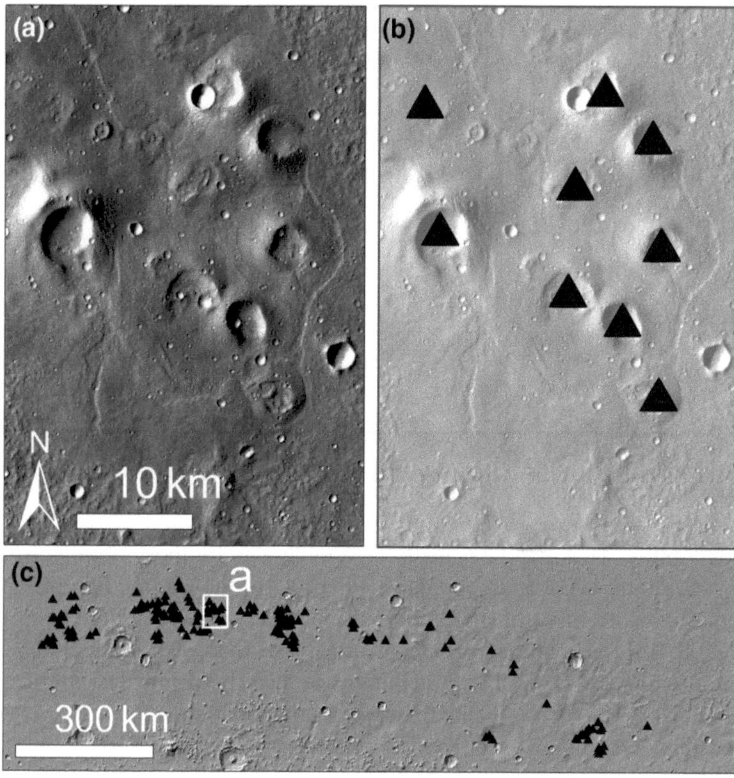

Fig. 1 Example of data preparation for point pattern analysis. Panel **a**, a satellite image obtained in the visible spectra shows a cluster of small conical edifices on the surface of Mars which represents object of our interest. The cones share a common mechanism of their formation based on morphological and morphometrical similarities. The position of those cones is mapped and marked **b**, both in local and regional scale **c**, and simplified from complex surface features to features with zero dimensionality represented by points. Panels a and b based on HRSC data obtained by Mars Express, and panel c based on shaded MOLA-relief obtained by Mars Global Surveyor

(Fig. 1c), or marks (e.g. Illian et al. 2008). While the points describe the locations of the objects in space only, the marks represent an attribute of the point and hence provide additional information about the characteristic of the study features (like shape, size, type, etc.). The rectangles, circles and triangles are mostly used to mark mapped objects of our interest, but other symbols are also used frequently.

Sometimes, it is not simply evident how to represent complex surface features by points. Faults, river beds, elongated impact craters or irregular volcanoes are good examples of such objects. In such cases, we may regard to determine the centre of these features by applying more sophisticated methods, like placing the point to the centre of the ellipse enclosing the feature of our interest. Simple manual marking in which the point would be drawn consistently to some parts of the elongated feature

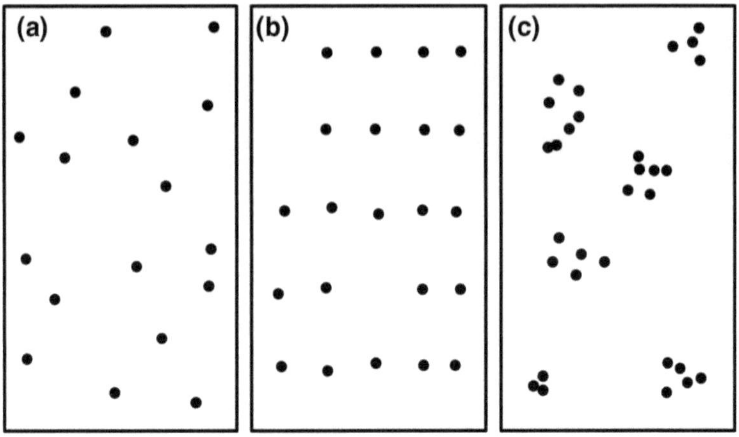

Fig. 2 Three examples of simulated point patterns: **a** random, **b** regular and **c** clustered sample

is also an option. This may be actually preferable way for those objects, where the geometrical centre is not as important as other part of the feature. For example, the position of the volcanic vent should be considered as the centre of the volcano (e.g. Cébria et al. 2011); regardless of whether such vent is located in the middle of the volcano or on its edge. Another possibility is to mark by points the intersections of the linear features, like the contact between polygonal structures. However, it is essential that such approach is applied consistently across the entire investigated area, so the resulting dataset is consistent.

Once the investigated objects are mapped and represented by points (Fig. 1), they can be shown on a point map from which a simple qualitative characterization of the type of the pattern can be revealed. In general, the position of each point can be independent from the other points; in such case, points are distributed randomly and hence points show random distribution. Or the distance between individual points can be maximized; in such case, points are evenly spaced, and hence they show regular distribution. Next possibility is that the distance between neighbouring points can be minimized; points are grouped, and hence they show clustered distribution (Fig. 2).

1.2 Choosing the Right Map Projection

It is also important to carefully select the proper spherical representation of the terrestrial body in the GIS environment and the appropriate map projection to minimize distortions in the geometries and positions of studied objects. In general, for global, pole-to-pole studies, equal-area projections such as global Sinusoidal or Mollweide projection might be the right choice. These projections retain areas; therefore, they are suitable for determination of densities of features per unit area. For polar regions,

Lambert equal-area projection with polar aspect is optimal, if the features analysed are concentrating near the poles and not near the equators. If we want to analyse the directions or orientation of the features, conformal projection should be used where local angles are preserved, but lengths and areas are not. If the distribution of features is expected to occur along orthodromes (great circles), gnomonic projection should be used where about one-third of the globe can be displayed at once.

For local studies, any equal-area projection can be used as they are comparable. For example, Sinusoidal projection retains areas and also lengths near parallels (circles of longitude) and angle distortions are also small. However, keep in mind that individual maps can not be pieced together at their edges. For polar areas, Lambert conformal or equidistant projections are recommended that also retain lengths at the centre parallel. For studying circular features locally, such as craters, stereographic projection is recommended as it retains circular forms.

1.3 Distribution: Implications

Information about the distribution of points may help to reveal the formation mechanism of the features of our interest as we know from Earth that many geological phenomena have tendency to show typical patterns in their distributions. Therefore by applying this knowledge on other planets, we can get a deeper insight about those processes that formed them. For example, it is known from terrestrial volcanic fields that volcanic vents, which may look similar to impact craters, have a tendency to cluster around underlying fractures. On the other hand, the distribution of primary impact craters is mainly random. Therefore, the pattern of the distribution may enable to distinguish between these two processes on other body of the solar system. This may be illustrated on rootless cones, a specific type of volcanic edifices formed due to the contact of extruded lava with water in liquid or solid state. The rootless cones show a non-random distribution both on Earth and on Mars. Such distribution is determined by preferred lava pathways and/or (sub)surface non-random volatile distributions (Bruno et al. 2004). By analysing their distribution, we can hence gain more information about the composition of the subsurface (e.g. variations in the presence of volatiles) or about the process of lava emplacement.

However, it would be difficult to reveal the point pattern just by naked eye. Therefore, it is appropriate to apply the ***point pattern analysis*** (for details, see, e.g., Fotheringham et al. 2000) represented by a wide variety of statistical methods allowing a precise quantification and distinction between different types of spatial behaviours (e.g. Illian et al. 2008; Hammer 2009). The methods of point pattern analysis have been successfully applied to identify and assess various spatial patterns both on the Earth and on other terrestrial bodies in the solar system and hence represent a useful tool for geologists. Next chapter focuses on the methods which are commonly used in planetary applications. However, keep in mind that there exist more methods and even more are being gradually developed because the point pattern analyses represent a dynamic field of statistics.

2 Methods

2.1 Cluster Analysis Methods

In general, point pattern analysis is a group of statistical approaches enabling to search for relationships and connections between investigated variables. Different approaches can be applied to fulfil this task; the *kernel density analysis*, *Gaussian kernel density analysis*, *power law distribution analysis or cluster analysis methods* (see, e.g., Corvec et al. 2013 and references therein). In planetary application, a prominent part of these approaches is represented by ***cluster analysis methods*** (e.g. Connor 1990), which enables to distinguish whether a set of objects occurs as separate entities or creates some sort of groups (clusters) in which objects share common characteristics. But without knowing if the groups are statistically significant or if the grouping is the best possible.

There are number of methods that can be used to perform a cluster analysis, but in general, they can be classified into two large groups, hierarchical and non-hierarchical. While non-hierarchical methods (also known as *k-means clustering methods*) require that the desired number of clusters is specified in advance and the appropriate solution is chosen, the hierarchical methods build the clusters incrementally. All the objects are hence linked together as a hierarchy, which is commonly showed as a tree diagram (dendogram) with individual elements at one end and a single cluster containing every element at the other. Such visualization allows the clusters to be read off. The number of clusters in non-hierarchical methods is a parameter of a sensitive choice which can strongly affect the desired output. On the other hand, the hierarchical methods are not limited to a pre-determined number of clusters—the choice may be based on different assumptions, and hence these methods can find similarity of samples across a wide range of scales. For this reason, they find their usefulness in many fields of the natural sciences (Romesburg 1984), including planetology.

In planetary applications, one of the most common hierarchical cluster analysis methods is the ***nearest neighbour technique*** (Clark and Evans 1954). This is a standard statistical technique which enables to determine the clustering or dispersing behaviour of investigated features by measuring the absolute distance from every point (i.e. surface feature) to its nearest neighbour. The method is based on testing the randomness in spatial distribution by calculating the ratio (R) between the observed mean distance and the expected mean distance for a random point distribution of our choice. The ratio is given as $R = 2\bar{d}(\sqrt{n}/a)$, where $\mathbf{\bar{d}}$ is the mean observed nearest neighbour distance, **n** is the total number of the points and **a** is the total area on which points are distributed. For more details and associated equations, readers are referred to Clark and Evans (1954). The one of the most widely used random distributions intends to distribute points randomly on the mathematical space is Poisson point process.

Randomly distributed objects would be expected to have $R \sim 1$ corresponding to the actual nearest neighbour distance approximates that expected of a Poisson distribution, a discrete frequency distribution which gives the probability of a number of independent events per unit time and a confidence estimate in the departure from randomness (c) <1.96 corresponding to no significant departure from randomness at the 0.05 significance level (i.e. c will exceed 1.96 in a Poisson distribution 5% of the time). This would suggest that features are independent of one another. In simply terms, if the ratio is <1, this suggests that features are clustered; the closer to zero, the more clustered (Clark and Evans 1954; Fig. 3). The advantage of this method is that there is no arbitrary selection of clusters by operator, so the choice which can significantly affect results is eliminated. This method is also directly implemented in the ArcGIS software package under the name *Average Nearest Neighbour* tool allowing easy usage for GIS users.

As many geological phenomena have tendency to cluster or show other typical pattern in distributions, the nearest neighbour technique is commonly used to reveal such behaviour of objects of our interest (e.g. Bruno et al. 2004; Baloga et al. 2007; Bleacher et al. 2009; Dutilleul et al. 2009; Brož and Hauber 2013; Hamilton et al. 2013; Richardson et al. 2013; Peters and Christensen 2017). For example, Bruno et al. (2004) applied this technique to assess the degree of spatial randomness within a cone field of rootless cones on Mars as the clustering pattern would suggest that those cones were formed as a surface signature of preferred lava pathways. Similarly, Richardson et al. (2013) applied this technique to investigate the clustering of volcanic vents within one field on Mars in the attempt to find whether those vents had been formed independently of each other as part of one larger magma production event or if multiple temporally distinct magma production events have occurred beneath this field.

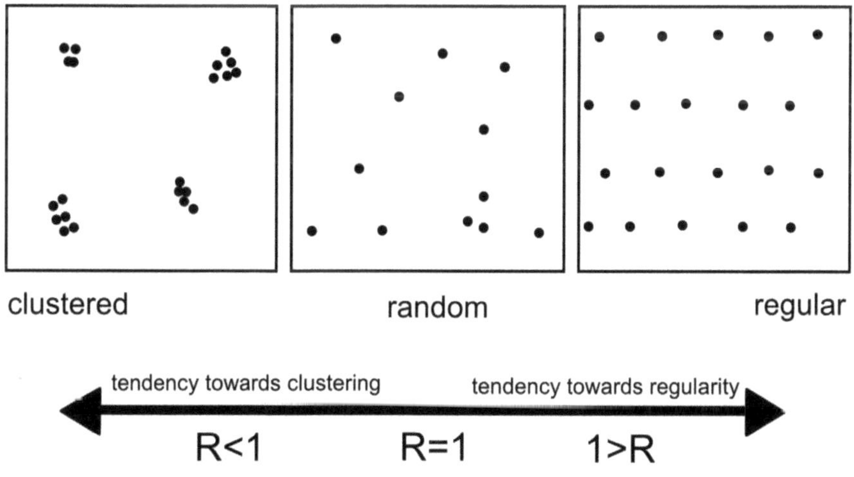

Fig. 3 Application of the nearest neighbour analysis enables to measure the extent to which particular pattern is clustered, random or regular. See the main text for details

2.2 Structural Alignment Methods

An application of cluster analysis methods has its limitation as these methods do not consider whether investigated points are aligned or not. Such information may be of considerable geologic interest as alignment may be reflecting underlying features, particularly faults and joins which are important prerequisite for many geological phenomena associated, for example, with the distribution of igneous or mud volcanoes. Therefore, to reveal such relationship, the ***structural alignment methods*** should be applied as these methods enable to find out if there is any structural alignment of the geologic significance between point-like features (e.g. Wadge and Cross 1988; Hammer 2009) exceeding reference distributions obtained from random probability distribution. A number of methods have been suggested, and they can be broadly divided into three classes: (a) *Spatial transformation methods*, (b) *Strip methods* and (c) *Azimuth methods* (Hammer 2009).

While *Spatial transform methods* are usually used in computer-aided image analysis or pattern recognition to transform the point positions into another domain where linear alignment can be more easily detected (Hammer 2009), the application of the *Strip methods* (Zhang and Lutz 1989) works with the idea that the entire investigated area would be swept by parallel strips at different positions and azimuths. This would enable the identification of the most significant strips those enclosing the higher density of points in comparison with densities derived from Monte Carlo simulations. In geological applications, one of the most widely used *Spatial transform method is* the ***Hough transform method*** (e.g. Duda and Hart 1972; Wadge and Cross 1988; Connor 1990).

Last group represent *Azimuth methods,* originally introduced by Lutz (1986), which enable quantitative analysis of the azimuth angles of lines connecting each point with all other points. In this sense, a point refers to the surface manifestation of the investigated objects, e.g. volcanic vent. The frequently used method from this class is the ***two-point azimuth technique*** (Lutz 1986) which is a probabilistic method looking for the existence of regional structural anisotropies by searching for a preferential alignment of points along certain orientations. This statistical method is based on the connection of all possible pairs of points in the investigated area by lines (for N points, the total number of lines is $N(N-1)/2$) and visualizing these results into a histogram or rose diagram, in which peaks in values indicate preferred alignments of features (Fig. 4). Straightforward application of this method leads to a relatively high number of non-significant data making it difficult to discriminate the most likely orientations. Therefore, some filtering should be applied. In the original study, Lutz (1986) used the Monte Carlo simulations of random patterns as a reference distribution from which confidence level were determined. Cébria et al. (2011) later offered simpler concept based on the idea that the volcanic centres in monogenetic fields, on which this method was successfully used, should be more closely related to the nearest vents, than those which are far away, and, very likely, this relationship should be a consequence of the presence of a fracture.

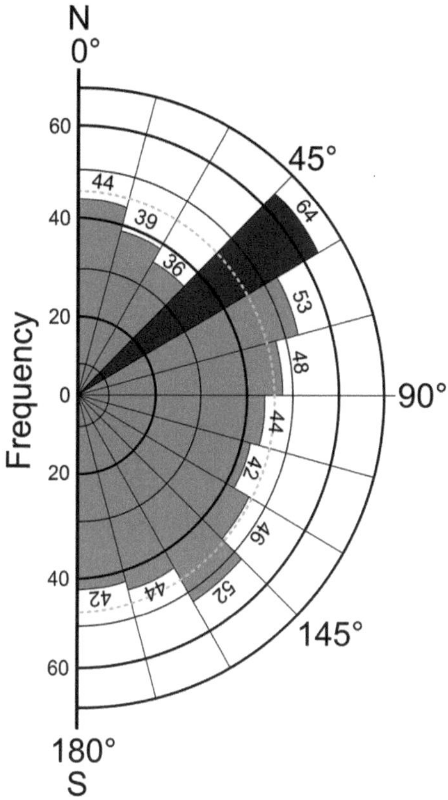

Fig. 4 Example of the output of the two-point azimuth technique (developed by of Lutz 1986 and modified by Cebriá et al. 2011) represented by a rose diagram with 15° bin intervals, containing the numbers of lines per bin for lines longer or equal to minimum significant distance (see the main text for explanation) long. The dotted line represents the arithmetic mean of frequency per bin, and the dark grey colour marks the bin where frequency is higher than one standard deviation above the mean suggesting the presence of the predominant orientation in the distribution of investigated phenomenon. Image from Brož and Hauber (2013)

Each of the above-mentioned methods has their own advantages and disadvantages. For example, the azimuth methods are sensitive to shape of the field of points and the heterogeneity of their distribution (Wadge and Cross 1988). A field that is elongated in a particular direction produces more azimuths with that orientation and this must be corrected. For this reason, Cebriá et al. (2011) defined a *minimum significant distance* between points to eliminate potential problems with a preferential alignment of points. The minimum significant distance (d) is defined as $d \leq (x - 1\sigma)/3$, where x is the mean of all distances between vents and σ is the standard deviation of the mean distance between vents. This modification enables to filter out any large amount of non-significant data corresponding to the most likely orientation caused by the shape of the investigated area (Fig. 5). But keep in mind

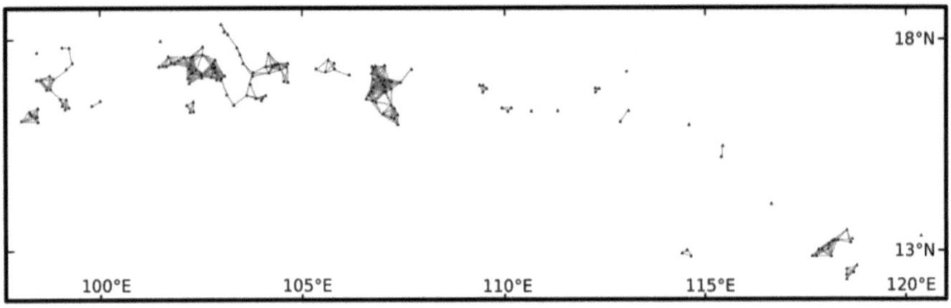

Fig. 5 Example of the output of the modified two-point azimuth technique (Cebriá et al. 2011) applied on the field of putative Martian kilometre-sized volcanic cones in the attempt to search for structural control of their distribution. Instead of a picture in which all points are connected together, the output show only those connections which are showing just those connecting equal or shorter than the value of minimum significant distance. Such picture makes it easier to see predominant alignments. Image from Brož and Hauber (2013)

that spatial transformation methods may also be affected by human visual capability and ability to select properly the sets of widely spread points on which the methods are based (Cebriá et al. 2011).

Despite these pitfalls, the structural alignment methods have been used on terrestrial bodies in the solar system in the attempt to quantify preferred orientations in clusters of point-like surface features (Lutz 1986; Wadge and Cross 1988; Connor 1990; Bleacher et al. 2009; Cebriá et al. 2011; Richardson et al. 2013; Brož and Hauber 2013; Thomson and Lang 2016), and hence provide additional information about the subsurface. For example, Bleacher et al. (2009) used two-point azimuth technique to show that the distribution of volcanic vents within one volcanic field on Mars displays alignment trends similar to orientations of older sets of faults and fractures observed in nearby area. Such match might suggest that the faults and fractures extend into the study area despite the fact that this extension is not directly detectable on the surface of Mars because of the later volcanic activity burying older surface by younger volcanic deposits. An interesting application of the two-point azimuth technique was also done by Thomson and Lang (2016) who tried to overcome the problem of the scarcity of impact craters on Venus making difficult to date the Venusian surface by inferring the relative timing between clusters of small shield volcanoes and sets of fractures in the surrounding terrains. Such application helped them to reveal the relative stratigraphy between different geologic units on Venus and hence refine the timing of their formation.

2.3 *Visualization of the Results*

The results of spatial distribution and alignment analysis can be visualized in many different ways in dependency on the applied technique and desired output. Different types of charts (e.g. line, column, bar, bubble or scatter charts), column histograms, rose diagrams, Voronoi diagrams or Delaunay tessellations are just some examples of those frequently used. Keep in mind that even during the visualization of your results care must be taken as the inappropriate method of visualization can make it difficult to read the results. For example, if the column histogram is selected to show large sets of data, it works well. But when applied to small sets, the result can be misleading as the results are grouped into class-intervals (bins) which smooth the distribution to some extent. And that smoothing is wholly arbitrary. Another option how to visualize the results is the usage of isolines or density maps which are connecting places that share a common value. Such outputs can be directly integrated into the final map products of the area of our interest as overlaying unit and hence easy to understand in their geological context.

3 Conclusion

An application of spatial distribution and alignment analyses represent a valuable tool how to get deeper insight into different geological phenomena shaping the surfaces of terrestrial bodies. Use of specific statistical techniques can reveal genetic relationships between surface edifices within investigated area but also find structural connections between their distribution and the structure of the subsurface layers. Before these methods can be applied, several simple steps need to be done; features of our interest have to be identified, localized and they have to be transformed from complex surface features to features with zero dimensionality represented by points which may be then a subject of point pattern analysis. Many different techniques of point pattern analysis are used in planetary applications, with the nearest neighbour technique, and two-point azimuth technique, as frequently used examples. These methods enable revealing the clustered or dispersed behaviour of investigated features, or to search for the existence of regional structural anisotropies, respectively, which can be diagnostic proofs of some geological processes and may hence help to reveal the mechanism of the formation of objects of our interest.

References

Baloga SM, Glaze LS, Bruno BC (2007) Nearest-neighbor analysis of small features on Mars: applications to tumuli and rootless cones. J Geophys Res 112

Bleacher JE, Glaze LS, Greeley R, Hauber E, Baloga SM, Sakimoto SEH et al (2009) Spatial and alignment analyses for a field of small volcanic vents south of Pavonis Mons and implications for the Tharsis province. Mars J Volcanol Geotherm Res 185:96–102

Bouley S, Baratoux D, Matsuyama I, Forget F, Séjourné A, Turbet M et al (2016) Late Tharsis formation and implications for early Mars. Nature 531:344–347

Brož P, Hauber E (2013) Hydrovolcanic tuff rings and cones as indicators for phreatomagmatic explosive eruptions on Mars. J. Geophys Res Planets 118:1656–1675

Bruno BC, Fagents SA, Thordarson T, Baloga SM, Pilger E (2004) Clustering within rootless cone groups on Iceland and Mars: effect of nonrandom processes. J Geophys Res 109:E07009

Byrne PK, Klimczak C, Sengör AMC, Solomon SC, Watters TR, Hauck SA II (2014) Mercury's global contraction much greater than earlier estimates. Nature Geosci 7:301–307

Cebriá JM, Martín-Escorza C, López-Ruiz J, Morán-Zenteno DJ, Martiny BM (2011) Numerical recognition of alignments in monogenetic volcanic areas: examples from the Michoacán-Guanajuato Volcanic Field in Mexico and Calatrava in Spain. J Volcanol Geotherm Res 201:73–82

Christensen PR et al (2004) The thermal emission imaging system (THEMIS) for the Mars 2001 odyssey mission. Space Sci Rev 110:85–130

Clark PJ, Evans FC (1954) Distance to nearest neighbor as a measure of spatial relationships in populations. Ecology 35(4):445–453

Connor CB (1990) Cinder cone clustering in the TranMexican volcanic belt: implications for structural and petrologic models. J Geophys Res 95:19395–19405

Duda RO, Hart PE (1972) Use of the hough transformation to detect lines and curves in pictures. Graphics and Image Process 15:11–15

Dutilleul P, Haltigin TW, Pollard WH (2009) Analysis of polygonal terrain landforms on Earth and Mars through spatial point patterns. Environmetrics 20:206–220

Fotheringham SA, Brunsdon C, Charlton M (2000) Quantitative geography: perspectives on spatial data analysis. Sage, London

Hammer Ø (2009) New statistical methods for detecting point alignments. Comput Geosci 3:659–666

Hamilton CW, Beggan CD, Still S, Beuthe M, Lopes RMC, Williams DA et al (2013) Spatial distribution of volcanoes on Io: Implications for tidal heating and magma ascent, Earth planet. Sci Lett 361:272–286

Illian J, Penttinen A, Stoyan H, Stoyan D (2008) Statistical analysis and modelling of spatial point patterns. Wiley, New Jersey ISBN 978-0-470-01491-2

Kneissl T, Reiss D, van Gasselt S, Neukum G (2010) Distribution and orientation of northern-hemisphere gullies on Mars from the evaluation of HRSC and MOCNA data. Earth Planet Sci Lett 294:357–367

Le Corvec N, Spörli KB, Rowland J, Lindsay J (2013) Spatial distribution and alignments of volcanic centers: clues to the formation of monogenetic volcanic fields. Earth Sci Rev 124:96–114

Lopes-Gautier R et al (1999) Active volcanism on Io: global distribution and variations in activity. Icarus 140:243–264

Lutz TM (1986) An analysis of the orientations of large-scale crustal structures: a statistical approach based on areal distributions of point-like features. J Geophys Res 91:421–434

Mueller N, Helbert, J, Hashimoto, GL, et al (2008) Venus surface thermal emission at 1 μm in VIRTIS imaging observations: evidence for variation of crust and mantle differentiation conditions. J Geophys Res 113

Peters SI, Christensen PR (2017) Flank vents and graben as indicators of Late Amazonian volcano-tectonic activity on Olympus Mons. JGR-Planets 122:501–523

Richardson JA, Bleacher JE, Glaze LS (2013) The volcanic history of Syria Planum. Mars J Volcanol Geotherm Res 252:1–13

Romesburg HC (1984) Cluster analysis for researchers. Lifetime Learning Publications, Belmont, p 334 ISBN 978-1411606173

Smrekar SE, Stofan ER, Mueller N et al (2010) Recent hotspot volcanism on Venus from VIRTIS emissivity data. Science 328:605–608

Thomson BJ, Lang NP (2016) Volcanic edifice alignment detection software in MATLAB: test data and preliminary results for shield fields on Venus. Comput Geosci 93:1–11

Wadge G, Cross A (1988) Quantitative methods for detecting aligned points: an application to the volcanic vents of the Michoacan-Guanajuato volcanic field, Mexico. Geology 16:815–818

Zhang D, Lutz TM (1989) Structural control of igneous complexes and kimberlites: a new statistical method. Tectonophysics 159:137–148

Topographic Roughness as Interquartile Range of the Second Derivatives: Calculation and Mapping

A. A. Kokhanov, I. P. Karachevtseva and Anastasia Zharkova

Abstract This paper describes methods of roughness calculation, based on different types of topography products: digital terrain models obtained by photogrammetric methods and results of laser altimetry observations. Implementation of processing results for global planetary mapping is presented.

Keywords Topographic roughness · DTM · The Moon · Mercury · Laser altimetry · LOLA · MLA

1 Introduction

Statistical characteristics of topography are useful tools for the morphological study of the surface. Topographic properties could be described by a set of parameters: slopes, aspect, curvature, roughness, etc. The study of statistical characteristics could reveal the surface structures that are obscured on the physical maps. Roughness is one of the most important characteristics, which shows the relative height variation within the studied area, and it can be defined using various methods that depend on the tasks. For example, *Area Ratio* characterizes the ratio of real 3D surface area to the area of its projection onto the 2D plane (Rashid 2010); *Dispersion Vector* effectively captures variability in slope and aspect into a single measure (Hobson 1972); *Relative Topographic Position* describes each pixel with respect to its local neighborhood (Jenness 2006). (For detailed definition see Grohmann and Hargitai 2014, for comparative analysis of the methods see Berti et al. 2013 and Grohmann et al. 2011).

Here we describe mapping of topographic roughness for two similar celestial bodies—the Moon and Mercury—where there is almost no atmosphere, and it does not affect the relief formation.

A. A. Kokhanov (✉) · I. P. Karachevtseva · A. Zharkova
MIIGAiK Extraterrestrial Laboratory (MExLab), Moscow State University of Geodesy and Cartography (MIIGAiK), Moscow, Russia
e-mail: a_kokhanov@miigaik.ru

© Springer Nature Switzerland AG 2019

H. Hargitai (ed.), *Planetary Cartography and GIS*, Lecture Notes in Geoinformation and Cartography, https://doi.org/10.1007/978-3-319-62849-3_18

2 Sources

The base source for the calculation of roughness for terrestrial planets is topographic information derived from remote sensing data, obtained by laser or radar altimeters, radar interferometry or from stereo image processing.

Modern data about the lunar surface is provided by the Lunar Reconnaissance Orbiter (LRO), which conducts frame and laser altimeter surveys of the Moon. The study of Mercurian topography was conducted by the MErcury Surface, Space ENvironment, GEochemistry, and Ranging (MESSENGER) spacecraft—also altimeter and space images.

2.1 The Moon

The whole surface of the Moon is covered by stereo images and laser altimeter measurements. These provide data about lunar topography with sub-kilometer spatial resolution. The digital terrain model (DTM) "Global Lunar DTM" or GLD100 (Scholten et al. 2012) was created from the photogrammetric processing of LRO WAC stereo images obtained by LRO (Robinson et al. 2010). The GLD100 is available in equidistant cylindrical projection for $\pm 60°$ latitudes and in stereographic projection for polar areas.[1] The spatial resolution of GLD100 is 100 m.

Another dataset is obtained by the instrument Lunar Orbiter Laser Altimeter (LOLA); its vertical resolution is 10 m (Smith et al. 2010). During one single measurement, LOLA gets five elevation points at the lunar surface. But for processing and roughness calculation, only one point inside each measurement was selected (Kreslavsky et al. 2013), so a single track consists of points located $\sim$60 m from each other.

2.2 Mercury

MESSENGER Laser Altimetry (MLA) (Cavanaugh et al. 2007) provided acceptable density tracks only for the northern subpolar area of Mercury between 65° and 84°N[2] due to the high eccentricity of the spacecraft orbit.

Mercury global DTM with resolution 665 m/pixel[3] was obtained recently (Becker et al. 2016). Some new DTMs with a resolution of about 200 m/pixel and vertical accuracy of 50 m (Preusker et al. 2017) can provide higher-quality data for a detailed study of Mercury topography.

[1]http://wms.lroc.asu.edu/lroc/global_product/256_ppd_DEM.

[2]http://pds-geosciences.wustl.edu/messenger/mess-e_v_h-mla-3_4-cdr_rdr-data-v1/.

[3]http://solarsystem.nasa.gov/news/2016/05/06/first-global-topographic-model-of-mercury.

3 Roughness Calculation

For all used data sources, we implemented the following general algorithm. Firstly, we computed the second derivatives of topography. For gridded DTM, it is Laplacian, calculated in sliding window, for altimeter single track—profile curvature. Secondly, we computed the interquartile range of the obtained values. We calculate interquartile range instead of the standard deviation of values, because this statistical characteristic does not depend on extreme values of morphometric parameters, calculated based on DTM (Kreslavsky et al. 2013).

As a laser altimeter track is in fact a topographic profile that contains precise height measurements, we suggest calculating roughness by laser altimeter tracks directly. Even with the possibility of creating an interpolated DTM, the true elevation points will be on track footprints only.

Figure 1 is an example of morphometric calculation results, obtained based on gridded DTM built by means of interpolation of laser altimeter tracks.

3.1 Gridded DTM Processing

To map surface topographic roughness, the interquartile range of Laplacian was chosen as a roughness quantitative value. (For detailed rationale of the choice of roughness measure see Kreslavsky et al. 2013.) The Laplacian as the sum of the second derivatives of topography could be calculated using Evans method (Evans 1979):

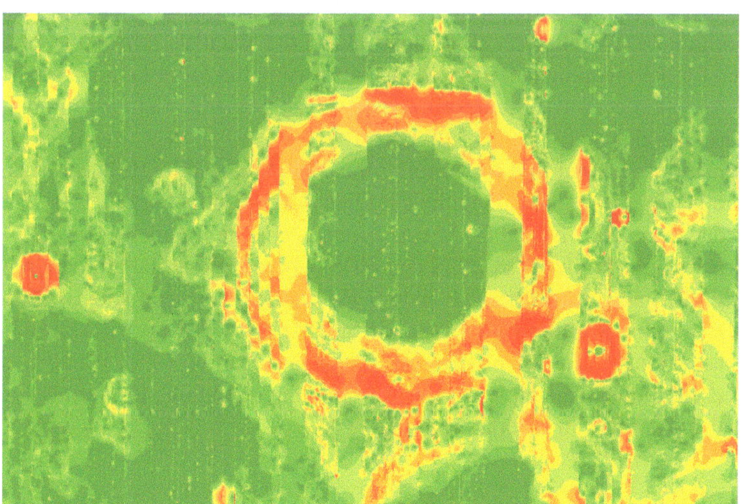

Fig. 1 Results of slope calculation by LDEM_1024_00N_15N_060_090 (http://imbrium.mit.edu/DATA/LOLA_GDR/CYLINDRICAL/JP2/)

$$L = r + t, \text{ where}$$

$$r = \frac{\partial^2 z}{\partial x^2}, \quad t = \frac{\partial^2 z}{\partial y^2}$$

These second derivatives are calculated in sliding window 3 × 3 with cell numbers:

Z_1	Z_2	Z_3
Z_5	Z_5	Z_6
Z_7	Z_8	Z_9

and L—distance between two cells.

$$r = \frac{Z_1 + Z_3 + Z_4 + Z_6 + z_7 + z_9 - 2(Z_2 + Z_5 + Z_8)}{3L^2}$$

$$t = \frac{Z_1 + Z_3 + Z_4 + Z_6 + z_7 + z_9 - 2(Z_4 + Z_5 + Z_6)}{3L^2}$$

Filtering algorithms were applied to the DTM derived by photogrammetric methods.

After that applying sliding circle window with radius 23 px in Laplacian raster, the first and third quartile and interquartile range were calculated.

To exclude the effect of map projection distortion when mapping the entire Moon surface, we introduced the following restriction: linear distance distortion along the parallel in the entire surface should not exceed 10%. To meet this condition, for the equatorial region (±60°) we used equidistant cylindrical projection from original DTM. Length distortions along the parallels (n) in the projection selected for the mapping surface are calculated using the formula: $n = \cos \varphi_k \sec \varphi$, where φ_k is the central latitude of the used projection and φ is the latitude, where distortion is calculated.

Estimations showed that for keeping the distortion within 10%, the size of the area should not exceed 40° along the meridian, when central latitude is 0°. Based on these conditions, the original DTM was cut into separate fragments (Fig. 2). For the two large regions bordered by 20°N and 20°S, where the original projection was not changed, the linear distortions are up to 6.4%. In the polar areas above 60° latitude, the stereographic azimuthal projection of the original DTM also was not changed, as its distortions up to 60° parallel are in the range up to 7.2%, which meets the imposed condition. This approach ensured that the conditions for permissible edge length distortion are in a certain range.

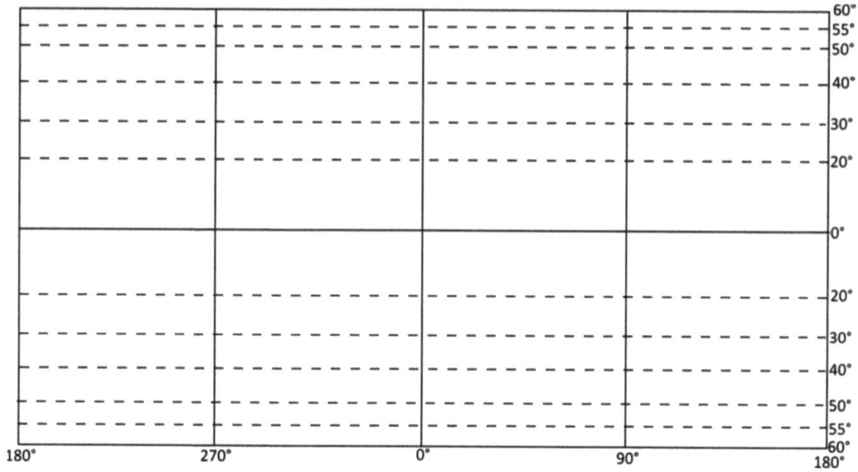

Fig. 2 Scheme of DTM fragments used for roughness calculation for the Moon and Mercury

3.2 *Laser Altimeter Tracks Processing*

Topographic roughness depends on spatial scale. To characterize this dependence, roughness calculations at different baselines were used for Mercury mapping.

The selection of minimum base length depends on typical distance between single measurements along altimeter track, which is 0.35 km for MESSENGER dataset. As the minimal dataset for calculation of roughness is 3 height measurements, the minimal basis is L_1 = 0.7 km. The longer baselines are suggested as 4- and 16-times more than minimal basis, L_2 = 2.8 km and L_3 = 11.2 km accordingly (Kreslavsky and Head 2002; Kreslavsky et al. 2013).

Quantitative value of roughness was calculated in two steps:

(1) For each altimeter measurement, we calculated curvature of topographic profile C along the orbit projection using formula:

$$C = (h_+ + h_- - 2h)/(l_+ + l_-)^2,$$

where h—single height measurement, h_+ and h_-—measurements of heights at the distance ½ of base length forward and backward along the altimeter track, l_+ and l_-—horizontal distance from h to h_+ and h, that means $l_+ + l_- \approx L$.

(2) For each created pixel with calculated value, the curvature values C were selected, which were inside the circle with defined radius R, centered in the pixel. For baselines 0.7 km and 2.8 km, the R was defined twice for two latitude belts: $R_{1,2}$ = 1.9 km (½ of diagonal of created pixel, 2.7 km size) for 70°N–84°N and $R_{1,2}$ = 3.8 km for 65°−70°N, where density of measurements is lower. For baseline 11.3 km, the one radius R_3 = 3.8 km was used.

For pixels with 20 and more, C values inside circle the 1st and 3rd quartile ($C_{1/4}$ and $C_{3/4}$) and interquartile range was calculated as

$$r = C_{3/4} - C_{1/4},$$

where r is roughness value, see for details (Kreslavsky et al. 2013).

Inside the mapped territory 65°N–84°N, there were 15% pixels without data, mostly on the lower latitudes. To fill them in we used the algorithm described in (Kreslavsky et al. 2014).

4 Map Design

The rasters with results of interquartile range of Laplacian or profile curvature calculating contain quantitative values, where the maximum value means the maximum roughness inside studied territory, the minimum—the smoothest. As there are no studies of clear interpretation of quantitative values, we use qualitative values “minimum” and “maximum” roughness inside map.

Roughness at different baselines shows different timescales: The larger baseline of roughness calculation is, the older relief features are characterized. So, at the hectometer baseline the roughness of young craters will be mapped and at kilometer baselines—bigger geological structures will be visualized.

Results of roughness calculation can be displayed in different ways: in grayscale mode (Fig. 3a, c) or as color images (Figs. 4 and 5). For many types of geological studies, visual presentation as a grayscale image with brighter shades representing a rougher surface is the most convenient as well as darker shades represent relatively smoother surface (Fig. 3a, c).

To demonstrate typical rough relief features on global level, three wall maps were compiled.

The *Map of Lunar Relief Parameters* (1:10,000,000) contains representation of heights and roughness on baseline 400 m. Comparing these two images, it is evident that the roughness shows old craters more smoothened (examples of craters

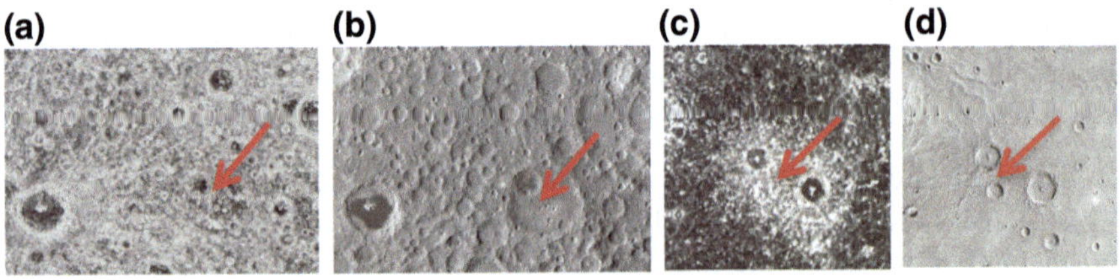

Fig. 3 Example of typical rough relief features in the grayscale rendition (**a**—Moon, **c**—Mercury) in comparison with image of planetary surface (**b**—Moon, **d**—Mercury). Arrows show “hidden” craters

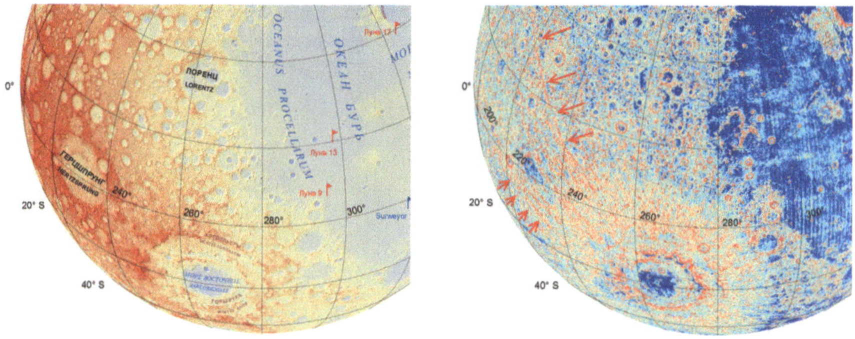

Fig. 4 Highlighting of ray structures on map of roughness (right) in comparison with topographic map (left)

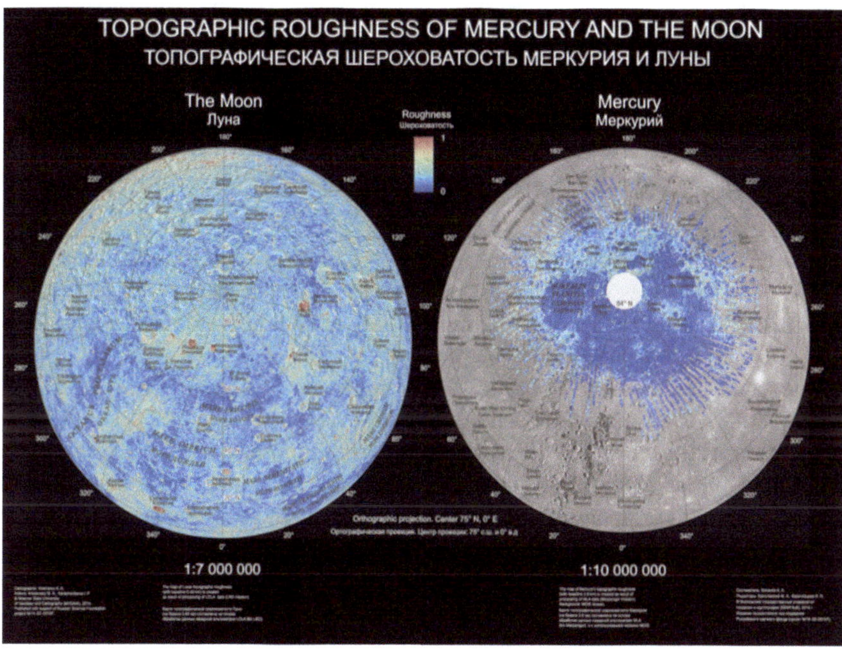

Fig. 5 Comparative map of the topographic roughness of Mercury and the Moon (MIIGAiK 2014)

"hiding" are shown by arrows on Fig. 3a–d), reflecting the general structure of the highland surface, and also reveals and visually emphasizes the ray structures of lunar maria (Fig. 4).

Color ramps for heights and roughness are designed as spectral gradient and close in colors. The quantitative values of roughness are not indicated, but its change from minimum to maximum values (from a smooth surface to a rough one)

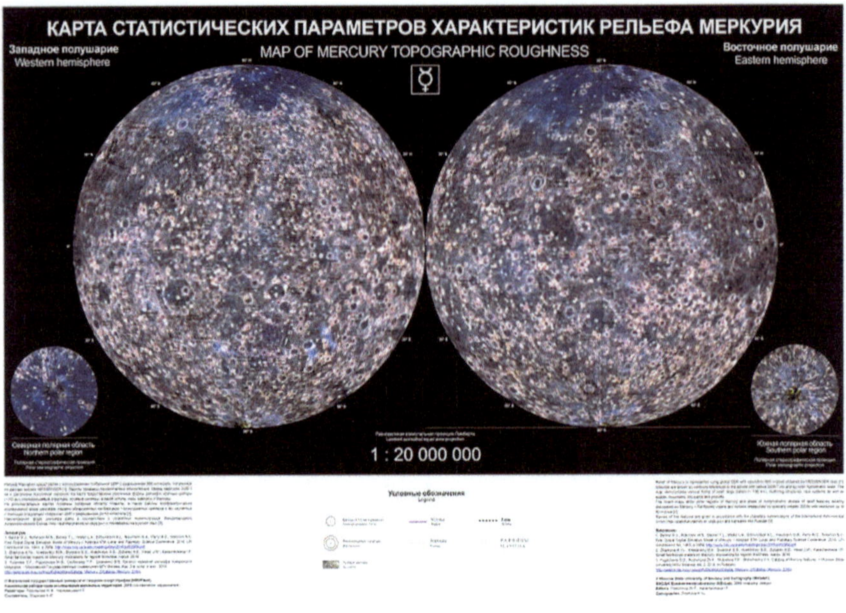

Fig. 6 Preliminary layout of new global map of Mercury roughness

is shown. At the hypsometric map, the relief is also shown by hillshading (Fig. 4). For navigation, names of the largest relief features (maria, montes, largest craters) are marked. Due to processing of origin DTM by sliding window, the final resolution of map is 3.8 km/px.

The layout of the map "*Topographic Roughness of Mercury and the Moon*" (Fig. 5) contains a map of the Mercury roughness on a baseline of 2.8 km (12 laser altimeter measurements) and a map of the lunar roughness on a baseline of 0.48 km (3 altimeter measurements). We compare these baselines on the single map, because only this way it appears the contrast between smooth and rough plains intrinsic for both bodies. Both maps are projected in oblique orthographic projection with their center at 75°N and 0°E for the best representation of the northern subpolar area. While the whole Moon surface is covered by the roughness data combined with hillshade, on Mercury only the northern subpolar area has roughness data, and MDIS mosaic[4] is used on other areas, which are not covered by MLA altimeter data.

To calculate for the first time roughness for the entire Mercury, recently obtained global gridded DTM[5] was implemented (Zharkova et al. 2016). For the new Mercury mapping, the layout design was based on the Multilingual Planetary Map Series (Shingareva et al. 2005) (Fig. 6).

[4]http://www.messenger-education.org.

[5]https://pdsimage2.wr.usgs.gov/data/mess-h-mdis-5-dem-elevation-v1.0/MESSDEM_1001/DEM/.

Here the roughness is visualized by RGB-composite image, with 42.6, 10.6, 2.7 km for the channels, respectively. At such map, the generally brighter area shows higher roughness and color variations show the scale dependence of roughness.

5 Summary

In this paper, we have described how to calculate topographic roughness as the interquartile range of the second derivatives of topography. Depending on the type data source, we calculate second derivatives in sliding window for DTMs, obtained by photogrammetric methods, and as profile curvature for altimeter tracks.

The provided results of calculation by this method intuitively correspond to the concept of roughness, do not depend on the tilt of the surface, and characterize a typical surface.

Maps of roughness could be designed in grayscale for combination with space orthoimages, in gradient color ramp for visualization of roughness on a defined scale or as synthetic RGB image for the visualization of roughness variation on different baselines.

References

Becker KJ, Robinson MS, Becker TL, Weller LA, Edmundson KL, Neumann GA, Perry ME, Solomon SC (2016) First global digital elevation model of mercury. In: Abstract 47th lunar and planetary science conference, LPI contribution no. 1903, p 2959. http://www.hou.usra.edu/meetings/lpsc2016/pdf/2959.pdf

Berti M, Corsini A, Daehne A (2013) Comparative analysis of surface roughness algorithms for the identification of active landslides. Geomorphology 182:15. https://doi.org/10.1016/j.geomorph.2012.10.022

Cavanaugh JF, Smith JC, Sun X, Bartels AE, Ramos-Izquierdo L, Krebs DJ, McGarry JF (2007) The Mercury laser altimeter instrument for the MESSENGER mission. SPACE Sci Rev 131:451–479

Evans IS (1979) Statistical characterization of altitude matrices by computer. An Integrated System of Terrain Analysis and Slope Mapping. The Final Report on Grant DA-ERO-591-73-G0040. Department of Geography, University of Durham, Durham, p 192

First Global Topographic Model of Mercury. http://solarsystem.nasa.gov/news/2016/05/06/first-global-topographic-model-of-mercury

Grohmann CH, Hargitai H (2014) Surface rourghness. Encyclopedia of planetary landforms. https://doi.org/10.1007/978-1-4614-9213-9_633-1

Grohmann CH, Smith MJ, Riccomini C (2011) Multiscale analysis of topographic surface roughness in the midland valley, Scotland. Geosci Remote Sens IEEE Trans 49:1200–1213. https://doi.org/10.1109/TGRS.2010.2053546

Hobson RD (1972) Surface roughness in topography: quantitative approach. In: Spatial analysis in geomorphology, pp 221–245

Jenness J (2006) Topographic Position Index (TPI) v 1.2. Extension for ArcView 3x. Jenness Enterprices. http://www.jennessent.com/arcview/tpi.htm

Kreslavsky MA, Head JW (2002) Mars: nature and evolution of young latitude-dependent water-ice-rich mantle. Geophys Res Lett 29:1719. https://doi.org/10.1029/2002GL015392

Kreslavsky MA, Head JW, Neumann GA, Rosenburg MA, Aharonson O, Smith DE, Zuber MT (2013) Lunar topographic roughness maps from Lunar Orbiter Laser Altimeter (LOLA) data: scale dependence and correlation with geologic features and units. Icarus 226:52–66

Kreslavsky MA, Head JW, Neumann GA, Zuber MT, Smith DE (2014) Kilometer-scale topographic roughness of Mercury: correlation with geologic features and units. Geophys Res Lett 41(23):8245–8251. https://doi.org/10.1002/2014GL062162

Mercury Dual Imaging System (MDIS). Messenger. http://www.messenger-education.org/instruments/mdis.php/

Messenger, PDS Geoscience Node. http://pds-geosciences.wustl.edu/messenger/mess-e_v_h-mla-3_4-cdr_rdr-data-v1

Preusker F, Oberst J, Stark A, Matz K-D, Gwinner K, Roatsch T (2017) High-resolution topography from MESSENGER orbital stereo imaging – The southern hemispehre. EPSC Abstracts, 11, EPSC2017-591

Rashid H (2010) 3-D surface-area computation of the state of Jammu & Kashmir using shuttle radar topographic mission (SRTM) data in geographical information system (GIS). J Geomat 4 (2):77–82

Robinson MS, Brylow SM, Tschimmel M, Humm D, Lawrence SJ, Thomas PC, Denevi BW, Bowman-Cisneros E, Zerr J, Ravine MA, Caplinger MA, Ghaemi FT, Schaffner JA, Malin MC, Mahanti P, Bartels A, Anderson J, Tran TN, Eliason EM, McEwen AS, Turtle E, Jolliff BL, Hiesinger H (2010) The Lunar Reconnaissance Orbiter Camera (LROC) instrument overview. Space Sci Rev 150:81–124

Scholten F, Oberst J, Matz K-D, Roatsch T, Wählisch M, Speyerer EJ, Robinson MS (2012) GLD100: the near-global lunar 100 m raster DTM from LROC WAC stereo image data. J Geophys Res: Planets 117:E00H17. https://doi.org/10.1029/2011JE003926

Shingareva KB, Zimbelman J, Buchroithner MF, Hargitai HI (2005) The realization of ICA commission projects on planetary cartography. Cartographica 40(4):105–114. https://doi.org/10.3138/3660-4078-55X1-3808

Smith DE, et al (2010) The lunar orbiter laser altimeter investigation on the lunar reconnaissance orbiter mission. Space Sci Rev 150:209–241

USGS PDS Imaging Node Server. https://pdsimage2.wr.usgs.gov/data/mess-h-mdis-5-dem-elevation-v1.0/MESSDEM_1001/DEM/

Viewing GLD100 WAC Global DTM. LROC. http://wms.lroc.asu.edu/lroc/global_product/256_ppd_DEM

Zharkova AY, Karachevtseva IP, Zubarev AE, Brusnikin ES, Kokhanov AA, Kreslavsky MA (2016) The study of Mercury's surface by cartography methods based on the newest topographic data derived from MESSENGER image processing. Sovremennye problemy distantsionnogo zondirovaniya Zemli iz kosmosa (Current problems in remote sensing of the Earth from space), 13(5):265–274. http://d33.infospace.ru/d33_conf/sb2016t5/265-274.pdf

Venus Topography and Boundary Conditions in 3D General Circulation Modeling

Michael J. Way and June Wang

Abstract We describe how one ingests 3D topographic data from NASA's Venus Magellan Spacecraft radar observations into the ROCKE-3D Planetary General Circulation Model. We also explain how boundary condition choices such as ocean/lake coverage/depth, rotation rate, atmospheric constituents, and other factors influence surface conditions in ROCKE-3D paleo-Venus simulations. Studies such as these should also be considered when examining liquid water habitability in similar exoplanet experiments.

Keywords General Circulation Model (GCM) · Topography · Magellan mission

1 Introduction

Given the similar size, density, probable composition (Chassefiere et al. 2012) and proximity of Venus to Earth planetary scientists have struggled to fully understand the differing climatic histories of these two worlds. For example, measurements of the deuterium to hydrogen ratio (D/H) by Pioneer Venus (Donahue et al. 1982, 1997) showed that the hot parched Venus of today had more water in previous epochs. Current estimates are that paleo-Venus could have had the equivalent of between 4 and 525 meters of liquid water if spread across a sphere of radius 6051.85 km (Donahue and Russell 1997). 6051.85 km is the mean radius from Venus as determined from the Magellan mission. How, when and at what rate it lost

M. J. Way (✉)
NASA Goddard Institute for Space Studies, New York, NY, USA
e-mail: michael.way@nasa.gov

M. J. Way
Theoretical Astrophysics, Department of Physics and Astronomy,
Uppsala University, Uppsala 75120, Sweden

J. Wang
Department of Earth and Planetary Science, Washington University
in Saint Louis, St. Louis, MO 63130, USA

© This is a U.S. government work and not under copyright protection in the U.S.; foreign copyright protection may apply 2019

H. Hargitai (ed.), *Planetary Cartography and GIS*, Lecture Notes in Geoinformation and Cartography, https://doi.org/10.1007/978-3-319-62849-3_19

its water is unclear (e.g., Kasting and Pollack 1983; Donahue 1999). Prior to 2016, most atmospheric modeling of paleo-Venus was via the 1D models of Grinspoon and Bullock (2007) who found that under the assumption of a 1 bar N_2 dominated atmosphere with 6 mb of H_2O at the surface and 100% cloud cover the surface air temperature would be around 27 °C. Recent three-dimensional General Circulation Model (3D GCM) work of Way et al. (2016) has shown that within the knowledge constraints of today a paleo-Venus could have had liquid water on its surface for well over 1 billion years of its early history. Below we describe the input Venus topographic data, the 3D GCM that utilized this data, and the choices made that revealed a paleo-Venus with a habitable climate in its early history.

2 Data

The most complete and highest resolution topographic maps of the Venus surface were derived from the NASA Magellan Radar altimetry observations taken from 1990 to 1992 (Ford and Pettengill 1992), together with previous Pioneer Venus (Pettengill et al. 1980) and Venera (Alexandrov et al. 1987; Barsukov et al. 1986) spacecraft observations. The grid resolution of the resulting topographic map is about $1° \times 1°$, and the vertical resolution of the data is approximately 80 meters. The topography data are in kilometers with the zero surface referenced to a sphere of 6051.85 km radius. Different formats of the maps are currently available at the PDS Geosciences node at Washington University in St. Louis.[1] The PDS topographic map of Venus covers a range from −120° to 240° in longitude and from −89.5° to 89.5° in latitude. The map is neither 'centered' at 0°, nor at 180° as is common. The map center was offset by 59.5° to see the Aphrodite Terra feature unobstructed by the edges of the display frame. Aphrodite Terra is located near the equator and includes four smaller highlands: Ovda, Thetis, Atla, and Ulfrun Regions (Young 1990).

3 Method

The topographic data from Sect. 2 were utilized in a 3D GCM known as ROCKE-3D (Way et al. 2017, hereafter Way17). Historically, 3D GCMs (also described as Global Climate Models) were constructed to study Earth climate over relatively long periods of time (100 s of years) in comparison with higher resolution models that studied weather phenomena on short timescales (less than 1 month). Today's Earth GCMs are coupled atmosphere–ocean models with detailed physics of radiative transfer, sea ice, land ice, ground hydrology, cloud dynamics and a host of other processes that govern the climate of Earth (e.g., Schmidt et al. 2014).

[1]http://pds-geosciences.wustl.edu/mgn/mgn-v-rss-5-gravity-l2-v1/mg_5201/.

They have been instrumental in understanding the possible impacts of climate change on Earth.[2] However, planetary scientists quickly realized that the same models could be used to model other planetary environments. Leovy and Mintz (1969) were the first to model another planet (Mars), and Joshi et al. (1997) made the first exoplanet GCM experiments. Presently, there are a host of GCMs modeling the solar system planets, dwarf planets, and moons with atmospheres in our solar system and hypothetical exoplanetary systems. Examples can be found in the Introduction section of Way17.

The topography data described in Sect. 2 above were ingested into a NetCDF[3] compatible format (NetCDF 2016) where the longitudinal offset mentioned above was taken into account so that zero longitude begins on the left-hand side as is standard in ROCKE-3D when displayed. In the specific case of the paleo-Venus simulations related to Way et al. 2016 (hereafter Way16), the data were down-sampled from the original resolution of 1° × 1° to 4° × 5° in latitude and longitude. Compared to modern climate simulations with the ROCKE-3D parent model GCM ModelE2 (Schmidt et al. 2014), resolution of 4° × 5° may seem quite coarse, but there are two reasons why Way16 chose this resolution. (1) The model run-time is a factor of 2–3 faster at this lower resolution compared to ModelE2 which runs at the resolution of 2° × 2.5° for the atmosphere and 1° × 1.25° for the ocean. (2) The lower resolution has little effect on the climate dynamics of a slowly rotating (day length greater than 64 Earth days long, see Way et al. 2015) world like Venus. This is because the Rossby radius of deformation is much larger than the 4° × 5° grid box size (Del Genio and Suozzo 1987).

In Way16, an ocean depth compatible with the D/H measurements mentioned above was desired for the simulations. The mean planetary radius of 6051.85 km was chosen as the ocean top. In the source topographic input files from the PDS, any down-sampled grid cell altitude with a negative 'height' would be assigned as an ocean cell. Any cell with a positive value would be assigned as a land grid cell. The resultant ocean had an average depth of 311 meters, if spread across a hypothetical flat surface sphere of radius 6051.85 km. This resulting ocean volume is within the estimated D/H range of possibilities in Donahue and Russell (1997). The resulting land-mask topography is shown in Fig. 1. Additionally, another step was taken: to prevent any possibility of issues related to particular ocean grid cells freezing to the bottom[4] and to simplify interpretation of land climatic effects, a number of small land grid cells with surrounding shallow ocean depths were transformed to sea grid cells with a depth set to the average of the deeper surrounding grid cells. One can see the elimination of these land grid cells when looking at Fig. 1a (with small islands) versus 1b (without small islands). The higher resolution map from the PDS was further

[2]http://www.ipcc.ch/.

[3]http://www.unidata.ucar.edu/software/netcdf.

[4]A known issue with ROCKE-3D for shallow ocean grid cells is that if they freeze to the bottom the GCM will crash as it is incapable of changing topographically defined surface types (e.g. ocean → land or vice versa).

utilized to calculate a roughness length in each 4° × 5° land grid cell based on the standard deviation of the twenty 1° × 1° cells used to construct each 4° × 5° grid cell. A water runoff direction was calculated based on the land topography so that lakes and rivers will form where necessary. For example, at points where there is a local minimum a lake will be allowed to form if precipitation outweighs evaporation on the same timescale. Lakes are given as fractional amounts of a given grid cell, from 0 to 100%. Bare soil of a 50/50 mix of clay/sand and an albedo of 0.2 was chosen following Yang et al. (2014).

It is believed that Venus has been largely resurfaced the past several hundred million years (McKinnon et al. 1997, Kreslavsky et al. 2015). Hence today's topographic relief may not be an accurate reflection of its ancient topography, although Way16 made the assumption that the resultant topography was the same over the epoch of the simulation: from 2.9 to 0.715 Gya. The 2.9 Gya was chosen to avoid early solar UV activity and geothermal heating effects that are hard to constrain with present-day knowledge. The 0.715 Gya was chosen as a period before the observed global resurfacing event. We were also able to obtain spectra from Mark Clare (Claire et al. 2012) for these particular times, hence the precision in the dates given. Typically, exoplanetary GCM work has relied upon either Earth topographic relief, idealized continental structures or assumed aquaplanet[5] surfaces (Yang et al. 2014). Way16 took the approach that using the present-day topography of Venus is the least-worst choice for a paleo-Venus world. However, comparable simulations in Way16 were performed with a present-day Earth-like topography. The main differences in these two topographies (as described in Way16) were the ratio of land-to-sea and ocean volume. The final paleo-Venus reconstruction used in Way16 has 10% of the water volume of modern Earth. The Venus topography also has more low-latitude land distribution versus the high-latitude land distribution of modern Earth. These differences played out in the final globally averaged surface temperatures (T_{surf}) where the Earth-like topography had a value of T_{surf} = 23 °C. This was 12 °C warmer than the Venus topography value of T_{surf} = 11 °C (see Table 1 in Way16) for the same incident solar flux. This is likely because of differing latitudinal distributions of land/sea and oceanic currents. Since Way16, we have explored other possible topographies. Aquaplanet simulations have been carried out assuming an ocean covered surface with a depth of 899 meters (to be published in a forthcoming paper). Those simulations have T_{surf} = 31 °C. This is higher than that of either the Venus or Earth topographic land reliefs mentioned above. This is due to the fact that no land ice is allowed to form at the poles, nor on the dark side of the slowly rotating planet at higher altitudes, as seen in the other topographic reliefs in Way16. These cold land surfaces tend to lower T_{surf}. Because of efficient ocean heat transport, there is no ocean ice anywhere in the aquaplanet simulation, hence the higher T_{surf} = 31 °C. One other topography attempted was a land planet simulation where 10 m equivalent depth water was spread over the Venus topography in lakes or small seas. This unpublished simulation has T_{surf} = 21 °C.

[5]A surface covered in 100% water.

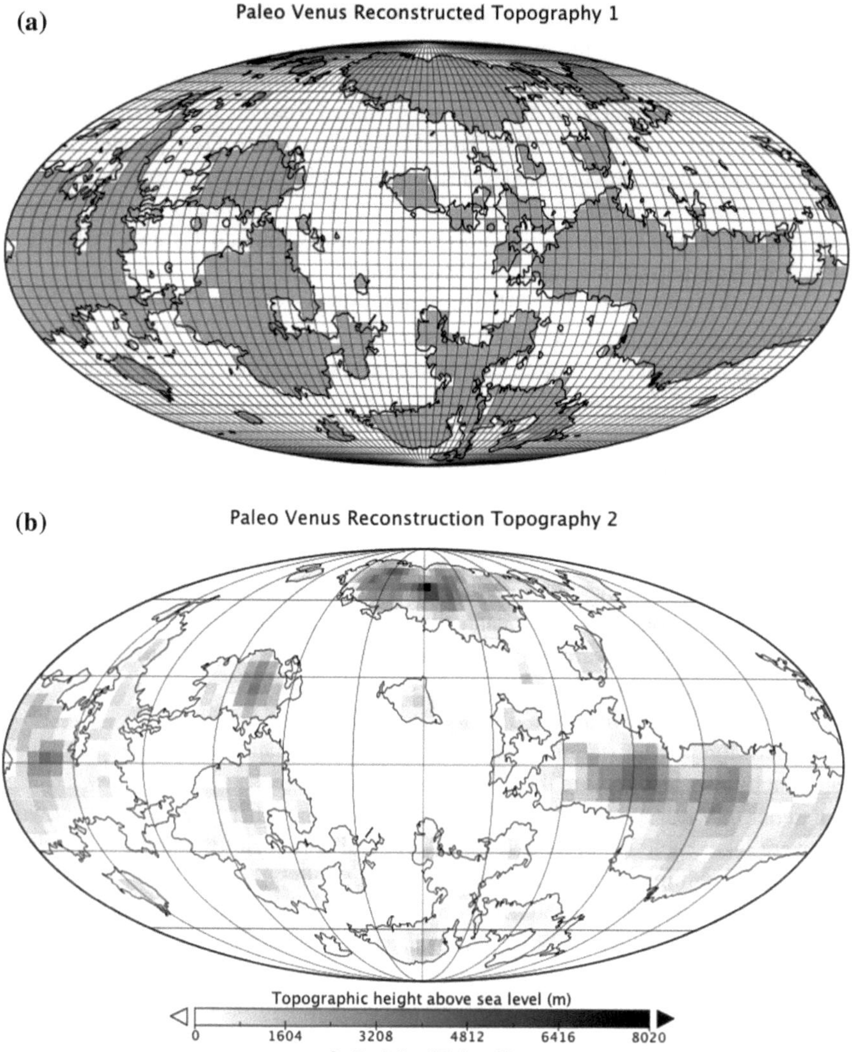

Fig. 1 Top **a**: Reconstructed topography from Magellan spacecraft data. Grid cells shown are equivalent to the 4° × 5° latitude–longitude resolution of the NetCDF ROCKE-3D GCM input file. Dark gray = land and light gray = ocean. Bottom **b**: Modified reconstructed topography from the Magellan spacecraft radar observations. Small islands have been removed. White = ocean and gray gradient = topographic height from 1 (light gray) to 8020 m (black). The center of the image's 30° spaced grid lines corresponds with 0° latitude and longitude. Each grid line to the north/south is ±30° of latitude and to the east/west is ±30° of longitude. West is to the left, and east is to the right

Several other key assumptions were made in the modeling efforts regardless of topography: (1) paleo-Venus would have the same orbital characteristics (rotation, obliquity, eccentricity) as present-day Venus, (2) the solar insolation and spectral distribution from several solar epochs were obtained from the models of Claire et al. (2012) and were ingested into ROCKE-3D: 2.9 Gya, 0.715 Gya, present day, and (3) the atmospheric pressure and constituents would be similar to modern Earth with a 1013mb N_2 dominated atmosphere with CO_2 = 400ppmv, CH_4 = 1ppmv and an H_2O profile similar to the 6mb one of Grinspoon and Bullock (2007) (see plate 1 in their paper), but no aerosols, O_2 nor O_3.

Input and output NetCDF (2016) formatted diagnostic data from the GCM are typically displayed using a widely available program called Panoply.[6] When viewing GCM diagnostic output, it is convenient to have an 'overlay' that outlines the boundaries of the continents, inland seas, and islands so that one knows if, for example, the surface humidity or temperature you are viewing is over land or sea. This is in contrast to contour lines that are often used to indicate the height of topographic features. The process to generate overlays involves loading the land/ocean topography map into NCAR Graphics[7] where one can then draw the desired boundaries with a mouse. A latitude x longitude sequence of points that comprises each polygon is then saved in a text file. For Panoply, each collection of latitude x longitude points that denote an island or continent is required by Panoply to be delimited by 999999 (an arbitrarily chosen number by the authors of Panoply). This file is then loaded into Panoply and shows up as an optional overlay. The Panoply Venus overlay is shown in the Mollweide projection of Fig. 1.

In addition to the definition of land and ocean mentioned above, the input topography files also require the definition of lakes if present, or land snow/ice (e.g., Antarctica). Way16's paleo-Venus simulations assumed there was no land snow/ice nor were lakes included at model start. A general approach to defining where lakes might be located involves the use of a 'river directions' file. This ROCKE-3D input file defines the directions where rivers would flow between grid cells based on their relative elevations in the topography file. The downstream flow is always to the grid cell with the greatest relative lower elevation. Most of the rivers will exit into the oceans. If a grid cell resides in a valley where all neighboring grid cells are at higher elevations, a lake may form. Lakes in ROCKE-3D can be variable quantities such that defined lakes in the topography input file can later evaporate, or from elsewhere as dictated by precipitation, evaporation, and topography. Figure 3a shows an example of this for the Venus land planet mentioned above that started with 10 m depth water equivalent if spread across the surface. Figure 3b shows what this hypothetical land Venus world would look like after running the model 4000 years. Looking at ground hydrology diagnostic outputs we can estimate when they have come into equilibrium. In this case, it was 4000 years of model run-time. There are also 'emergency' river directions. These may be invoked when a given grid cell has

[6]http://www.giss.nasa.gov/tools/panoply.

[7]http://ngwww.ucar.edu/.

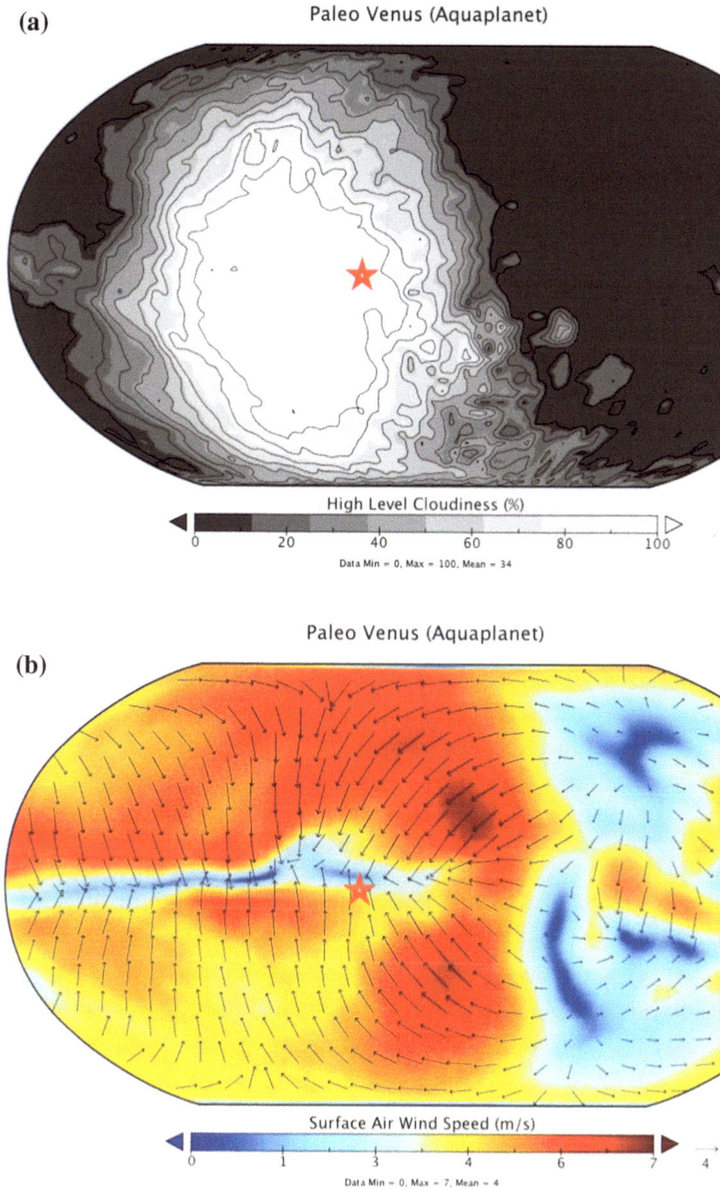

Fig. 2 ROCKE-3D aquaplanet simulations of paleo-Venus at 0.715Gya with a constant 899-m-depth fully coupled ocean. Red star in each is the location of the substellar point. Each image is an average over 1/6 of a diurnal cycle (19 Earth days in length), and they are taken from the same exact time in the simulation. Top **a**: Black = 0% PCLDH and White = 100% PCLDH (high level cloudiness—see main text). Note that the substellar point is not at the center of the high albedo clouds because the planet is rotating and the clouds are an average over 1/6th of a diurnal cycle. Bottom **b**: Surface wind speeds where the longest vectors represent speeds up to 7 m/s. These maps were generated in Panoply from post-processed ROCKE-3D data

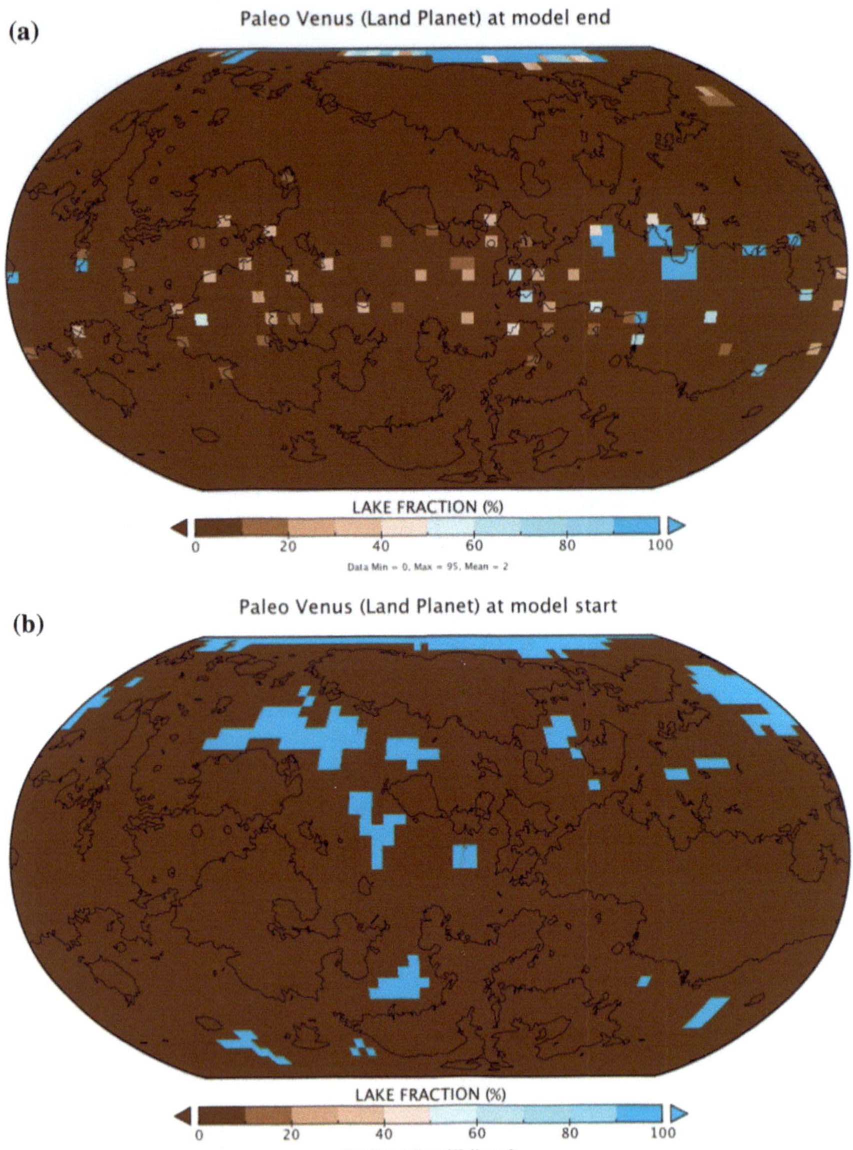

Fig. 3 ROCKE-3D land planet simulations of paleo-Venus at 2.9 Gya with 10 m equivalent depth of water if spread across the surface of a sphere of radius 6051.85 km. Top **a** shows the lakes residing at the lowest points in the topography at model start. Bottom **b** shows where the lakes are after running 4000 model years. One can see that precipitation and evaporation patterns in the atmosphere have tended to consolidate the lakes into larger contiguous regions. This simulation also used a 50/50 sand/clay soil type

no defined river direction because it was surrounded by higher elevation grid cells, but the lake height in this grid cell increases (due to precipitation and a lack of evaporation) to an elevation higher than one of the surrounding grid cells. When this occurs, the emergency river direction is invoked and water is allowed to flow into the (now) lower neighboring grid cell.

Completed GCM runs are normally analyzed or post-processed by examining quantities such as surface temperature, wind speed, and albedo. Using Panoply, a large selection of projections are also available in addition to the Mollweide used in Fig. 1. In Fig. 2, we present results from one of the paleo-Venus aquaplanet runs using a Robinson projection. The figures are the results of averaging over 1/6 of a diurnal cycle (approximately 19 Earth days in length). Figure 2a is a typical GCM diagnostic output called high-level cloudiness (known as PCLDH within ROCKE-3D). Because of the very slow rotation rate of this world, the atmospheric dynamics dictate that it has a single large Hadley-type cell in the northern hemisphere and a corresponding one in the south. On Earth (a fast spinner), the Coriolis force acts to break up the circulation pattern into several northern and southern Hadley cells. In Fig. 2a, the substellar point is offset from the center of the maximum of PCLDH. These high-level clouds also have a very high albedo, and hence, this map is not only a PCLDH map, but would also mirror the albedo map if we had shown it. It is these clouds, their high albedo and their contiguous nature that prevent a paleo-Venus-type world at 0.715 Gya from having extremely high surface temperatures even though it sees almost 1.7 times as much solar flux as modern-day Earth (see Table 1 in Way16). Figure 2b shows the ground level wind speeds. One can see that the ground level winds below the peak of the high-level clouds near the equator have very high relative velocities. These winds drive the large convective updraft that creates the high-level clouds and borders each hemisphere's Hadley cell.

4 Conclusions

The process to transform Magellan spacecraft topographic data into useful input files at the default resolution of $4° \times 5°$ for the ROCKE-3D GCM is described. The horizontal and vertical resolution of the Magellan topographic data are sufficient for exploring Venus' climatic history with modern-day GCMs (Way et al. 2016), but higher resolution data are desired as modern GCMs can utilize such data. Unfortunately, higher resolution data can only be obtained via another mission to Venus, and currently no space agency has such a mission presently funded. The ROCKE-3D GCM topographic NETCDF input and output files may be obtained from the authors upon request.

Acknowledgements This research was supported by the NASA Astrobiology Program through our participation in the Nexus for Exoplanet System Science, and by the NASA Planetary Atmospheres Program, Exobiology Program, and Habitable Worlds Program. M.J.W. acknowledges support from GSFC Sellers Exoplanet Environments Collaboration (SEEC), which is funded by the NASA Planetary Science Division's Internal Scientist Funding Model. Special thanks to Jeff Jonas, Robert Schmunk, and Linda Sohl for advice related to the creation of the topographic overlay file.

References

Alexandrov YN, DubrovinVM, Zakharov AI, Kotel'nikov VA, Krimov AA, Petrov GM, Rzhiga ON, Sidorenko AI, Sinilo VP, Sokolov GA (1987) The creation of a Venus radar map. In Problems of today's radio-engineering and electronics. Nauka, Moscow, pp 44–69

Barsukov VL, Sukhanov AL et al (1986) Geomorphological characteristics of the northern hemisphere of Venus. Geotektonica 4:4–25 NASA TM-88506, Translation of "geologo-morfologicheskaya kharacteristika severnogo polushariya Venery

Chassefiere E, Wieler R, Marty B, Leblanc F (2012) The evolution of Venus: present state of knowledge and future exploration. Planet Space Sci 63–64:15–23

Claire MW, Sheets J, Cohen M, Ribas I, Meadows VS, Catling DC (2012) The evolution of solar flux from 0.1 nm to 160 um: quantitative estimates for planetary studies. Astrophys J 757:95. https://doi.org/10.1088/0004-637x/757/1/95 (12pp)

Del Genio AD, Suozzo RJ (1987) A comparative study of rapidly and slowly rotating dynamical regimes in a terrestrial general circulation model. J Amos Sci 44:973–986

Donahue TM (1999) New analysis of hydrogen and deuterium escape from Venus. Icarus 141:226–235

Donahue TM, Russell CT (1997) The Venus atmosphere and ionosphere and their interaction with the solar wind: an overview. In: Baugher SW, Hunten DM, Phillips RJ (eds) Venus II geology, geophysics, atmosphere and solar wind environment. University of Ariz Press, Tucson, pp 3–31

Donahue TM, Hoffman JJ, Hodges RR Jr, Watson AJ (1982) Venus was wet: a measurement of the ratio of deuterium to hydrogen. Science 216:630

Donahue TM, Grinspoon DH, Hartle RE, Hodges RR Jr (1997) Ion/neutral escape of hydrogen and deuterium: evolution of water. In: Bougher SW, Hunten DM, Phillips RJ (eds) Venus II geology, geophysics, atmosphere and solar wind environment. University of Ariz Press, Tucson, pp 414–585

Ford PG, Pettengill GH (1992) Venus topography and kilometer-scale slopes. J Geophys Res 97 (E8):13103–13114

Grinspoon DH, Bullock MA (2007) Astrobiology and Venus exploration. In: Esposito LW, Stofan ER, Cravens TE (eds) Exploring Venus as a terrestrial planet. AGU, Washington, pp 191–206

Joshi MM, Haberle RM, Reynolds RT (1997) Simulations of the atmospheres of synchronously rotating terrestrial planets orbiting M dwarfs: conditions for atmospheric collapse and the implications for habitability. Icar 129:450

Kasting JF, Pollack JB (1983) Loss of water from Venus. I. Hydrodynamic escape of hydrogen. Icarus 53:479

Kreslavsky M, Ivanov A, Head JW (2015) The resurfacing history of Venus: constraints from buffered crater densities. Icarus 250:438

Leovy C, Mintz Y (1969) Numerical simulation of the atmospheric circulation and climate of Mars. JAtS 26:1167

McKinnon WB, Zahnle KJ, Ivanov BA, Melosh HJ (1997) Cratering on Venus: modeling and observations. In: Bougher SW, Hunten DM, Phillips RJ (eds) Venus II geology, geophysics, atmosphere and solar wind environment. University of Ariz Press, Tucson, pp 414–585

NetCDF (2016) NETwork, Common Data Format (NetCDF). http://doi.org/10.5065/D6H70CW6

Pettengill GH, Eliason E, Ford PG, Loriot GB, Masursky H, McGill GE (1980) Pioneer Venus radar results: altimetry and surface properties. J Geophys Res 85:8261–8270
Schmidt GA et al (2014) Configuration and assessment of the GISS ModelE2 contributions to the CMIP5 archive. J Adv Model Earth Syst 6(1):141–184. https://doi.org/10.1002/2013MS000265
Way MJ, Del Genio AD, Kelley M, Aleinov I, Clune T (2015) Exploring the inner edge of the habitable zone with fully coupled oceans. arXiv:1511.07283
Way MJ et al (2016) Was Venus the first habitable world of our solar system? Geophys Res Lett 43(16):8376–8383. https://doi.org/10.1002/2016GL069790
Way MJ, Aleinov I, Amundsen DS, Chandler M, Clune T, Del Genio AD, Fujii Y, Kelley M, Kiang NY, Sohl L, Tsigaridis K (2017) Resolving orbital and climate keys of earth and extraterrestrial environments with dynamics (ROCKE-3D) 1.0: a general circulation model for simulating the climates of rocky planets. Astrophys J Suppl Ser 231:12. arXiv:1701.02360
Yang J, Gweanel B, Fabrycky DC, Abbot DS (2014) Strong dependence of the inner edge of the habitable zone on planetary rotation rate. Astrophys J 787:2
Young C (1990) The Magellan Venus explorer's guide. Jet Propulsion Laboratory, Pasadena, 90–24 JPL Publication

Exoplanet *Terra Incognita*

Svetlana V. Berdyugina, Jeff R. Kuhn, Ruslan Belikov and Slava G. Turyshev

Abstract Exoplanet surface imaging, cartography, and the search for exolife are the next frontiers of planetology and astrophysics. Here we present an overview of ideas and techniques to resolve albedo features on exoplanetary surfaces. Albedo maps obtained in various spectral bands (similar to true-color images) may reveal exoplanet terrains, geological history, life colonies, and even artificial mega-structures or planet-engineering projects of advanced civilizations.

Keywords Exoplanets · Surface imaging · Exolife · Biosignatures · Technosignatures · Civilizations

1 Introduction

Extrasolar life (exolife) detection by SETI techniques is ongoing for decades (Tarter et al. 2010), while exoplanet habitability and biosignatures is a recent and very active subject of study (e.g., Grenfell 2017). Here we explore how unambiguous exolife detection will be improved with direct or indirect imaging techniques that are capable of resolving the surfaces of exoplanets. The Earth seen as an unresolved planet shows a very marginal spectral signature of the chlorophyll red edge—a sign

S. V. Berdyugina (✉)
Kiepenheuer Institut für Sonnenphysik, Freiburg, Germany
e-mail: sveta@leibniz-kis.de

S. V. Berdyugina · J. R. Kuhn
PLANETS Foundation, Kula, HI, USA

J. R. Kuhn
Institute for Astronomy, University of Hawaii, Pukalani, Honolulu, HI, USA

R. Belikov
Space Science and Astrobiology, NASA Ames, Mountain View, CA, USA

S. G. Turyshev
Jet Propulsion Laboratory, Pasadena, CA, USA

© Springer Nature Switzerland AG 2019
H. Hargitai (ed.), *Planetary Cartography and GIS*, Lecture Notes in Geoinformation and Cartography, https://doi.org/10.1007/978-3-319-62849-3_20

of the dominant photosynthetic life on present-day Earth, as well as concomitant abundant oxygen, water, and methane in the atmosphere (e.g., Sterzik et al. 2012). Modeling signals from distant Earth-like planets indicate that life-forms should cover more than 50% of the visible planet surface to be detected spectrally from unresolved planet (Berdyugina et al. 2016). In spatially resolved Earth's albedo maps, however, one can clearly see terrestrial global life colonies, such as vegetation (on a subcontinent scale), and our civilization footprints (at a higher resolution). Remote sensing of resolved surface structures on Earth can unambiguously identify their nature (e.g., Greenberg et al. 1993).

Here we discuss a similar strategy applied to exoplanets. Atmospheric biosignatures, such as the presence of biogenic gases and their disequilibrium (Krissansen-Totton et al. 2018), will help to identify exoplanets which possibly host life (perhaps similar to ours, since we focus our search on planets in a water-based habitable zone (WHZ) while other HZ can be defined for non-Earth-like life based on different solvents and for different temperature ranges). There is, however, a risk of false-positive detections, since these gases may also form due to geochemical and photochemical processes affecting the composition of the planetary atmosphere (e.g., Meadows 2017). Therefore, obtaining resolved albedo maps of such potentially inhabited exoplanets (PIEs) will be crucial for identifying sources of biogenic gases and studying their distribution and evolution. In fact, both life and planetary activities are of great interest for planetology, astrophysics, and astrobiology, as we will expand the realm of planetary and life morphology, which is now still limited to the Solar system.

While much of the current efforts in exoplanetary science are focused on identifying PIEs using atmospheric biosignatures, here we review emerging efforts to develop techniques and technologies which will provide resolved maps of exoplanets and open the door toward unambiguous detection of exolife. A few model solutions for albedo maps and orbital parameters based on reflected flux or polarization light curves have been demonstrated for exoplanets (e.g., Cowan et al. 2009; Fluri and Berdyugina 2010; Kawahara and Fujii 2011; Fujii and Kawahara 2012; Schwartz et al. 2016). More recently, an independent light-curve inversion formalism ExoPlanet Surface Imaging (EPSI) elaborated applications for planets with and without clouds, with seasonal variations, photosynthetic organism colonies, and artificial structures built by advanced civilizations (Berdyugina and Kuhn 2017, hereafter BK17). It was demonstrated that inversions of multi-wavelength observations of such planets may enable detection of primitive and even advanced exolife with high confidence.

Here we distinguish between two approaches: *direct* and *indirect* Exoplanet Surface Imaging (EPSI). Direct EPSI requires km- to mm-size telescopes or interferometers on or near Earth, or alternatively targeted space missions beyond the Solar system. Such technologies are being developed and may be implemented in several decades if optimistic hopes for continued exponential technology growth are realized. Indirect EPSI is significantly less costly and in some cases comparably powerful: it can be done from the ground with 20-m to 100-m telescopes optimized for low scattered light. The technology to build such telescopes is within reach, such that investments now could yield an exolife telescope within a decade (Kuhn et al. 2018a).

2 Indirect Exoplanet Surface Imaging

2.1 Cartography of Unresolved Solar System Bodies

Early cartography of some of the distant and small bodies in the Solar system, before space missions delivered their high-resolution images, was based on indirect imaging techniques. Historically important methods include occultations, limb, shadow, and terminator measurements, as well as inversion of light curves measured at different rotational and orbital phases (see other reviews in this volume).

A light-curve analysis to reveal surface features of unresolved moons and asteroids was considered more than a hundred years ago (Russell 1906). Much later, such light-curve inversions have revealed the first surface maps of Iapetus (Morrison et al. 1975) and Pluto and Charon (see review by Buie et al. 1997). This allowed a study of their surfaces well before we saw images taken by spacecraft, which confirmed the main albedo features on planet surfaces recovered indirectly. Studying asteroid shapes and morphology from light-curve inversion (including polarimetry) is currently an active research field, which is enhanced by high-precision photometry from ground and space telescopes (e.g., Kaasalainen et al. 1992; Carbognani et al. 2012).

2.2 Gas Giant Exoplanets

The first light curves of exoplanets were obtained for hot Jupiters, since they are large and very close to the host star, so they are not spatially resolved from stars. Infrared light curves of hot Jupiters were obtained with the *Spitzer* space telescope (e.g., Harrington et al. 2006; Knutson 2007; Crossfield et al. 2010). Optical light curves were obtained using ground-based polarimetry (Berdyugina et al. 2008, 2011) and space telescopes COROT (e.g., Snellen et al. 2009) and *Kepler* (Demory et al. 2011). From 2018, high-precision photometry of stars with exoplanets will be delivered by TESS in the optical and JWST in the infrared space telescopes. A combined analysis of optical and infrared light curves provides stronger constraints on their atmospheres.

Hot Jupiters may be tidally locked to their orbital motion, with the rotational axis perpendicular to the orbital plane. In this case, their light curves can resolve features only in longitude (no resolution in latitude). Infrared light curves of hot Jupiters often show maximum radiation at orbital phases away from the exterior conjunction phase, which is when the planet is farther from the observer than the star. This was interpreted as the presence of hot spots and heat transport to the "dark side" of the planet (e.g., Harrington et al. 2006; Knutson 2007; Cowan and Agol 2008). Optical light curves provide a disk-average geometrical albedo of the planetary atmosphere, its dominant color, and may indicate the presence of clouds or hazes (Berdyugina et al. 2011; Evans et al. 2013; Pont et al. 2013).

Jupiter-, Saturn-, and Neptune-like gas giant planets at wider orbits (further from the host star) may rotate significantly faster than the orbital motion and be inclined to the orbital plane. Hence, the terminator misaligned with the rotational axis breaks the symmetry and allows spatial resolution in both longitude and latitude, when reflected light is sampled during several rotational periods at various orbital phases (preferably during the entire orbit). Since each flux measurement contains information on different parts of the partially illuminated planetary disk, one can employ a numerical inversion technique to "invert" such a 1D time series of flux measurements (light curve) into a time-average 2D surface map (BK17). For gas giants, this helps resolving global circulation patterns imprinted into the cloud structure. Even when the planet is "rolling" on its orbit (like Uranus), there is still useful information to map its surface. This was demonstrated using Solar system's Jupiter and Neptune images as inputs for EPSI inversions (BK17). Their Great Red and Great Dark Spots as well as zones and belts on Jupiter were successfully reconstructed (Fig. 1). This holds promise for studying global circulation on gas giant exoplanets on wider orbits in other planetary systems, which may be observed already now with large telescopes.

Inversions of thermal emission light curves are similar to stellar and brown-dwarf light-curve inversions to recover brightness inhomogeneities due to magnetic spots or clouds (e.g., Berdyugina et al. 2002; Crossfield et al. 2014).

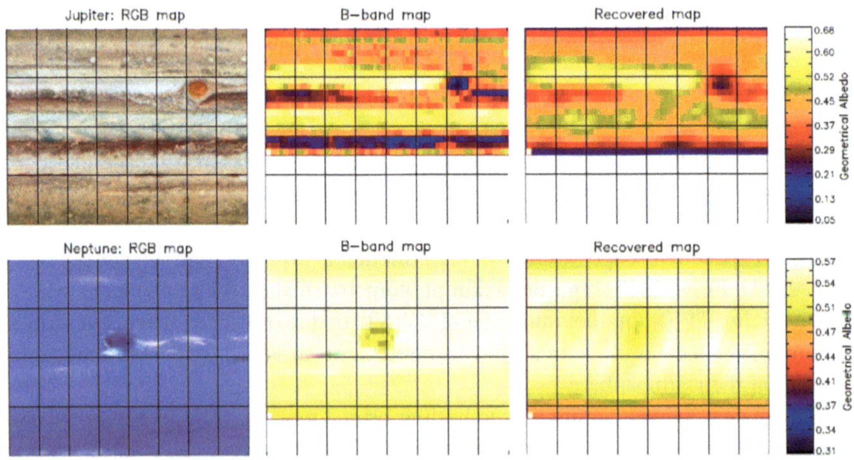

Fig. 1 EPSI inversion applied to the Solar system Jupiter and Neptune images as inputs (BK17). In the first column are original RGB maps of higher resolution, in the second column are original albedo maps rebinned to a coarser grid in a particular band, and in the third column are recovered albedo maps in the same band

2.3 Rocky Exoplanets

Rocky planets may have a variety of interesting surface features, if not fully covered by clouds, such as Venus. If they are covered in clouds, they would then be similar to the case of gas giant planets discussed in Sect. 2.2. In the Solar system, we see oceans, continents, deserts and forests on the Earth, ice caps, canyons and fluvial networks on Mars, volcanos on Io, heavily cratered terrains on the Moon, Mercury and moons of gas giants, glaciers on Pluto, and so on. Mapping albedo of such geological features on rocky exoplanets would provide hints to their geological history and internal structure and composition. Obviously, only large-scale and high-contrast features can be distinguished and analyzed.

To resolve (indirectly) such surface features on Earth-size exoplanets requires high-precision light curves (signal-to-noise ratio SNR > 20) sampled at a range of orbital and rotational phases, especially if planets are intermittently covered by clouds, like the Earth. The interpretation of recovered albedo features in terms of geological, biological, or artificial structures becomes possible. Here, EPSI techniques applied to the Solar system planet maps as models for rocky exoplanets can help in identifying such structures. In Fig. 2, inversions for several solar planetary

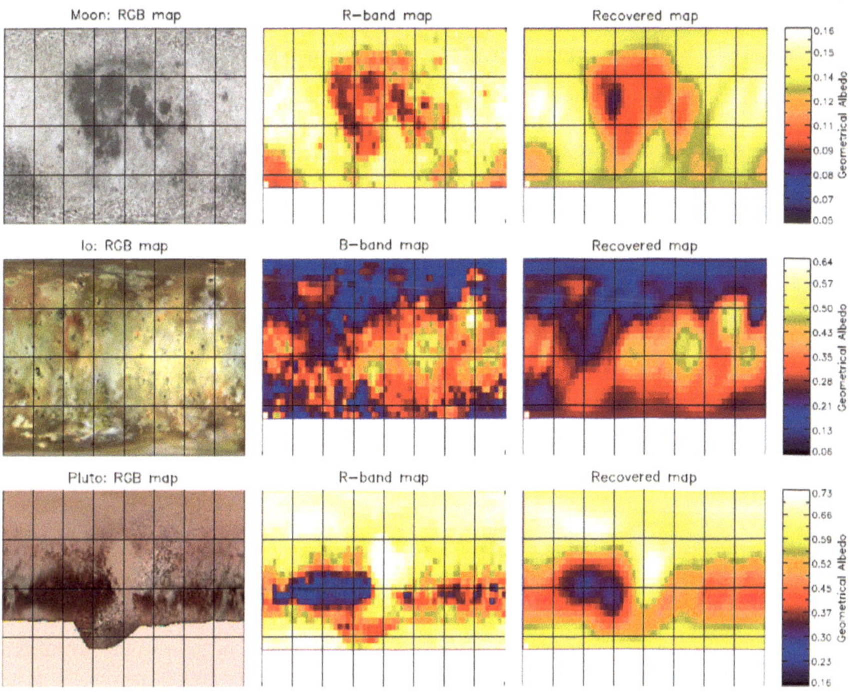

Fig. 2 Same as Fig. 1 but for the Solar system rocky planetary bodies (BK17)

bodies are shown as examples. They remarkably recover large-scale terrains, continents, volcanoes (plums), and polar caps. Albedo maps obtained in various spectral bands can provide information on chemical composition of terrains (see Sect. 2.4).

2.4 *Biosignatures*

Detecting biosignatures and life on exo-Earths requires an analysis of the spectral content of the reflected, emitted or/and absorbed flux from an exoplanet. Light curves measured in several spectral bands (e.g., blue, green, red, infrared) can provide color maps of exoplanets using the EPSI technique. This is a powerful tool for detecting living organisms on the surface, in particular photosynthetic organisms which have dominated the Earth for billions years.

A test inversion for an exoplanet with the Earth's surface features, i.e., ice polar caps, ocean, deserts, and forest, is shown in Fig. 3 (BK17). Now, by resolving surface features, spectra of various areas can be extracted and explored for life signatures. An example average spectrum (in four broad spectral passbands) of a small green area at mid-latitudes in the eastern part of the planet reproduces well the red edge chlorophyll signature near 700 nm: a deep absorption in the visible wavelengths and high albedo in the near-infrared. Since photosynthetic organisms have a variety of biopigments absorbing light in broad bands, it may possible to

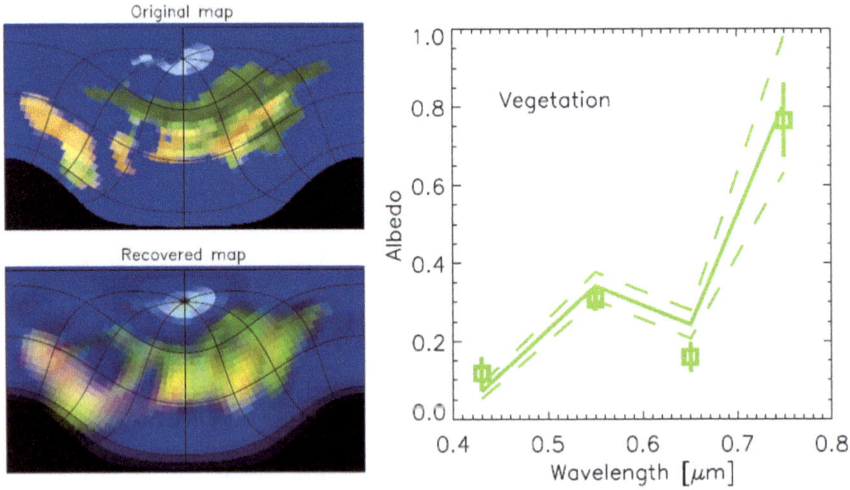

Fig. 3 Test inversion for an exoplanet with Earth's surface samples in "true"-color (left) and an average spectrum (right) of a small green area at mid-latitudes in the eastern part of the planet image with the red edge chlorophyll signature near 700 nm. "Measurements" from the recovered image in four broad spectral passbands are shown with symbols, while original spectra are shown with lines (BK17)

distinguish them using such low-resolution spectra if they cluster in relatively large areas on the planetary surface. Also, measuring polarized light curves helps to identify biopigments, minerals, and water reservoirs (Berdyugina et al. 2016). In particular, polarized light can measure glint from bodies of water (or other liquids, in general), which may be a good way to detect oceans (Karalidi and Stam 2012). Thus, spatially resolved images of exoplanets dramatically increase our chances to detect exolife, because we can carry out a spectral analysis of areas with high concentrations of living organisms (BK17).

2.5 Technosignatures

Although this is at present rather speculative, the EPSI technique allows also for detecting artificial mega-structures (AMS) constructed by advanced civilizations either on the surface or in the near-space of an exoplanet (circumplanetary space) (BK17). AMS could be of some regular shape and/or homogeneous albedo, or even above the surface as "geostationary" technology (e.g., for communications).

Low-albedo installations similar to our photovoltaic systems can be on the planet surface and in space. High-albedo installations, also on the surface or in the near-space, can redirect the incident stellar light, e.g., for heat mitigation by reflecting the light back to space. Such AMS may also efficiently absorb/reflect only a particular part of the spectrum, similar to photosynthetic organisms having specific spectral edges. An inversion example for a low-albedo installation in space (above clouds) is shown in Fig. 4.

These models show that recognizing AMS in inferred images requires large areas covered by them. Also, higher contrast structures with respect to the natural environment are easier to recover. Also, analyzing the spectral content of the reflected light can be useful.

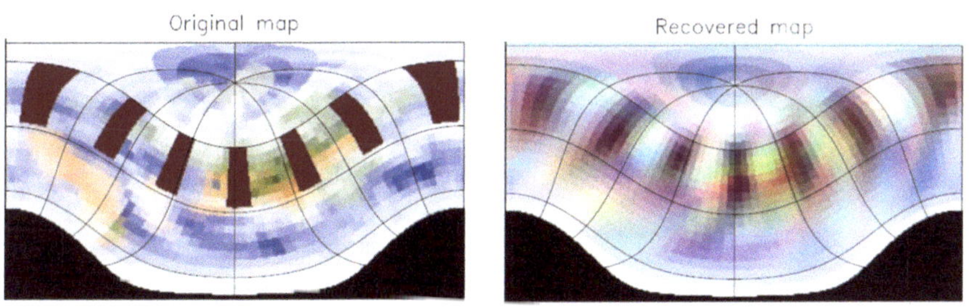

Fig. 4 A test inversion for an exoplanet with artificial structures in "true"-color (BK17)

2.6 Targets and Telescopes for Indirect EPSI

The Alpha Centauri system stars A, B, and Proxima Centauri are the closest stars to the Sun (4.3 ly). The only exoplanet discovered so far in this system is Proxima b (Anglada-Escudé et al. 2016), possibly a rocky planet in the WHZ of the M5 red dwarf. Other nearby possibly rocky planets in a WHZ are Ross 128 b (11 ly) orbiting an M4 red dwarf (Bonfils et al. 2017) and Tau Ceti planets e and f orbiting an orange dwarf G8 V (12 ly) (Feng et al. 2017). These nearby exoplanets are excellent candidates for first-time exoplanet surface imaging. In addition, more planets are statistically expected to exist around Alpha Centauri and other nearby stars (Burke et al. 2015).

Before the indirect EPSI technique can be applied, one has to first directly image an exoplanet, i.e., obtain an image with the star and planet sufficiently separated, so that the planet flux can be measured with sufficient signal to noise against the background of scattered light from the star and other background light sources. This is a challenge because of (1) a high star-to-planet contrast ratio and (2) small star-planet angular separation, especially for Earth-size planets. These challenges imply stringent constraints on the telescope size and performance.

The telescope must be large enough to collect enough photons from the faint exoplanet to make it detectable. We consider the nearest exoplanet Proxima b in an example calculation. The bright stellar background generally implies that the integration time to achieve a given signal-to-noise (SNR) ratio here scales more steeply than even the inverse fourth power of the telescope aperture diameter. In Fig. 5 (left panel) realistic SNR are computed for 1 h exposure time in different

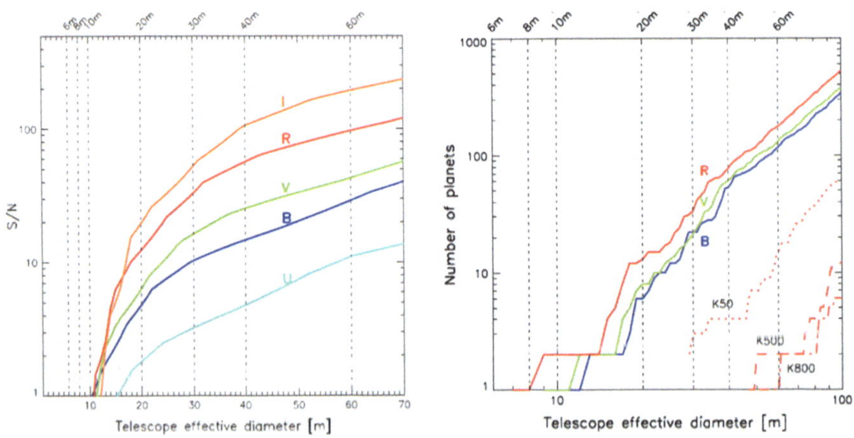

Fig. 5 *Left:* Proxima b light-curve SNR depending on the size of the telescope, assuming 0.2 geometrical albedo and 1 h exposure time. *Right:* Number of Earth-size exoplanets to be detected with the SNR ≥ 5 depending on the size of the telescope, assuming one planet per nearby star, 0.2 geometrical albedo, and 4 h exposure time. Solid lines are for high-contrast imaging telescopes (such as VLT and ELF), while dotted and dashed lines are for highly segmented Keck-style telescopes with a high scattered light background (BK17)

passbands depending on the size of a telescope (BK17). Here the size of Proxima b is assumed 1.15 Earth radii, and its average geometrical albedo 0.2. This calculation employs also a "state-of-the-art" point spread function including scattered stellar light and speckles. We conclude that a low scattered light telescope with the effective diameter of ≥ 12 m is necessary to obtain meaningful images of Proxima b. Since the assumed telescope is effectively diffraction limited, a space telescope would not significantly improve the Proxima b detectability (except in wavelengths where the telescope thermal emission or the atmosphere absorption dominates).

To achieve high-contrast measurements, one approach is to block the star with a coronagraph, starshade, and/or post-processing techniques, and directly image the planet, assuming it is spatially resolved from the blocked star. This has been achieved for stars with young, self-luminous, massive planets at orbits of tens of AU using such instruments as SPHERE at 8-m VLT (e.g., Bonnefoy et al. 2016) and GPI on 8-m Gemini (e.g., Ingraham et al. 2014). Currently deployed instruments are not yet capable of imaging WHZ exoplanets, but there are several concepts that may do it within the next two decades, and some as early as ~ 2 to 5 years from now. For example, imaging Alpha Centauri AB potential planets may be possible in the next 3–5 years in the visible light with a space telescope as small as 30 cm, with a powerful enough coronagraph (Belikov et al. 2015; Bendek et al. 2015). Also, a 1.5-Earth-radius WHZ planet around Alpha Centauri B may be within the capability of WFIRST using Multi-Star Wavefront Control (Sirbu et al. 2017), especially if post-processing methods can benefit from the greater brightness of Alpha Centauri.

Recently, a new concept for a ground-based high-contrast imaging system was proposed. The ExoLife Finder (ELF: Kuhn et al. 2018b) is an array of off-axis telescopes with their primary mirrors being segments of the same parabola and served by individual adaptive secondary mirrors. It combines advantages of high-contrast off-axis optical telescopes with interferometric options to control wavefront. For instance, by changing phases of primary mirrors one can achieve nulling interferometry within a limited field of view (FOV), i.e., a low scattered light area, or dark spot, without a physical coronagraph. This spot can be moved within FOV for high-contrast imaging of an exoplanet at different orbital phases. In this respect, ELF's optical system is a hybrid interferometric telescope optimized for low scattered light and exoplanet observations.

A 20-m ELF can continuously image at least a dozen of Earth-size planets in WHZ (Fig. 5, right panel). It will provide unique information on seasonal variations of their surfaces, atmospheres, and possibly life as well as possible geological activity (e.g., volcanoes). Its estimated cost is only about 100 M$, and it can be built within 5 years. A 70-m ELF (Kuhn et al. 2014; Moretto et al. 2016) could obtain images of all kinds of exoplanets within about 20pc, and its cost would be about 500 M$.

Such large diffraction-limited optics will not soon be deployed in space because of extremely high cost (>10B$) and technological challenges. Note that comparable aperture space and ground instruments can differ in cost by factors of 100–1000. Current space telescope concepts, such as NASA's LUVOIR or HabEx, may be implemented sometime in the late 2030s, and they may be able to image same PIEs around nearby stars as a 20-m ELF but in infrared bands which are not accessible from the ground. In the mid 2020s, also rather expensive (>1B$), ground-based Keck-style extremely large telescopes (ELTs) promise to directly image exoplanets (Guyon et al. 2014), but it becomes recognized that additional significant efforts (e.g., high-dispersion coronagraphy, Wang et al. 2017) are needed to overcome large scattered light caused by their highly segmented mirrors.

Yet another approach is to image planetary systems in the mid-infrared, where planet-to-star contrast ratio increases due to thermal radiation of the planet (e.g., Defrère et al. 2015). The ESO VLT 8-m telescope instrumentation in the Southern hemisphere is being upgraded to employ this technique for imaging Alpha Centauri AB at 10 microns and search for exoplanets in this system. At this wavelength, we expect planet flux contributions from both reflection of the stellar light and own radiation of the planet. If planet radiation dominates, indirect EPSI will result in very low-resolution maps (mainly resolved only in longitude). Also, to be sensitive to Earth-radius planets, thermal infrared imaging may require larger than 8-m telescopes.

3 Direct Exoplanet Surface Imaging

3.1 Direct EPSI from the Earth Distance

High-resolution direct surface imaging is the most expensive and most challenging for studying exoplanets. For example, a 2-km-size optical telescope/interferometer is needed to theoretically resolve the disk of an Earth-size exoplanets (1 pixel image) at the distance of 1.3 pc (4.3 ly) from the Earth (Alpha Cen distance). Under the same conditions, a Jupiter-size planet can be resolved into an image of about 10×10 pixels, and a Sun-size star in a 100×100 pixel image. Correspondingly, to resolve directly the Proxima b planet into at least 10×10 pixel image requires a 20-km-scale imaging system (possibly an array of smaller telescopes). This is a factor of 1000 larger than the ELF telescope which can provide indirect images of Proxima b of higher surface resolution (BK17). Also, building a 20-km optical system is truly a technological challenge, and its cost may be forbidding.

3.2 Direct EPSI with the Solar Gravitational Lens Telescope

The 1.3 pc to the nearest exoplanet is about 27,0000 AU. At about 550 AU from the Sun, a focal line of the solar gravitational lens (SGL) begins, i.e., the solar gravitational field can focus light from a faint, distant source along that line by bending photon trajectories (Fig. 6). The remarkable optical properties of the SGL include major brightness amplification ($\sim 10^{11}$ at the near-infrared wavelength of $\lambda = 1$ μm) and extreme angular resolution ($\sim 10^{-10}$ arcsec) within a narrow field of view (Turyshev and Toth 2017). It was proposed to position a telescope beyond 550 AU and use the SGL to magnify light from distant objects on the opposite side of the Sun, e.g., Proxima b (Eshleman 1979; Maccone 2009; Turyshev and Toth 2017; Turyshev et al. 2018).

The angular diameter of an Earth-like exoplanet at 30 pc is 1.4×10^{-11} rad. To resolve the disk of this planet as a single pixel, a telescope array with a baseline of ~ 74.6 km would be needed. Resolving the planet with 10^3 linear pixels would require a baseline $\sim 1 \times 10^5$ km ($\sim 12R_\oplus$), which is not feasible. In contrast, a 1-m telescope, placed at the focal line of the SGL at 750 AU from the Sun, has a collecting area equivalent to a diffraction-limited telescope with diameter of ~ 80 km and the angular resolution of an optical interferometer with a baseline of $12R_\oplus$. While building an instrument with the SGL's capabilities is far beyond our technological reach, we can use the SGL's unique capabilities to capture megapixel images of exoplanets.

A modest telescope at the SGL could be used for direct imaging of an exoplanet. The entire image of an exo-Earth is projected by the SGL into an instantaneous cylindrical volume with a diameter of ~ 1.3 km. Moving outward while staying within this volume, the telescope will take photometric data of the Einstein ring around the Sun, formed by the light from the exoplanet. The collected data will be processed to reconstruct the desirable high-resolution image by relying on modern deconvolution techniques.

To sample a planet image within the Einstein ring, an SGL telescope will have to operate on a pixel-by-pixel basis by moving the spacecraft in the image plane. A $10^3 \times 10^3$ pixel image will therefore require 10^6 pointings with the spacecraft step of ~ 1 m. Since each pointing corresponds to a different impact parameter (see Fig. 6) with respect to the Sun, one has to account for light from adjacent surface areas on the planet. Also, to place and keep the telescope on the focal line, high-precision astrometry of the planetary orbit is needed, as the telescope spacecraft will have to compensate stellar and planetary motion. The parent star will be completely resolved from the planet, and its light will be amplified many AU away from the optical axis, making the parent star contamination issue negligible. For example, an exo-Earth at 30pc can be imaged by a 1-m SGL telescope operating at

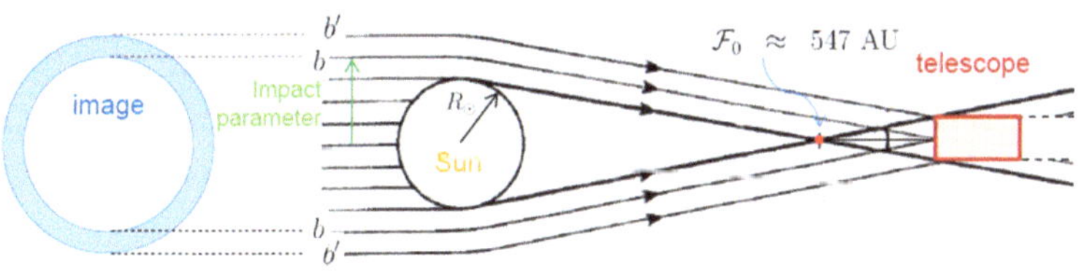

Fig. 6 Imaging of an exo-Earth with solar gravitational Lens. The exo-Earth occupies about 1 km × 1 km area at the image plane. Using a 1-m telescope as a 1-pixel detector provides a 1000 × 1000-pixel image (Color figure online)

650AU with SNR of 7 in ~1 week of integration time, when accounting for zodi background, the solar corona brightness, coronagraphic suppression of 10^{-6}, spacecraft jitter, and other realistic losses (Turyshev and Toth 2017; Turyshev et al. 2018).

The SGL telescope is a single-target instrument and, therefore, a powerful follow up mission for a target with signatures of life revealed by EPSI mapping and spectral analysis. Given the recent progress in finding exoplanets and in their characterization with indirect techniques (Sect. 2), it is a matter of a relatively short time that we will have plenty of exciting targets to explore with the SGL. It would take, however, ~30 years of "cruise" before the spacecraft would reach 600AU with the current technology yielding escape velocity of 15–20 AU/yr, see (Turyshev et al. 2018). In comparison, the farthest man-made object Voyager 1 moves at the speed of 3.6 AU/yr (17 km/s), and after 40 years it is at 141 AU from the Sun. A mission to the SGL enables the unique possibility of direct high-resolution imaging and spectroscopy of a habitable Earth-like exoplanet.

3.3 Direct EPSI with a Flyby Probe

The Alpha Centauri system may in some distant future also be explored by interstellar probes. The 4.3 ly distance implies a 43-year mission, if a tiny probe can be accelerated to 10% speed of light (~3,0000 km/s, or 6300 AU/yr). In comparison, the largest geocentric velocity achieved by the DLR/NASA Helios 1 spacecraft was ~100 km/s, and the largest heliocentric velocity by the NASA Juno mission was ~60 km/s. Thus, achieving 10% speed of light is yet quite challenging, but significantly reducing the size of the probe may help.

It was recently proposed to develop a light-sail spacecraft of only 1 g or less to fly to the Alpha Centauri system at 20% of the speed of light (Lubin 2016). The spacecraft *StarChip* is being developed by the Starshot project of Breakthrough Initiatives and to be propelled by a 100 GW/km^2 laser array. A thousand units would be released to compensate for possible interplanetary and interstellar dust collisions. Planet images will be obtained during a flyby with a high speed, so deconvolution will be still needed to achieve higher spatial resolution, but a slow-down maneuver seems possible (Heller and Hippke 2017; Heller et al. 2017). It is desirable that onboard miniature cameras can deliver multi-passband images for a spectral, polarimetric, and morphological analysis, so that the chemical, structural, and hopefully biological composition of these planets can be studied directly. Another interesting opportunity is to arrange a direct probe of the planetary atmosphere and even surface, if some of the probes can intercept the planet.

4 Conclusions

In just a few years, we can discover *terra incognita* on exoplanets and start their exploration in terms of surface composition, morphology, diversity, frequency, life/ civilization detection, etc. Similar to the last century indirect imaging of small and distant Solar system bodies, we can start practicing exoplanetology and exogeology, and even exobiology and exosociology of extraterrestrial life and civilizations.

Albedo maps in optical and infrared bands can be obtained using light-curve inversions for a dozen of Earth-size planets with a 20-m telescope like ExoLife Finder and for almost all exoplanets within about 20 pc with a 70-m ELF. These telescopes are optimized for low scattered light and exoplanet observations. The technology to build such telescopes is within reach, such that they could become operational within a decade.

In a more distant future, concurring space outside the Solar system and sending probes to selected exoplanets could be possible. Such missions are technology drivers and with the help of Moore's law may become reality in a couple of decades. Meanwhile, we should get busy with the indirect EPSI in order to select the most promising targets for direct probes.

References

Anglada-Escudé G, Amado PJ, Barnes J et al (2016) A terrestrial planet candidate in a temperate orbit around Proxima Centauri. Nature 536:437

Belikov R, Bendek E, Thomas SJ, Males J, Lozi J (2015) How to directly image a habitable planet around Alpha Centauri with a ~30–45cm space telescope. In: SPIE, vol 9605, id. 960517, p 12

Bendek E, Belikov R, Lozi J, Thomas SJ, Males J, Weston S, McElwain M (2015) How to directly image a habitable planet around Alpha Centauri with a ~30–45cm space telescope. In: SPIE, vol 9605, id. 960516, p 15

Berdyugina SV, Kuhn JR (2017) Surface imaging of proxima b and other exoplanets: topography, biosignatures, and artificial mega-structures. Astrophys Planet J, arXiv:1711.00185 (submitted)

Berdyugina SV, Pelt J, Tuominen I (2002) Magnetic activity in the young solar analog LQ Hydrae. I. Active longitudes and cycles. Astron Astrophys 394:505

Berdyugina SV, Berdyugin AV, Fluri DM, Piirola V (2008) First detection of polarized scattered light from an exoplanetary atmosphere. Astrophys J Lett 673:L83

Berdyugina SV, Berdyugin AV, Fluri DM, Piirola V (2011) Polarized reflected light from the exoplanet HD189733b: first multicolor observations and confirmation of detection. Astrophys J Lett 728:L6

Berdyugina SV, Kuhn JR, Harrington DM, Šantl-Temkiv T, Messersmith EJ (2016) Remote sensing of life: polarimetric signatures of photosynthetic pigments as sensitive biomarkers. Int J Astrobiol 15:45

Bonfils X, Astudillo-Defru N, Díaz R et al (2017) A temperate exo-Earth around a quiet M dwarf at 3.4 parsecs. A&A, arXiv:1711.06177 (in press)

Bonnefoy M, Zurlo A, Baudino JL et al (2016) First light of the VLT planet finder SPHERE. IV. Physical and chemical properties of the planets around HR8799. Astron Astrophys 587:58

Buie MW, Young EF, Binzel RP (1997) Surface appearance of Pluto and Charon. In: Alan Stern S, Tholen DJ (eds) Pluto and Charon. University of Arizona Press, Tucson, p 269

Burke C, Christiansen JL, Mullally F et al (2015) Terrestrial planet occurrence rates for the Kepler GK dwarf sample. Astrophys J 809(8):19

Carbognani A, Tanga P, Cellino A, Delbo M, Mottola S, Marchese E (2012) The representation of asteroid shapes: a test for the inversion of Gaia photometry. Planet Space Sci 73:80

Cowan NB, Agol E (2008) Inverting phase functions to map exoplanets. Astrophys J Lett 678: L129

Cowan NB, Agol E, Meadows VS et al (2009) Alien maps of an ocean-bearing world. Astrophys J 700:915

Crossfield IJM, Hansen BMS, Harrington J et al (2010) A new 24 μm phase curve for υ andromedae b. Astrophys J 723:1436

Crossfield IJM, Biller B, Schlieder JE et al (2014) A global cloud map of the nearest known brown dwarf. Nature 505:654

Defrère D, Hinz P, Skemer A et al (2015) Exoplanet science with the LBTI: instrument status and plans. In: SPIE, vol 9605, 96051G, p 12

Demory B-O, Seager S, Madhusudhan N et al (2011) The high albedo of the hot Jupiter Kepler-7 b. Astrophys J Lett 735:L12

Eshleman VR (1979) Gravitational lens of the sun—its potential for observations and communications over interstellar distances. Science 205:1133

Evans TM, Pont F, Sing DK et al (2013) The deep blue color of HD 189733b: albedo measurements with hubble space telescope/space telescope imaging spectrograph at visible wavelengths. Astrophys J Lett 772:L16

Feng F, Tuomi M, Jones HRA et al (2017) Color difference makes a difference: four planet candidates around τ Ceti. Astron J 154(135):23

Fluri DM, Berdyugina SV (2010) Orbital parameters of extrasolar planets derived from polarimetry. Astron Astrophys 512:12 id. A59

Fujii Y, Kawahara H (2012) Mapping earth analogs from photometric variability: spin-orbit tomography for planets in inclined orbits. Astrophys J 755:14 id. 101

Greenberg R, Belton M, Dejong E et al (1993) Earth imaging results from Galileo's second encounter. In: Lunar and planetary science conference, part 2: G-M, p 569

Grenfell JL (2017) A review of exoplanetary biosignatures. Phys Rep 713:1

Guyon O, Hayano Y, Tamura M et al (2014) Adaptive optics at the Subaru telescope: current capabilities and development. In: SPIE, vol 9148, id. 91481R, p 10

Harrington J, Hansen BM, Luszcz SH et al (2006) The phase-dependent infrared brightness of the extrasolar planet υ Andromedae b. Science 314:623

Heller R, Hippke M (2017) Deceleration of high-velocity interstellar photon sails into bound orbits at α Centauri. Astrophys J Lett 835:L32

Heller R, Hippke M, Kervella P (2017) Optimized trajectories to the nearest stars using lightweight high-velocity photon sails. Astron J 154(115):11

Ingraham P, Marley MS, Saumon D et al (2014) Gemini planet imager spectroscopy of the HR 8799 planets c and d. Astrophys J Lett 794:L15

Kaasalainen M, Lamberg L, Lumme K, Bowell E (1992) Interpretation of lightcurves of atmosphereless bodies. I—general theory and new inversion schemes. Astron Astrophys 259:318

Karalidi T, Stam DM (2012) Modeled flux and polarization signals of horizontally inhomogeneous exoplanets applied to Earth-like planets. Astron Astrophys 546:56

Kawahara H, Fujii Y (2011) Mapping clouds and terrain of Earth-like planets from photometric variability: demonstration with planets in face-on orbits. Astrophys J Lett 739:6 id. L62

Knutson HA, Charbonneau D, Allen LE et al (2007) A map of the day-night contrast of the extrasolar planet HD 189733b. Nature 447:183

Krissansen-Totton J, Olson S, Catling DC (2018) Disequilibrium biosignatures over Earth history and implications for detecting exoplanet life. Sci Adv 4:eaao5747

Kuhn JR, Berdyugina SV, Langlois M, Moretto G, Thiébaut E, Harlingten C, Halliday D (2014) Looking beyond 30m-class telescopes: the Colossus project. In: SPIE, vol 9145, id. 91451G, p 8
Kuhn JR, Cunnyngham I, Moretto G, Berdyugina SV, Langlois M, Thiébaut E (2018a) The Exo-Life Finder (ELF) telescope: design and synthetic beam concepts. In: SPIE, AS103-146
Kuhn JR, Berdyugina SV, Apai D et al (2018b) Exo-Life Finder (ELF): a hybrid optical telescope for imaging exo-Earths, astrobiology science strategy for the search of life in the universe, NAS, white paper, p 6. http://surveygizmoresponseuploads.s3.amazonaws.com/fileuploads/15647/3949545/254-4d6977cde1e82c73294e42111c7f904e_JeffKuhn.pdf
Lubin P (2016) A roadmap to interstellar flight. arXiv:1604.01356
Maccone C (2009) Deep space flight and communications exploiting the sun as a gravitational lens. Springer, Berlin, Heidelberg
Meadows VS (2017) Reflections on O_2 as a biosignature in exoplanetary atmospheres. Astrobiology 17:1022
Moretto G, Kuhn JR, Berdyugina SV, Langlois M, Tallon M, Thiébaut E, Halliday D (2016) Partially filled aperture interferometric telescopes: achieving large aperture and coronagraphic performance. In: SPIE, vol 9906, id. 99062B, p 12
Morrison D, Jones TJ, Cruikshank DP, Murphy RE (1975) The two faces of Iapetus. Icarus 24:157
Pont F, Sing DK, Gibson NP, Aigrain S, Henry G, Husnoo N (2013) The prevalence of dust on the exoplanet HD 189733b from Hubble and Spitzer observations. MNRAS 432:2917
Russell HN (1906) On the light variations of asteroids and satellites. Astrophys J 24:1
Schwartz JC, Sekowski C, Haggard HM, Palle E, Cowan NB (2016) Inferring planetary obliquity using rotational and orbital photometry. MNRAS 457:926
Sirbu D, Thomas S, Belikov R, Bendek E (2017) Techniques for high-contrast imaging in multi-star systems. II. Multi-star wavefront control. Astrophys J 849(142):13
Snellen IAG, de Mooij EJW, Albrecht S (2009) The changing phases of extrasolar planet CoRoT-1b. Nature 459:543
Sterzik MF, Bagnulo S, Palle E (2012) Biosignatures as revealed by spectropolarimetry of Earthshine. Nature 483:64
Tarter JC, Agrawal A, Ackermann R et al (2010) SETI turns 50: five decades of progress in the search for extraterrestrial intelligence. In: SPIE, vol 7819, id. 781902
Turyshev SG, Toth VT (2017) Diffraction of electromagnetic waves in the gravitational field of the Sun. Phys Rev D, 96, id. 024008
Turyshev SG, Shao M, Alkalai L et al (2018) Direct multipixel imaging & spectroscopy of an ex-oplanet with a solar gravity lens mission. The final report. A NASA innovative advanced concepts (NIAC) phase I grant, arXiv:1802.08421 [gr-qc]
Wang J, Mawet D, Ruane G, Hu R, Benneke B (2017) Observing exoplanets with high dispersion coronagraphy. I. The scientific potential of current and next-generation large ground and space telescopes. Astron J, arXiv:1703.00582 (in press)

Part V
The Community

Participants and Initiatives in Planetary Cartography

Andrea Naß and Henrik Hargitai

Abstract Planetary cartography forms an integral part of cartography and has found its manifestation in governmental activities, community efforts, professional organizations, and commercial activities. These activities substantiated with the revival of planetary exploration in the early 2000s when Europe visited the Moon for the first time, and the USA launched a number of exploration missions. With the success of Asian spacecraft missions to the Moon and Mars joining the global planetary exploration endeavor, planetary mapping is increasingly becoming a global collaborative effort, and planetary cartography being one of its main tools to accomplish the goals. This chapter introduces institutes and groups working in the field of planetary cartography and mapping. Some of them have a long history in planetary cartography, while others represent more recent efforts. Their activities are usually organized on a national level, but they are internationally related to each other through research cooperation and collaborative projects. This overview should provide a cross section covering main institutions as well as groups and initiatives. However, it does not qualify to be complete in listing all active organizations and groups.

Keywords Planetary cartography · Mapping centers · Organizational background · Active mapping · Astrogeology

1 Introduction

For centuries, planetary cartography was limited to be the profession or hobby of few astronomers. In the twentieth century, International Astronomical Union (IAU) began actively shaping this field, while during the space race, its centers

A. Naß (✉)
German Aerospace Center (DLR), Rutherfordstr. 2, 12489 Berlin, Germany
e-mail: andrea.nass@dlr.de

H. Hargitai
Eötvös Loránd University, Múzeum krt 6-8, Budapest 1088, Hungary
e-mail: hargitaih@caesar.elte.hu

© Springer Nature Switzerland AG 2019

H. Hargitai (ed.), *Planetary Cartography and GIS*, Lecture Notes in Geoinformation and Cartography, https://doi.org/10.1007/978-3-319-62849-3_21

moved to military and then to American and Soviet government agencies, universities, research institutions, and other international organizations. Today, when maps and map components can be shared globally, planetary cartographic work is done by national and international organizations in the USA and several countries in Europe and Asia. But mapping projects are still initiated by individual scientists devoted to this field. They may closely collaborate with mission scientists either to support landing site selection or to utilize returned data (see Chapter "Cartography: Its Role and Interdisciplinary Character in Planetary Science"). In this chapter, we introduce entities that work on planetary cartography, especially map production and planetary research that involves mapping. We acknowledge that there are many more participants involved in planetary spatial infrastructure, from developing software to defining standards, handling geodetic issues, producing image and elevation models, refining control networks, planning and operating mission instruments, validating data, maintaining online databases and archives, coordinating collaborations, organizing workshops, just to name a few. The chapter could therefore not provide a complete and global listing, but rather an overview that highlights some of the efforts, to inform the reader who may seek to find the most appropriate research environment that best fits their short- or long-term research goals.

2 Agencies, Institutes, Universities, and Facilities Across the World

2.1 USA

2.1.1 National Aeronautics and Space Administration—NASA

Several NASA centers support active planetary mapping groups in many ways. NASA's *Planetary Science Research Program* finances mapping proposals submitted by any US-based researcher, within the appropriate *Data Analysis Program* for science-driven proposals, and the *Planetary Data, Archiving, & Tools Program* for reference mapping projects. NASA's global system of *Regional Planetary Image Facilities* (RPIFs), established in 1977, stores and makes accessible photographic and digital planetary images and also maps (Hagerty et al. 2017) and is present in several US institutions, in Canada, the UK, France, Germany, Italy, Israel, and Japan.

2.1.2 Astrogeology Science Center—ASC/USGS

The *Astrogeology Science Center (ASC)* was established in Flagstaff, Arizona, on July 1, 1963, as a research facility of the *United State Geological Survey (USGS)* (Schaber 2005) through the efforts and requests of several geoscientists and cartographers, perhaps most notably geologist Shoemaker and Hackman (1961). The USGS and NASA agreed on the benefits of a research center that focused on compiling planetary maps, developing observational instrumentation, and training both astronauts and fellow researchers. Therein, a location in Flagstaff proved to be advantageous as it built on the existing resident planetary research community, proximity to Moon observing telescopes, as well as a geologically diverse, lunar-like yet highly accessible terrain. The USGS ASC has evolved over the past five decades in response not only to the changing needs of NASA and the planetary science community but also to the increased volume and diversity of modern, technologically advanced data sets acquired for planetary bodies. By doing so, within the *Planetary Geologic Mapping Program*[1] (founded by NASA and coordinated by the USGS ACS) planetary maps and cartographic products were produced that reveal topography, geology, geomorphology, image mosaics, and more. The aim of this program is to support the international research community with high-quality peer-reviewed geologic maps of planetary bodies. To accomplish this and to have comparable map results, the mapping process has been standardized and is coordinated from its beginning (input data) up to the final map layout, printing, publishing, and archiving (Tanaka et al. 2011; Skinner 2015). All the resulted products are available to the international scientific community and the general public.[2]

In order to handle these tasks effectively, i.e., to ensure that unnecessary duplication is reduced to a minimum, cooperation is critical, wherein multiple institutions and organizational bodies must cross-collaborate. Thus, the USGS ACS established cooperation with institutions including NASA (at multiple programmatic levels), *European Space Agency* (ESA), and *Jet Propulsion Laboratory* (JPL) but also community organizations such as the *International Astronomical Union* (IAU) and MAPSIT.

The USGS ASC is active in developing GIS tools[3] and tools for standards that ensure the interoperability between planetary and terrestrial GIS systems (Hare et al. 2015a, 2017). The Astrogeology *Mapping, Remote-Sensing, Cartography, Technology, and Research* (MRCTR, pronounced "Mercator") GIS Lab[4] produced a number of WebGIS tools (Hare and Tanaka 2001; Hare et al. 2014). The *Planetary Image Locator Tool (*PILOT, Bailen et al. 2013)[5] provides user-friendly access to raw spacecraft images from the *Planetary Data System (PDS)—Cartography and Imaging Science Node* and uses the *Unified Planetary Coordinate*

[1]For more information, see http://planetarymapping.wr.usgs.gov/.

[2]http://astrogeology.usgs.gov/.

[3]https://astrogeology.usgs.gov/facilities/mrctr/gis-tools.

[4]https://astrogeology.usgs.gov/facilities/mrctr-gis-lab.

[5]http://pilot.wr.usgs.gov/.

(UPC, Akins et al. 2009) database with improved geometric and positional information; the *Map Projection On the Web* (POW, Hare et al. 2013)[6] produces calibrated, science-ready images; and in *Map-a-Planet 2*[7] users can browse and work with global image mosaics. The USGS ASC also organizes workshops (Planetary GIS[8] and Planetary Data Workshop[9]) to train mappers at the USGS/NASA Planetary Photogrammetry Guest Facility (Kirk et al. 2009).

Furthermore, USGS publishes outreach maps for the general public (Hare et al. 2015b), and nomenclature reference maps in collaboration with the IAU WGPSN.

2.1.3 Other Research Facilities and Universities in the USA

In addition to the already mentioned institutions, planetary maps are produced in the following places: *Arizona State University* (ASU), Tempe, AZ; *Northern Arizona University* (NAU), Flagstaff, AZ; *Lunar Planetary Laboratory* (LPL), University of Arizona (UA), Tucson, AZ; *Planetary Science Institute* (PSI), Tucson, AZ; *Jet Propulsion Laboratory* (JPL), Pasadena, CA; *SETI Institute*, Mountain View, CA; *University of Colorado Boulder* (CU Boulder), Boulder, CO; *Southwest Research Institute* (SwRI), Boulder, CO; *Smithsonian Institution*, Washington, DC; *University Hawaii* (UH), Honolulu, HI; *University of Minnesota —Duluth* (UMD), Duluth, MN; *Cornell University*, Ithaca, NY; *Lunar and Planetary Institute* (LPI), Houston, TX.

WebGIS platforms that enable viewing, editing, and analyzing spatial data play an important role internationally. These tools include *JMARS/JVesta/JMoon,*[10] and *LunaServ,*[11] both projects led by Arizona State University—ASU, and the Web Map Tiling Service (WMTS) *Trek* portals (*Vesta/Moon/MarsTrek),*[12] developed at NASA's Jet Propulsion Laboratory (JPL) and NASA Ames Center. *ACT-REACT Quickmap* by ACT Corporation is used in ASU's *Lunar LRO WebGIS*[13] and the Mercury Quickmap[14] developed by the MESSENGER team at the Johns Hopkins University Applied Physics Laboratory (APL) in Laurel, MD. USGS and ASU also maintain their multilayer *Web Map Services* (WMS) that give access to georeferenced data sets directly from GIS applications.

[6]https://astrocloud.wr.usgs.gov/.

[7]https://astrogeology.usgs.gov/tools/map-a-planet-2.

[8]https://astrogeology.usgs.gov/facilities/mrctr/gis-tutorials.

[9]https://astrogeology.usgs.gov/groups/planetary-data-workshop.

[10]https://jmars.asu.edu/.

[11]http://lunaserv.lroc.asu.edu/.

[12]https://marstrek.jpl.nasa.gov/.

[13]http://target.lroc.asu.edu/q3/.

[14]https://messenger.quickmap.io/.

2.2 *Europe*

2.2.1 European Space Agency—ESA

The European Space Agency (ESA) only supports missions, and it does not maintain a dedicated group of facility responsible for coordinating or supporting direct planetary geologic map production and publication in Europe. ESA maintains an online data archive for its mission data called *Planetary Science Archive* (PSA) (Besse et al. 2017), which is compatible and comparable to NASA's Planetary Data System. ESA carried out a planetary mapping pilot program for its Mars Express mission (ESA 2010). *ESA's Space Astronomy Centre* (ESAC) supported its first Planetary GIS workshop in 2015 (Manaud et al. 2016a).

2.2.2 German Aerospace Center—DLR

In Germany, the *Institute of Planetary Research* at the *German Aerospace Center (DLR)* in Berlin—among other groups in Germany—has been involved in a number of planetary cartographic projects since its establishment in the early 1990s. A series of Mars maps were compiled in the early 1990s (Hiller et al. 1993; Neugebauer and Dorrer 1996) in preparation of the Russian *Mars-96* mission. Since that time, the main focus of DLR's cartography has been put on the generation of image mosaics and topographic maps. Maps based on the *High-Resolution Stereo Camera* (HRSC) onboard *Mars Express* (Jaumann et al. 2007) have been produced in close collaboration with groups at universities such as TU of Berlin, FU Berlin, TU Dresden, TU Munich, or University Hannover. Recent products include those based on the integration of data from multiple orbits (Gwinner et al. 2016), controlled images and ortho-image mosaics for Phobos (Wählisch et al. 2010, 2014; Willner et al. 2010, 2014), the icy Saturnian satellites[15] (Roatsch et al. 2009) and global mosaics of Ceres and Vesta that serve as base map for different atlas collections[16] (e.g., Roatsch et al. 2013, 2016).

During the last ten years, DLR has also been focusing on GIS-based mapping and tasks of processing, analyzing, archiving, and visualizing scientific results (e.g., Deuchler et al. 2004; Saiger et al. 2005). Topics of interests also include standardized cartographic visualization of scientific map results in order to create homogenous and comparable maps and data archiving products (e.g., van Gasselt and Nass 2011; Nass et al. 2011). To provide sophisticated user experiences, which satisfies scientific as well as public outreach purposes, first steps have been undertaken to set up *WebMap Services* and *Web Graphics Library* (WebGL) applications by using image data and maps from the *Apollo-17* and *LRO* missions (Clever 2014; Maslonka 2014).

---

[15]http://ciclops.org/maps/ and http://pds-imaging.jpl.nasa.gov/.

[16]http://dawngis.dlr.de/atlas (Dawn GIS web page).

2.2.3 Other Research Facilities and Universities in Europe

In Germany, besides DLR, work in planetary mapping is done at the *Technische Universität* (TU) and the *Freie Universität* (FU) of Berlin, the Institute for Cartography of the *Technische Univerität* (TU) Dresden, *Technische Univerität* (TU) Munich, and the *University Hannover*. This involves, e.g., production of image and topographic maps, definition of a large-scale quadrangle schema for cartographic representation of Mars (Lehmann et al. 1997; Albertz et al. 2005), or automatization of map generation processes (*PIMap*, Gehrke et al. 2006).

In the United Kingdom, *University College London's* Centre for Planetary Sciences and Mullard Space Science Laboratory[17] is active in planetary mapping.

With Italian key instruments onboard BepiColombo for observing Mercury, members of several Italian scientific centers and universities joined the planetary geologic mapper's community (Galluzzi et al. 2016) including those in Rome, Perugia, Padova, and Napoli. Research activities at the *International Research School of Planetary Sciences* (IRSPS, located in the Università G.D'Annunzio, Pescara, Italy) include mapping of Mars, Venus, and Titan.

The project *Planetary Geographic Information System* (PAGIS), run by an Italian consortium that includes IRSPS and the Italian Space Agency (ASI), aims "to provide coordinated data analysis and data dissemination"[18] of planetary geological maps. The geomorphic map of Ares Vallis, subtitled "ASI Planetary Map Series—Map No. 1," was published in the Italian Journal of Geoscience (Pacifici 2008), without follow-up.

The Italian–Chinese (*Italian Space Agency*–ASI; *Chinese National Remote Sensing Center of China*–NRSCC) project "Moon Mapping" (Scaioni et al. 2016) was established in 2014 and promotes cooperation and exchange between undergraduate students.

2.2.4 Inter-organizational Groups

Several Lunar and Mars mapping projects involved groups of scientists from multiple institutes over the last decades of the twentieth century. Some projects are international. The grid-based mapping project runs at the *International Space Science Institute* (ISSI) as a collaborative project of nine European countries and the USA. This project is led by *The Open University*, UK (Ramsdale et al. 2017). Other European institutions include: the *Arctic Planetary Science Institute* (APSI), Finland[19]; the *University of Münster*, Germany (e.g., Hiesinger et al. 2013); the *Universidad Rey Juan Carlos,* Madrid, Spain (e.g., López and Hansen 2015); the

[17]https://www.ucl.ac.uk/mssl/imaging/research.

[18]http://www.planetek.it/eng/projects/pagis.

[19]http://planetaryscience.fi/.

Cosmic Materials Space Research Group of the Eötvös Loránd University, Hungary, with a number of outreach map projects (Hargitai et al. 2008); and the *Wroclaw Group for Planetology Advance* (WROONA) of the Polish Academy of Sciences, Wroclaw, Poland[20] (e.g., Mège et al. 2014) with a geomorphic mapping project of Valles Marineris (Dębniak et al. 2017).

2.3 Russia

2.3.1 Moscow State University of Geodesy and Cartography—MIIGAiK

MIIGAiK established its Fundamental Research Laboratory at the Department of Aerial Photo Surveying in 1961. The laboratory focused on photographic lunar image processing and mapping through the 1970s and the production of the Atlas of the Terrestrial Planets and their Moons in the 1980s, published in 1992. Since 1999 MIIGAiK, Dresden Technical University (Germany), Eötvös Loránd University (Hungary), and the University of Western Ontario (Canada) have participated in the project Multilingual Maps of Terrestrial Planets and their Moons (Shingareva et al. 2005).

Projects at the *Extraterrestrial Laboratory (MExLab)* of MIIGAiK focus on planetary geodesy and GIS-based cartographic methods for Solar System bodies (Karachevtseva et al. 2016; Shevchenko et al. 2016). Research topics include the determination of fundamental parameters of celestial bodies such as shape, rotational parameters, and forced librations as well as planetary coordinate systems. Recent products include the Atlas of Phobos (Savinykh et al. 2015; Karachevtseva et al. 2015; Oberst et al. 2014), a map and a globe of Mercury and wall maps. One of the research branches is devoted to the cartographic support of future lunar missions (landing site selection for Luna-25, 27–28) as well as planning of a future orbital mission (Luna-26). The results of these studies are stored in MExLab's planetary geodatabase that is accessible through MExLab's Planetary Geoportal.[21]

2.3.2 Institutes of the Academy of Sciences—IKI, GEOKhI

The *Space Research Institute* (IKI) of the *Academy of Sciences of the USSR*, named the *Russian Academy of Sciences* (RAS) since 1991, was involved in the production of the hypsometric globes of the Moon and Venus. The *Geological Institute* of the Academy of Sciences compiled geological–morphological maps of the Moon (Rodionova 2000).

[20]http://dmzone.org/wroona/.

[21]http://cartsrv.mexlab.ru/geoportal.

The *Vernadsky Institute of Geochemistry and Analytical Chemistry* (GEOKhI) of the Academy of Sciences produced the maps of the surface of Venus based on Venera 15 and 16 data in 1989, published by the *General Directorate of Geodesy and Cartography* (GUGK 1988), and continues to be involved in the USGS-coordinated geological mapping of Venus (e.g., Ivanov and Head 2011). The Russian localization of the planetary nomenclature has been coordinated by G. A. Burba (e.g., Burba 1984) at GEOKhI.

A team of scientists from the *Institute of Geography GIS Research Laboratory* of RAS and *MSU* are involved in the development of cartographic projections for triaxial ellipsoid, which can be applied as a reference surface not only for small Solar System bodies but also for satellites like Phobos and larger celestial bodies. (Nyrtsov et al. 2015; Fleis et al. 2013).

2.3.3 Sternberg Astronomical Institute—SAI/GAISh

The Department of Lunar and Planetary Investigations of the *Sternberg Astronomical Institute (SAI or GAISh) of the Lomonosov Moscow State University* (MSU/MGU) was founded by S. Korolev. The first major planetary project here was the production of the first map and atlas of the far side of the Moon in 1961 led by Yu. Lipsky (Lipsky 1967; Rodionova et al. 2014; Shevchenko et al. 2016). The Institute recently published a number of hypsometric planetary maps using NASA's data (Rodionova et al. 2014). Further projects include the morphologic study of craters and the Russian language localization of the planetary nomenclature (e.g., Shevchenko et al. 2016). The *Department of Geography* of MSU has a special course in "The Mapping of Extraterrestrial Objects" where students are involved in the production of planetary maps, together with SAI (Rodionova et al. 2014).

2.4 Asia (without Russia) and Middle East

2.4.1 Chinese Academy of Sciences—CAS

Driven by China's Lunar exploration missions Chang'E-1, Chang'E-2, and Chang'E-3, lunar cartography work has been undertaken by several institutions of the *Chinese Academy of Sciences (CAS)*, including the *National Astronomical Observatories of China* (NAOC), *Institute of Geochemistry* (IGCAS), and the *Institute of Remote Sensing and Digital Earth* (RADI). NOAC established the *Science and Application Center for Lunar and Deep Space Exploration* in 2003. One of its major responsibilities is to produce global cartographic products from data acquired by the Chang'E-1 and Chang'E-2. NAOC released several lunar maps: a global ortho-image and DEM mosaic using orbiter imagery data of

Chang'E-1 (Li et al. 2010a); a global lunar digital elevation model (DEM) using altimetry data of Chang'E-1 (Ping et al. 2009; Li et al. 2010b); a high-resolution global DEM and ortho-image using stereo imagery data from Chang'E-2 (Li et al. 2015); and high-precision DEMs and ortho-images of the Chang'E-3 landing site using data from the lander (descent images) and the rover (stereo images) (Liu et al. 2014). IGCAS established the *Center for Lunar and Planetary Sciences* in 2005 with focus on geological mapping using the same data sets. RADI established the *Planetary Mapping and Remote Sensing Laboratory* in 2008. This laboratory works on the development of high-precision planetary mapping methods using multi-source data and has produced high-resolution topographic products of the Chang'E-3 landing site using orbital and rover images.

2.4.2 Other Universities and Other Research Facilities in Asia and Middle East

In China, universities have also been involved in the analysis of Chinese lunar data, including the *Hong Kong Polytechnic University* (high-resolution DEM and ortho-image map of Chang'E-3 landing area by Chang'E-2 stereo imagery and LOLA data, Wu et al. 2014) and the *Space Science Institute/Lunar and Planetary Science Laboratory of the Macau University of Science and Technology* (Chang'E-1 mapping) (Zheng et al. 2012).

Countries with developing planetary research infrastructure are planning planetary missions with mapping capabilities, such as the *Indian Space and Research Organization* (ISRO) (ISRO 2015) and the *Japanese Aerospace Exploration Agency* (JAXA) (Demura et al. 2006). Others are in the stage of developing future missions, such as a mission to the Moon by the *Korea Aerospace Research Institute* (KARI) in South Korea (Kim et al. 2016), and future missions to Mars by the *Mohammed bin Rashid Space Centre* and the *UAE Space Agency* in the United Arab Emirates (Sharaf et al. 2017).

3 Organizations and Initiatives

Planetary cartography has found its manifestation in governmental activities, community efforts, professional organizations, and, in recent years, private activities. That activity substantiated with the revival of planetary exploration in the early 2000s when Europe visited the Moon for the first time, and the USA launched a number of exploration missions. With the success of Asian spacecraft missions, planetary cartography is increasingly becoming a global collaborative effort.

3.1 International Astronomical Union—IAU

IAU was founded in 1919 partly to oversee assigning names for stars, planets, moons, asteroids, comets, and surface features on them (Blaauw 1994). The first goals were to normalize various systems used in Lunar and Martian nomenclatures across different countries (Blagg and Müller 1935). The updated nomenclature database and the Gazetteer of Planetary Nomenclature are maintained by the USGS ASC on behalf of the IAU *Working Group for Planetary System Nomenclature* (WGPSN). The map archive of WGPSN is also maintained at USGS ASC in Flagstaff, AZ.

In 1976, the IAU established a working group on the *Cartographic Coordinates and Rotational Elements of Planets and Satellites* that reports triennially on the preferred parameters of the rotation rate, spin axis, prime meridian, and reference surface for planets and satellites (Archinal et al. 2011) to allow consistent data usage. However, the IAU's oversight does not cover other standards essential for digital planetary mapping including common feature attributions, feature symbols, recommended mapping scales and metadata or the documentation of the data. For the USA, this role has been filled by the *Federal Geographic Data Committee (FGDC)* that defines feature attributes and their assigned symbols for terrestrial and planetary maps (FGDC 2006; Hare et al. 2017). Planetary map symbols are primarily based on the same set of attributes and symbols as those used in regurlar terrestrial mapping (see, e.g., Nass et al. 2011; Hargitai and Shingareva 2011; Hargitai et al. 2014).

3.2 International Society for Photogrammetry and Remote Sensing—ISPRS

The ISPRS inter-commission working group III/II *Planetary Remote Sensing and Mapping*[22] was transformed from the previous working group *IV/8—Planetary Mapping and Spatial Databases* in 2016. This group is built by the community to provide a platform for those involved in all topics of planetary cartography, such as data acquisition, processing, and information extraction from planetary remote sensing data. This also includes the evaluation and refinement of reference systems, coordinate systems, control networks, map sheet definitions, etc., and their standardization. The group organizes workshops and symposia on planetary mapping, cartography, and remote sensing and promotes international cooperation since 1998.

[22]http://www2.isprs.org/commissions/comm3/icwg-3-2.html.

3.3 International Cartographic Association—ICA

The *Commission on Planetary Cartography (CPC) of* ICA was established with the goal of disseminating products and initiating outreach and professional projects in countries where planetary cartographic materials are scarcely available or altogether absent. The commission also focuses on supporting planetary mapping with various databases, and providing information on international activities.

Shortly after the establishment of the ICA *Working Group on Planetary Cartography* in 1995 (the forerunner of the ICA CPC, founded in 1999), upon request of Kira Shingareva, Manfred Buchroithner, and Jim Zimbelman (joined as cofounder in 1999) at the 17th International Cartographic Conference 1995 in Barcelona both European home institutions, the Moscow State University of Surveying and Mapping (MIIGAiK) and TU Dresden, began to produce and publish a multilingual series of planetary maps. Explanatory notes on the back page and legends on the front page were given in English, French, German, Spanish, and Russian. This first-ever multilingual planetary map series covering Mars, Venus, Moon, Mercury, Io as well as Phobos and Deimos (Buchroithner; Krasnopevtseva et al. 2007). The commission members produced three multilingual outreach map series: the series edited in Dresden (Buchroithner), a Central European edition (Shingareva et al. 2005), and a special series for children (Hargitai et al. 2015). Other projects include the Multilingual Glossary of Planetary Cartography (Shingareva and Krasnopevsteva 2011); the GIS-ready Integrated Database of Planetary Features (Hargitai 2016); an EPO planetary cartographic application that can be used to compare sizes of planetary features to countries and states (Gede and Hargitai 2015, 2017); the Digital Museum of Planetary Mapping[23]; and the International Catalog of Planetary Maps (Hargitai and Pitura 2018).

3.4 Other Initiatives

In addition to institutional and organizational efforts, a number of initiatives have arisen in recent years motivated not only by individuals but also by commercial entities that specialize in combining planetary spatial data with Web technologies. Today, different startups and organizations offer platforms and preexisting cartographic databases which often feature open-source mapping technologies at their core. It has made it even easier for non-GIS specialists, researchers, or data enthusiasts to manipulate, visualize, and share their data and maps on the Web (Zastrow 2015).

[23]http://planetarymapping.wordpress.com.

3.4.1 European Initiatives

Europlanet is a European initiative to connect and support the European planetary scientist community. It operates since 2005 and receives funding every four years from the European Union. Its consortium consists of European research institutions, organizations, and universities coordinated by the Open University, UK, with the aim "to overcome fragmentation and to share resources in Europe's planetary science community."[24]

In 2017, the first Workshop for *Planetary Mapping and Virtual Observatory* was organized within the *Virtual European Solar and Planetary Access (VESPA)* and *Europlanet* programs to bring together the geologic, geospatial, and *Virtual Observatory (VO)* communities for focusing on forward knowledge, tools, and standards for mapping the Solar System.

A number of projects funded by the *European Commission* are directly or indirectly relevant to planetary mapping. The largest, and one of the most long-lasting efforts, is *VESPA* (Erard et al. 2014, 2016), the Europlanet H2020 Research Infrastructure component that deals with accessibility and distribution of planetary science data from very diverse scientific domains, including a specific surface mapping task (Rossi et al. 2015). A three-year-long H2020 project (2018–2021) finances the geological mapping of Mars, Mercury, and the Moon. This project is coordinated by the University of Padova with participants in Italy, the UK, Germany, and France, (Cordis 2018), following USGS protocol (Open University 2018). E-Infrastructure projects with broad Earth Science focus, such as EarthServer 2, include a planetary science component, *PlanetServer* (Baumann et al. 2015; Rossi et al. 2016). Rather than focusing on data searches and discovery or on-demand processing, *PlanetServer* primarily uses the OGC WCPS standard (Baumann 2010) to perform real-time data analytics (Rossi et al. 2016; Marco Figuera et al. 2015).

The *OpenPlanetary*[25] initiative, created in 2015 (Manaud et al. 2016b), provides an online framework to help collaborate on common planetary mapping and data analysis problems. Furthermore, it focused on building the first Open Planetary Mapping and social platform for researchers, educators, storytellers, and the general public (Manaud et al. 2017).

CARTO, a company focused on Web-based geospatial data visualization and analysis, collaborated with ESA to support an open-source outreach project intended to raise public awareness of ESA's ExoMars rover mission through an interactive map of the candidate landing sites (*Where On Mars?,*[26] Manaud et al. 2015).

[24]http://www.europlanet-2020-ri.eu/.

[25]openplanetary.co.

[26]whereonmars.co.

3.4.2 American Strategic Planning

The NASA-recognized *Mapping and Planetary Spatial Infrastructure Team (MAPSIT)*[27] was established as a result of NASA and USGS efforts in 2014 to serve as a planetary cartography strategic planning group to oversee issues related to the "planetary spatial data infrastructure" (PSDI) (NSF 2012; Lawrence et al. 2016; Laura et al. 2017; Gaddis 2017), within NASA's Planetary Science Division. Spatial data infrastructure covers "the technology, policies, standards, and human resources necessary to acquire, process, store, distribute, and improve utilization of geospatial data" (Federal Register 1994). MAPSIT's goals include enabling the best possible and most useful science return from missions; planning data acquisition, data processing, and cartographic products suitable for planetary exploration; lowering costs. Since 2017, its *Steering Committee* assists the USGS in issues related to planetary mapping planning and implementation (e.g., Archinal 2014; Williams 2016).

3.4.3 Citizen Scientist Projects

Citizen scientist projects with clear planetary mapping aims also have been developed starting with *Clickworkers* where users analyzed Mars image data (Kanefsky et al. 2001). Some of the citizen scientist projects are embedded in a broader context, such as *iMars* (Muller et al. 2016); others originate from experiment-driven effort, involving users in the detection, measurement, and classification of surface features, such as in the *CosmoQuest* (Robbins et al. 2012) and Zooniverse projects that include *PlanetFour* (Aye et al. 2016) and *Moon Zoo* (Joy et al. 2011).

3.5 Private/Commercial Companies and Individuals

The *National Geographic Society* has produced, and from time to time updated, maps of Mars and the Moon, as supplements to the National Geographic magazine.

Google's Google Earth Mars/Moon modules are popular planetary geospatial browsers, which were created in cooperation between NASA Ames Research Center and Google, Inc. In addition to provide a platform for public outreach and education, Google's planetary geobrowser is designed to allow scientists to rapidly browse and locate features and data sets, and share data within the community (Hancher et al. 2009). Google's Moon was released in 2007 (Mewhinney 2007), and Google Mars was released in 2009 and updated in 2012. In 2016, Google added 16 new planetary bodies to explore (Marquardt 2017).

[27]http://www.lpi.usra.edu/mapsit/.

Map manufacturer companies that produce planetary and celestial maps or globes include *ZES Brno* (maps, Czechia), *Replogle* (globes, USA), and *Global Imagination* (digital video globes, USA).

In addition to organization-curated maps, individuals, including gamers, designers, artists, amateur astronomers, are increasingly active in creating planetary maps that they usually publish online (Hargitai and Pitura 2018).

4 Meetings and Journals

Members of the planetary cartographic community present their results in several meetings during a year. Active, NASA-funded planetary geologic mappers are required to present their project status in the annual Planetary Mappers' Meeting[28] (Parker et al. 2001). Meetings with planetary cartographic presentations include the *Lunar and Planetary Science Conference* (LPSC, Houston, TX, 1970–), Europlanet's the *European Planetary Science Congress* (EPSC, 2006–), the *American Geophysical Union* (AGU) Fall Meeting (usually in San Francisco, CA), the *American Astronomical Society Division of Planetary Sciences* (AAS DPS) Meeting, the *European Geosciences Union* (EGU) General Assembly, IAU, ISPRS, and ICA meetings and commission symposia.

Major peer-reviewed research journals have slightly different aims and impacts (e.g., Knight and Steinback 2008) and accept manuscripts about different aspects of planetary cartography. Peer-reviewed specialized planetary geologic maps are published in the *US Geological Survey Scientific Investigations Map* series.

Journals focusing on cartographic research include *The International Journal of Cartography* (IJC, 2015–, the official journal of the International Cartographic Association), *The Cartographic Journal* (1964–, British Cartographic Society), and *Cartographica* (1964–, Canadian Cartographic Association). The fully open-access *Journal of Maps* (JoM, 2005–, Taylor and Francis) publishes complete maps with accompanying short papers.

Other journals that welcome manuscripts in the field of planetary mapping include[29] the folowings. *Space Science Reviews'* (SSRv, 1962–, Springer) topics focus on the physics of planetary systems, magnetospheres, etc. The *Earth and Planetary Science Letters* (EPSL, 1966–, Elsevier) publishes concise "letters" about planetary processes and evolution. The *Geophysical Research Letters* (GRL, 1974–, AGU) accepts high-impact short articles with rapid publication. The *Journal of Geophysical Research Planets* (JGR–Planets, 1991–, AGU) is more oriented toward scientific analysis of data, including those obtained through science-driven planetary geologic mapping. *Icarus* (1963–, AAS DPS) accepts manuscripts on

---

[28]https://planetarymapping.wr.usgs.gov/Page/view/Meetings.

[29]In decreasing order of their impact factor in 2016.

observational, experimental, or theoretical research in Solar System studies, including geology and mapping. *Geomorphology* (1987–, Elsevier) welcomes planetary geomorphology-related papers. *Planetary and Space Science* (PSS, 1959–, Elsevier) topics include the history of planetary research, exoplanets, surface morphology, and celestial mechanics and issues concerning the control of planetary cartographic products. *Acta Astronautica* (1974–, International Academy of Astronautics) contributions are in the fields of engineering, life, and social space sciences and include education and public outreach-related subjects. *Astrobiology* (2001–, Mary Ann Liebert, Inc.) topics include mapping of planetary surfaces with and astrobiological potential. *Science China: Earth Sciences* (Science China Press/ Springer) is an English language platform for Chinese planetary research papers, supervised by the Chinese Academy of Sciences. *Photogrammetric Engineering and Remote Sensing* (PE&RS, American Society of Photogrammetry and Remote Sensing) is a journal for imaging and geospatial information science and technology. *Earth, Moon, and Planets* (EMP, 1970–, Springer) focuses on research that includes subjects like planetary systems, small bodies, and impact hazards. *Solar System Research* (2000– Springer) is the English language version of the Russian language *Astronomicheskii Vestnik* (1967–, Pleiades Publishing, Moscow). The open-access *Earth and Space Science* (ESS, 2004–, American Geophysical Union) welcomes technical papers that are focused on useful data sets, observations, mapping/cataloging, modeling and highlights methodology.

Acknowledgements The authors are grateful to Trent Hare, Irina Karachevtseva, Lingli Mu, Maxim V. Nyrtsov, Jürgen Oberst, Teemu Öhmann, Zhanna Rodionova, and Marita Wählisch for their help in improving the manuscript.

References

Akins SW, Gaddis L et al (2009) Status of the PDS unified planetary coordinates database and the planetary image locator tool (pilot). In: 40th Lunar and planetary science conference (LPSC), LPI No. 1903, #2002

Albertz J, Attwenger M et al (2005) HRSC on Mars express—photogrammetric and cartographic research. Photogramm Eng Remote Sens 71:1153–1160

Archinal BA (2014) The Need to Restart NASA Planetary Cartography Planning. NASA Planetary Science Division Presentation 21 October 2014

Archinal BA, A'Hearn MF et al (2011) Report of the IAU working group on cartographic coordinates and rotational elements: 2009. Celest Mech Dyn Astron 109(2):101–135. https://doi.org/10.1007/s10569-010-9320-4

Aye K-M, Schwamb ME et al (2016) Analysis pipeline and results from the PlanetFour Citizen Science Project. In: 47th Lunar and planetary science conference (LPSC), LPI No. 1903, #3056

Bailen MS, Sucharski RM et al (2013) Using the PDS planetary image locator tool (pilot) to identify and download spacecraft data for research. In: 44th Lunar and planetary science conference (LPSC), LPI No. 1903, #2246

Baumann P (2010) The OGC web coverage processing service (WCPS) standard. Geoinformatica 14(4):447–479

Baumann P, Mazzetti P et al (2015) Big data analytics for earth sciences: the earthserver approach. Int J Digit Earth https://doi.org/10.1080/17538947.2014.1003106

Besse S, Vallat C et al (2017) ESA's planetary science archive: preserve and present reliable scientific data sets. Planet Space Sci. https://doi.org/10.1016/j.pss.2017.07.013

Blaauw A (1994) History of the IAU: the birth and first half-century of the IAU. Kluwer Academic Publishers, Dordrecht. ISBN 0-7923-2979-1

Blagg MA, Müller K (1935) Named Lunar Formations, 196 p. International Astronomical Union, Percy Lund, Humphries and Co. Ltd., London

Buchroithner MF (1999) Mars map: the first of the series of multilingual relief maps of terrestrial planets and their moon. In: Proceedings 19th International cartographic conference, Ottawa, Canada, 14–21 August 1999

Burba GA (1984) Nomenklatura Detaley Relyefa Galileevych Sputnikov Jupitera. Nauka, Moskva

Clever S (2014) LROC—Anaglyphen vom Mond: Erstellung und Visualisierung in einem Webmapping Projekt. Ba. thesis, Beuth Hochschule für Technik Berlin

Cordis (2018) PLANMAP Project ID: 776276. https://cordis.europa.eu/project/rcn/212904_en.html

Dębniak K, Mège D, Gurgurewicz J (2017) Geomorphology of Ius Chasma, Valles Marineris, Mars. J Maps 13(2):260–269. https://doi.org/10.1080/17445647.2017.1296790

Demura H, Kobayashi S et al (2006) Pole and global shape of 25143 Itokawa. Sci 312. https://doi.org/10.1126/science.1126574

Deuchler C, Wählisch M et al (2004) Combining Mars data in GRASS GIS for geological mapping. In: International Society for Photogrammetry and Remote Sensing conference (ISPRS), Istanbul, Turkey, #455

Erard E, Cecconi B et al (2014) Planetary science virtual observatory architecture. Astron Comput 7:71–80

Erard S, Cecconi B et al (2016) VESPA: developing the planetary science virtual observatory in H2020. In: European geoscience union conference (EGU), vol 18, #2016-17527

ESA (2010) The Mars Planetary Mapping Pilot Project. Final report. http://www.gispla.net/docs/Final_PMPP_Report_Issued.pdf

Federal Register (1994) Executive Order 12906 of April 11, 1994, Coordinating Geographic Data Acquisition and Access: The National Spatial Data Infrastructure. Federal Register 59(71)

FGDC (2006) Digital Cartographic Standard for Geologic Map Symbolization, Federal Geographic Data Committee, Document # FGDC-STD-013-2006

Fleis ME, Nyrtsov MV, Borisov MM (2013) Cylindrical projection conformality of triaxial ellipsoid. Dokl Earth Sci 451(1):787–789

Gaddis L (2017) The Role of NASA's Planetary Data System in the Planetary Spatial Data Infrastructure Initiative. In: American Geophysical Union (AGU) Fall Meeting, #225497

Galluzzi V, Guzzetta L et al (2016) Geology of the Victoria quadrangle (H02). Mercury J Maps 12:227–238. https://doi.org/10.1080/17445647.2016.1193777

Gede M, Hargitai H (2015) Country movers—an extraterrestrial geographical application. In: Reyes NJ (eds) Cartography beyond the ordinary world, joint ICA symposium, pp 178–183

Gede M, Hargitai H (2017) An online planetary exploration tool: "country movers". Acta Astronaut. https://doi.org/10.1016/j.actaastro.2017.04.028

Gehrke S, Wählisch M et al (2006) Generation of topographic and thematic planetary maps using the software system "PIMap". In: 37th Lunar and planetary science conference (LPSC), LPI No. 1903, #1322

GUGK (1988) Map series of Venus based on Venera 16 and Venera 17 observations. GUGK SSSR

Gwinner K, Jaumann R et al (2016) The high resolution stereo camera (HRSC) of Mars express and its approach to science analysis and mapping for Mars and its satellites. Planet Space Sci (PSS) 126:93–138

Hagerty JJ, Mouginis-Mark et al (2017) The NASA regional planetary image facility network: a globally distributed resource for the planetary science community. In: Planetary science vision 2050 workshop 2017, LPI Contribution No. 1989, #8063

Hancher MD, Beyer R et al (2009) Visualizing Mars data and imagery with Google earth. In: 40th Lunar and planetary science conference (LPSC), LPI No. 1903, #2308

Hare TM, Tanaka KL (2001) Planetary Interactive GIS-on-the-Web Analyzable Database (PIGWAD). In: 20th International Cartography conference (ICC), Beijing, China

Hare TM, Akins SW et al (2013) Map projection web service for PDS images. In: 44th Lunar and planetary science conference (LPSC), LPI No. 1903, #2068

Hare TM, Keszthelyi L et al (2014) Online planetary data and services at USGS astrogeology. In: 45th Lunar and planetary science conference (LPSC), LPI No. 1903, #2487

Hare TM, Skinner JA et al (2015a) Planetary GIS at the U.S. geological survey astrogeology science center, 2nd planetary data workshop, held Flagstaff, Arizona, 8–11 June 2015. LPI No. 1846, 2015, #7005

Hare TM, Hayward RK et al (2015b) Image mosaic and topographic map of the moon: U.S. geological survey scientific investigations map 3316, 2 sheets. https://doi.org/10.3133/sim3316

Hare TM, Pio Rossi A et al (2017) Interoperability in planetary research for geospatial data analysis. Planet Space Sci (PSS). https://doi.org/10.1016/j.pss.2017.04.004

Hargitai H (2016) Metacatalog of planetary surface features for multicriteria evaluation of surface evolution: the integrated planetary feature database. In: 48th Devision of Planetary Science (DPS)/11th European Planetary Science conference (EPSC), #426.23

Hargitai H, Pitura M (2018) International catalog of planetary maps 1600–2017. In: 49th Lunar and planetary science conference (LPSC), LPI No. 1903, #2083

Hargitai H, Shingareva KB (2011) Planetary nomenclature: a representation of human culture and alien landscapes. Advances in Cartography and Giscience. Lecture notes in geoinformation and cartography, 6, pp 275–288, https://doi.org/10.1007/978-3-642-19214-2_18

Hargitai H, Gucsik A, Okumura T (2008) The new bilingual photomap of Itokawa. In: 3rd planetology seminar, Budapest, 4–5 September 2008. #C2

Hargitai H, Li C et al (2014) Chinese and Russian language equivalents of the IAU gazetteer of planetary nomenclature: an overview of planetary toponym localization methods. Cartogr J. https://doi.org/10.1179/1743277413

Hargitai H, Gede M et al (2015) Multilingual narrative planetary maps for children. In: Robbi SC, Madureira Cruz CB, Leal de Menezes PM (eds) Cartography—Maps Connecting the World. Lecture notes in geoinformation and cartography, pp 17–30, Springer, Cham

Hiesinger H, Ruesch O et al (2013) Geologic map of the northern hemisphere of Vesta based on DAWN FC images. In: 44th Lunar and planetary science conference (LPSC), LPI No. 1903, #2582

Hiller K, Hauber E et al (1993) Digitale Kartenherstellung der Planetenbildkarten Olympus Mons/ Planet Mars. *Kartograph.* Nachrichten, Heft 2 Kirschbaum Verlag, Bonn

ISRO (2015) Mars orbiter mission (MOM) Mars atlas. Space Applications Centre. Indian Research Organisation (ISRO), Ahmedabad, India. https://www.issdc.gov.in/docs/mr1/Mars-atlas-MOM.pdf

Ivanov MA, Head JW (2011) Global geological map of Venus. Planet Space Sci (PSS), 59:1559–1600

Jaumann R, Neukum G et al (2007) The high-resolution stereo camera (HRSC) experiment on Mars express: instrument aspects and experiment conduct from interplanetary cruise through the nominal mission. Planet Space Sci (PSS), 55:928–952

Joy K, Crawford I et al (2011) Moon Zoo: citizen science in lunar exploration. Astron Geophys 52 (2):2–10. https://doi.org/10.1111/j.1468-4004.2011.52210.x

Kanefsky B, Barlow NG, Gulick VC (2001) Can distributed volunteers accomplish massive data analysis tasks? In: 32th Lunar and planetary science conference (LPSC), LPI No. 1903, #1272

Karachevtseva I, Kokhanov A et al (2015) Development of a new Phobos atlas based on Mars express image data. Planet Space Sci (PSS) 108:24–30

Karachevtseva IP, Kokhanov AA et al (2016) Modern methodology and new tools for planetary mapping. In: Gartner G, Jobst M, Huang H (eds) Progress in Cartography, pp 207–227, 480 p. Springer. https://doi.org/10.1007/978-3-319-19602-2

Kim KJ, van Gasselt S et al (2016) Framework of Lunar landing site selection and resource analysis for the 2020 Korean Lunar mission. In: 47th Lunar and planetary science conference (LPSC), LPI No. 1903, #1706

Kirk RL, Howington-Kraus E, Rosiek MR (2009) Build your own topographic model: a photogrammetry guest facility for planetary researchers. In: 40th Lunar and planetary science conference (LPSC), LPI No. 1903, #1414

Knight LV, Steinbach TA (2008) Selecting an appropriate publication outlet: a comprehensive model of journal selection criteria for researchers in a broad range of academic disciplines. Int J Dr Stud 3:59–79

Krasnopevtseva BV, Buchroithner MF, Shingareva KB, Leonenko SM, Fleis ME, Stoke Ph. (2007) The project of multilingual maps series on celestial bodies: current state and near future. In: Proceedings 23rd international cartographic conference (ICC), Moscow, 4–10 August 2007, 6 p. CD-ROM

Laura JR, Hare TM et al (2017) Towards a planetary spatial data infrastructure. ISPRS Int J Geo-Inf 6:181

Lawrence SJ et al (2016) The mapping and planetary spatial infrastructure team (MAPSIT): addressing strategic planning needs for planetary cartography. In: 47th Lunar and planetary science conference (LPSC), LPI No. 1903, #1710

Lehmann H, Scholten F et al (1997) Mapping a whole planet—the new topographic series 1:200 000 for Mars. In: 18th International cartographic conference (ICC), Stockholm, Sweden

Li C, Liu J et al (2010a) The global image of the moon obtained by the Chang'E-1: data processing and lunar cartography. Sci China Earth Sci 53(8):1091–1102

Li C, Ren X et al (2010b) Laser altimetry data of Chang'E-1 and the global lunar DEM model. Sci China Earth Sci 53(11):1582–1593

Li C, Ren X et al (2015) A new global and high resolution topographic map product of the Moon from Chang'E-2 image data. In: 46th Lunar and planetary science conference (LPSC), LPI No. 1903, #1638

Lipsky Yu N (ed) (1967) Atlas of the Far side of the Moon I-II. Nauka, Moskva

Liu Z, Di K et al (2014) High precision landing site mapping and rover localization for Chang'3 mission. Sci China Phys Mech Astron 58(1):19601

López I, Hansen VL (2015) Progress report on the geologic mapping of the 1:10 m Niobe map area, Venus. In: Abstracts of the annual meeting of planetary geologic mappers, HI 2015, Honolulu, 9–10

Manaud N, Boix O et al (2015) Where on Mars? A web map visualisation of the ExoMars 2018 rover candidate landing sites. In: 10th European planetary science conference (EPSC), #2015-228

Manaud N, Rossi AP et al (2016a) Summary and recommendations from the 2015 ESAC planetary GIS workshop. In: 47th Lunar and planetary science conference (LPSC), LPI No. 1903, #1387

Manaud N, Rossi AP et al (2016b) The OpenPlanetary initiative. In: 48th Devision of Planetary Science (DPS)/11th European Planetary Science conference (EPSC), #2587836

Manaud N, Rossi AP et al (2017) OpenPlanetaryMap: building the first open planetary mapping and Social platform for researchers, educators, storytellers, and the general public. In: 3rd planetary data workshop 2017, LPI Contribution No. 1986, #7024

Marco Figuera R, Rossi AP et al (2015) Analyzing Lunar DTMs through web services with EarthServer/PlanetServer-2. In: ISPRS commission VI, WG VI/4 meeting, Berlin

Marquardt S (2017) Space out with planets in Google Maps. https://www.blog.google/products/maps/space-out-planets-google-maps/

Maslonka C (2014) Bestimmung und Kartierung von Stationspunkten der Traverse der Apollo-17-Landestelle auf Basis von Lunar Reconnaissance Orbiter Camera Daten. Ma. thesis, Beuth Hochschule für Technik Berlin

Mège D, Chicarro A et al (eds) (2014) MPSE 2014, Warsaw, 3–5 June 2014. http://dmzone.org/wroona/MPSE2014/AbstractBook_MPSE2014.pdf

Mewhinney M (2007) NASA maps the moon with Google. Astrogram, September https://www.nasa.gov/centers/ames/pdf/190041main_07_09astrogram.pdf

Muller J-P, Tao Y et al (2016) EU-FP7-iMARS: analysis of Mars multi-resolution images using auto-coregistration data mining and crowd source techniques: processed results—a first look. In: International archives of the photogrammetry, remote sensing and spatial information sciences, pp 453–458. https://doi.org/10.5194/isprs-archives-xli-b4-453-2016

Naß A, Hargitai H (this book) Cartography—Its role and interdisciplinary character as intersection in planetary science. In: Hargitai H, Wolak J (eds) Planetary cartography—concepts, tools, and methods. Lecture notes in geoinformation and cartography. Springer

Nass A, van Gasselt S et al (2011) Implementation of cartographic symbols for planetary mapping in geographic information systems. *Planet Space Sci* 59:1255–1264. Special Issue: Planetary Mapping, Elsevier Ltd. https://doi.org/10.1016/j.pss.2010.08.022

Neugebauer G, Dorrer E (1996) Experimentelle Untersuchungen zur kartographischen Darstellung der Marsoberfläche. *Kartographische Nachrichten*, Heft 2 Kirschbaum Verlag, Bonn

NSF (2012) A vision and strategy for science, engineering, and education: Cyberinfrastructure framework for the 21st century. NSF 12-113

Nyrtsov MV, Fleis MM et al (2015) Equal-area projections of the triaxial ellipsoid: first time derivation and implementation of cylindrical and azimuthal projections for small solar system bodies. Cartogr J 52(2):114–124

Oberst J, Zakharov A, Schulz R (eds) (2014) Phobos (special issue). In: Planet Space Sci (PSS), vol. 102, 182 p. Elsevier

Open University 2018 Geological mapping in Mercury's southern hemisphere. https://goo.gl/SZ36fE

Pacifici A (2008) Geomorphological map of Ares Vallis, Mars; ASI planetary map series; map no. 1. Ital J Geosci 127(1):75–92 http://www.neogeo.unisi.it/IWSViewer/socgeol.php?id=B0404

Parker T, Tanaka KL, Hirsch DD (eds.) (2001) Abstracts of the annual planetary geologic mappers meeting, 22–24 June 2000, with a section on field trip notes; geologic controls on select springs near Flagstaff, Arizona, by Bills, D.J.: U.S. Geological Survey Open-File Report 01–018, 53 p. https://pubs.usgs.gov/of/2001/of01-018/

Ping J, Huang Q et al (2009) Lunar topographic model CLTM-s01 from Chang'E-1 laser altimeter. Sci China Ser G: Phys Mech Astron 52(7):1105–1114

Ramsdale JD, Balme MR et al (2017) Grid-based mapping: a method for rapidly determining the spatial distributions of small features over very large areas. Planet Space Sci 140:49–61. https://doi.org/10.1016/j.pss.2017.04.002

Roatsch T, Jaumann R et al (2009) Cartographic mapping of the Icy satellites using ISS and VIMS data. In: Dougherty M, Esposito L, Krimigis S (eds) Saturn from Cassini-Huygens, pp 763–781. Springer

Roatsch T, Kersten E et al (2013) High-resolution Vesta low altitude mapping orbit atlas derived from dawn framing camera images. Planet Space Sci (PSS) 85:293–298. https://doi.org/10.1016/j.pss.2013.06.024

Roatsch T, Kersten E et al (2016) High-resolution Ceres high altitude mapping orbit atlas derived from Dawn framing camera images. Planet Space Sci (PSS) 129:103–107. https://doi.org/10.1016/j.pss.2016.05.011

Robbins SJ, Antonenko I et al (2012) Cataloging the Moon with the CosmoQuest Moon mappers citizen science project. In: 43th Lunar and planetary science conference (LPSC), LPI No. 1903, #2901

Rodionova JF (2000) Lunar maps and space achievements. In: Proceedings of the Fourth International Conference on the Exploration and Utilisation of the Moon. (ESA SP-462, September 2000), pp 11–14

Rodionova Z, Berlyant A et al (2014) Maps of Mars compiled by students at Lomonosov Moscow State University In: Bandrova et al (eds) Thematic cartography for the society. Lecture notes in geoinformation and cartography. Springer International Publishing Switzerland. https://doi.org/10.1007/978-3-319-08180-9

Rossi AP, Cecconi B et al (2015) Planetary GIS and EuroPlanet-RI H2020. In: 10th European planetary science conference (EPSC), #2015-178

Rossi AP, Marco Figuera R et al (2016) Remote sensing data analytics for planetary science with planet/earthserver. In: European geoscience union (EGU), #2016-3996
Saiger P, Wählisch M et al (2005) ArcGIS and GRASS GIS for planetary data. In: 1st Mars Express Science Conference
Savinykh VP, Karachevtseva IP, Konopikhin AA (eds) (2015) The Phobos atlas Moscow. MIIGAiK, 220 p, 43 maps (In Russian)
Scaioni M, Giommi P et al (2016) The 'Moon Mapping' project to promote cooperation between students of Italy and China. In: The International Archives of the Photogrammetry, Remote Sensing and Spatial Information Sciences, vol XLI-B6. https://doi.org/10.5194/isprsarchives-xli-b6-71-2016
Schaber GG (2005) The U.S. geological survey branch of astrogeology—a chronology of activities from conception through the end of project Apollo (1960–1973), USGS Open Report 2005-1190
Sharaf O, Amiri S et al (2017) Emirates Mars mission (EMM) science overview. In: 48th Lunar and planetary science conference (LPSC), LPI No. 1903, #1992
Shevchenko V, Rodionova Z, Michael G (2016) Lunar and Planetary Cartography in Russia. Springer. https://doi.org/10.1007/978-3-319-21039-1
Shingareva K, Krasnopevsteva B (2011) A new version of the multilingual glossary of planetary cartography. In: Advances in cartography and GIScience. Lecture notes in geoinformation and cartography, vol 2, pp 289–295
Shingareva KB, Zimbelman J et al (2005) The realization of ICA commission projects on planetary cartography. Cartographica 40(4):105–114. https://doi.org/10.3138/3660-4078-55x1-3808
Shoemaker E, Hackman RJ (1961) Lunar Photogeologic Chart LPC 58. Copernicus, Prototype Chart, USGS, unpublished
Skinner JA Jr (2015) The Challenges of Standardized Planetary Geologic Mapping, 2nd Planetary Data Workshop, #7071
Tanaka KL, Skinner JA Jr, Hare TM (2011) Planetary Geologic Mapping Handbook—2011. USGS
van Gasselt S, Nass A (2011) Planetary mapping: the datamodel's perspective and GIS framework. Planet Space Sci (PSS) 59:1231–1242. Special Issue: Planetary Mapping, Elsevier Ltd., https://doi.org/10.1016/j.pss.2010.09.012
Wählisch M, Willner K et al (2010) A new topographic image atlas of Phobos. Earth Planet Sci Lett 294:547–553
Wählisch M, Stooke PJ et al (2014) Phobos and Deimos cartography. Planet Space Sci (PSS) 102:60–73
Williams D (2016) NASA's planetary geologic mapping program: overview. In: The international archives of the photogrammetry, remote sensing and spatial information sciences, vol XLI-B4, pp 519–520. https://doi.org/10.5194/isprs-archives-xli-b4-519-2016
Willner K, Oberst J et al (2010) Phobos control point network, rotation, and shape. Earth Planet Sci Lett 294:541–546
Willner K, Shi X, Oberst J (2014) Phobos' shape and topography models. Planet Space Sci (PSS) 102:51–59
Wu B, Li F et al (2014) Topographic modelling and analysis of the landing site of Chang'E-3 on the Moon. Earth Planet Sci Lett 405:257–273
Zastrow M (2015) Data visualization: science on the map. Nature 519:120
Zheng Y-C, Tsang KT et al (2012) First microwave map of the Moon with Chang'E-1 data: the role of local time in global imaging. Icarus 219:194–210. https://doi.org/10.1016/j.icarus.2012.02.017

MIX
Papier aus verantwortungsvollen Quellen
Paper from responsible sources
FSC® C105338

If you have any concerns about our products,
you can contact us on
ProductSafety@springernature.com

In case Publisher is established outside the EU,
the EU authorized representative is:
Springer Nature Customer Service Center GmbH
Europaplatz 3, 69115 Heidelberg, Germany

Printed by Libri Plureos GmbH
in Hamburg, Germany